AF577018

C M P

Mc
Graw
Hill

Cover illustration by courtesy of
British Aerospace plc.

Boundary Elements
An Introductory Course

C. A. Brebbia
J. Dominguez

Computational Mechanics Publications
Southampton Boston

Co-published with

McGraw-Hill Book Company
New York St. Louis San Francisco Colorado Springs
Mexico Montreal Oklahoma City San Juan Toronto

C. A. BREBBIA
Computational Mechanics Institute
Ashurst Lodge, Ashurst
Southampton, SO4 2AA
U.K.

J. DOMINGUEZ
Escuela Tecnica Superior de Ingenieros Industriales
University of Seville
Av. Reina Mercedes, s/n.
41012, Sevilla
SPAIN

British Library Cataloguing in Publication Data

Brebbia, C.A. (Carlos Alberto)
Boundary Elements An Introductory Course
1. Mathematics. Boundary element methods
I. Title II. Dominguez, Jose
515. 3′53

ISBN 0-905451-76-7

Library of Congress Catalog Card Number 88-72042

ISBN 0-905451-76-7 Computational Mechanics Publications, Southampton
ISBN 0-931215-66-8 Computational Mechanics Publications, Boston
ISBN 0-07-007414-3 McGraw-Hill Book Company New York

Second impression November 1989

Printed in Great Britain by The Bath Press, Avon

to

Alexander, Beatriz, Isabel and Pelayo

DISKETTE

The diskette for all the codes in this book is available in two formats, $3\frac{1}{2}$ inch or $5\frac{1}{4}$ inch, for IBM PC, PS2, or compatible machines.

Please apply to:

The Computational Mechanics Institute
Ashurst Lodge
Ashurst
Southampton
SO4 2AA
U.K.

Please send a cheque or money order for £18 (or the equivalent in dollars) and state diskette format required.

Table of Contents

Why Boundary Elements?

Engineers who have been exposed to finite elements may ask themselves why it is necessary to produce yet another computational technique. The answer is that finite elements have been proved to be inadequate or inefficient in many engineering applications and what is perhaps more important is in many cases cumbersome to use and hence difficult to implement in Computer Aided Engineering systems. Finite Element Analysis is still a comparatively slow process due to the need to define or redefine meshes in the piece or domain under study.

Boundary elements [1] have emerged as a powerful alternative to finite elements particularly in cases where better accuracy is required due to problems such as stress concentration or where the domain extends to infinity. The most important features of boundary elements however is that it only requires discretization of the surface rather than the volume. Hence boundary element codes are easier to use with existing solid modellers and mesh generators. This advantage is particularly important for designing as the process usually involves a series of modifications which are more difficult to carry out using finite elements. Meshes can easily be generated and design changes do not require a complete remeshing.

This point is illustrated in figure 1 by two views of a turbine blade section, one discretized using a finite element code and the other with boundary elements. Notice the presence of a series of cooling ducts in the blade whose size, position and number have to be reviewed during the design process. Such a variation creates difficulties for finite elements as some elements may easily become distorted or have bad dimension ratios. The boundary element mesh instead is easy to modify. Figure 1 describes a two dimensional application; these problems are of course compounded for finite elements when working in three dimensions.

Boundary element meshes, especially three dimensional ones can easily be linked to CAE systems as the structure is defined using only the boundary. The discretization process is even simpler when using discontinuous elements, which are not admissible in finite elements. The mesh shown in figure 2 represents the surface discretization of one eighth of a problem, i.e. a cylinder with a cylindrical perforation across. Notice that the use of elements which sometimes do not meet at corners and are consequently discontinuous in terms of their variables, facilitates the meshing. In addition there is no need to use elements on the planes of symmetry.

Figure 3 describes part of a bearing cap using discontinuous elements and taking full advantage of symmetry. These views can be easily rotated to check that elements on the surface are not missing. Notice that discontinuous elements allow for a simple mesh grading. The reason why these elements are possible in boundary elements is explained in some of the chapters in this book. From the user's point

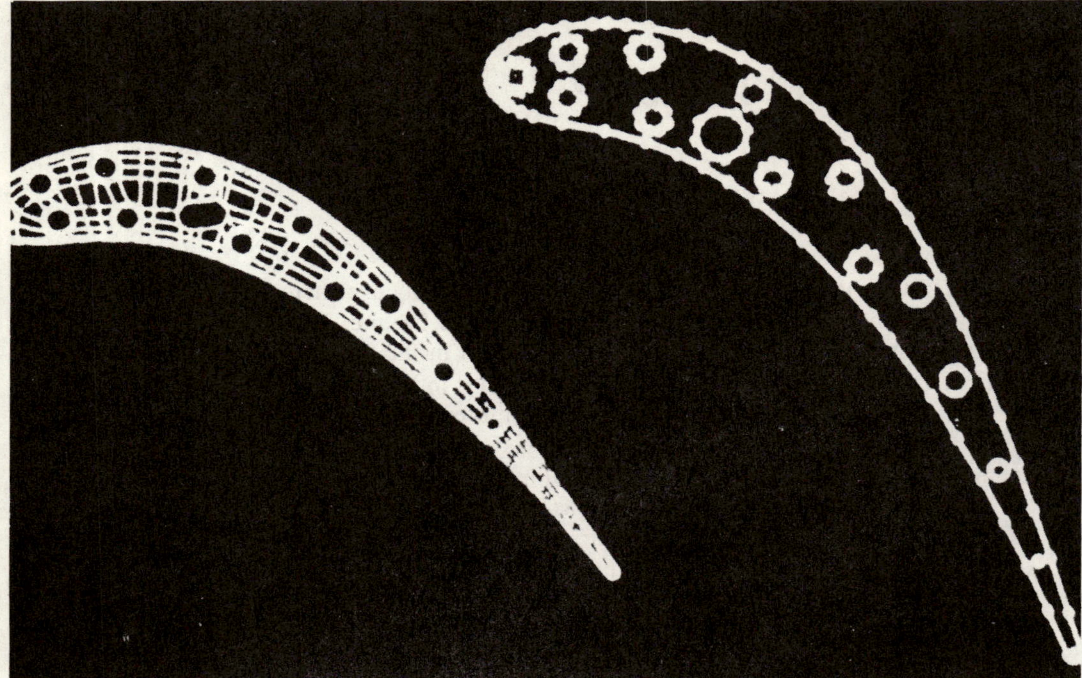

Figure 1. Analysis of a turbine blade using FEM and BEM. Notice that a variation in the configuration of cooling elements creates difficulties for the FE code (from a colour original)

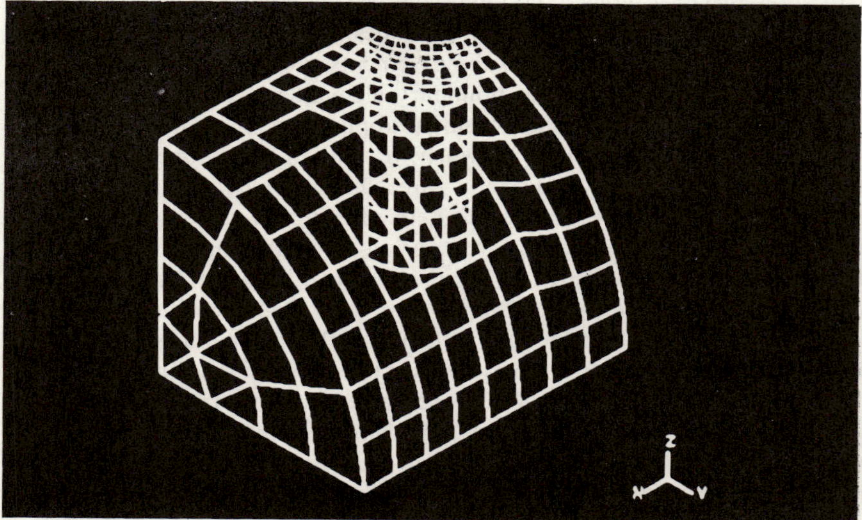

Figure 2. Cylinder with a cylindrical perforation. The boundary element mesh represents discretization of one eighth of the problem

of view they offer many advantages in terms of alterations of the meshes and general versatility. Figure 4 shows some displacements contours obtained for this cap bearing which are plotted using well known post processing systems.

More complex three dimensional structures such as part of the piston shown in figure 5 can be discretized relatively easily using a combination of continuous and discontinuous elements [2].

Boundary element codes can be used to analyse rapidly stresses or temperatures in different types of components. Figure 6 describes the mesh used to analyse part of a crankshaft. Results for the Von Mises stresses on the surface are also plotted. Figure 7 describes a section with a cylindrical hole at an angle which is analysed to determine its temperature distribution which is plotted in figure 8.

It is evident from these examples that boundary elements are an ideal tool for CAD mainly because it is easy to generate the data required to run a problem and carry out the modifications needed to achieve an optimum design. With computer costs declining while engineers' time becomes (or should become!) more expensive, the saving in engineers' time is of primary importance. (Also, engineers need relief from the dreary task of preparing finite element data.) More important still, any tool that can shorten the 'turn around' time taken by the analysis and design can bring forward the completion date of a project.

The future of BEM in engineering is promising and will continue to be so as long as the developers do not alienate the users by producing codes which are unreliable or cumbersome to use. Most of the advantages of BEM are related to

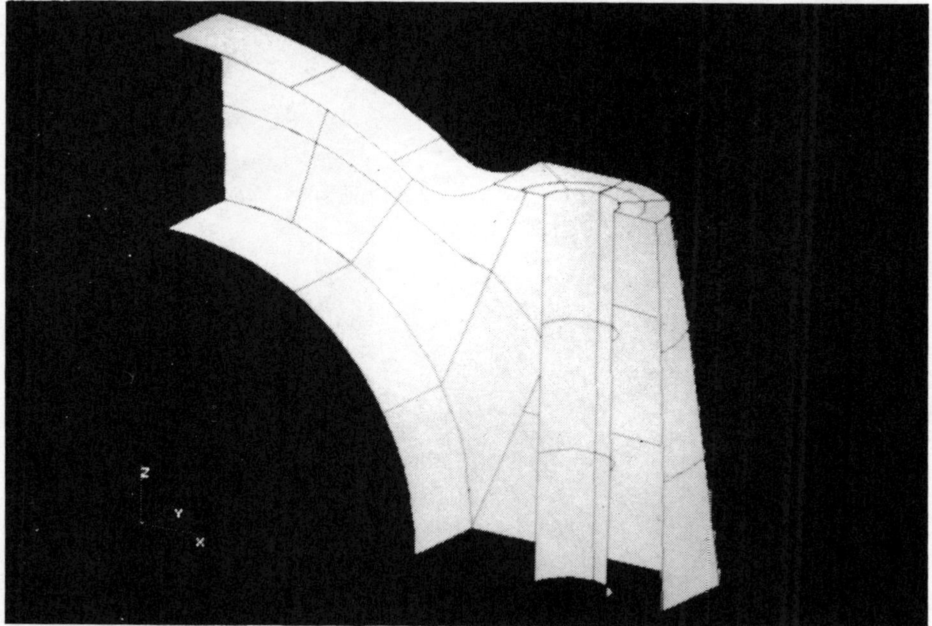

Figure 3. Discretization of the bearing cap into continuous and discontinuous elements (from a colour original)

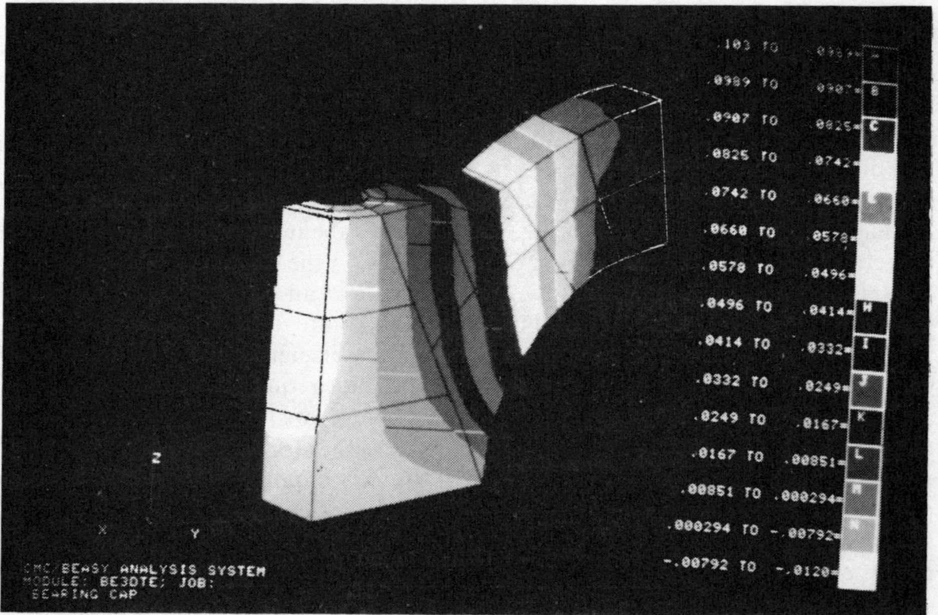

Figure 4. Contours for displacements in the cap bearing obtained using PATRAN for post processing (from a colour original)

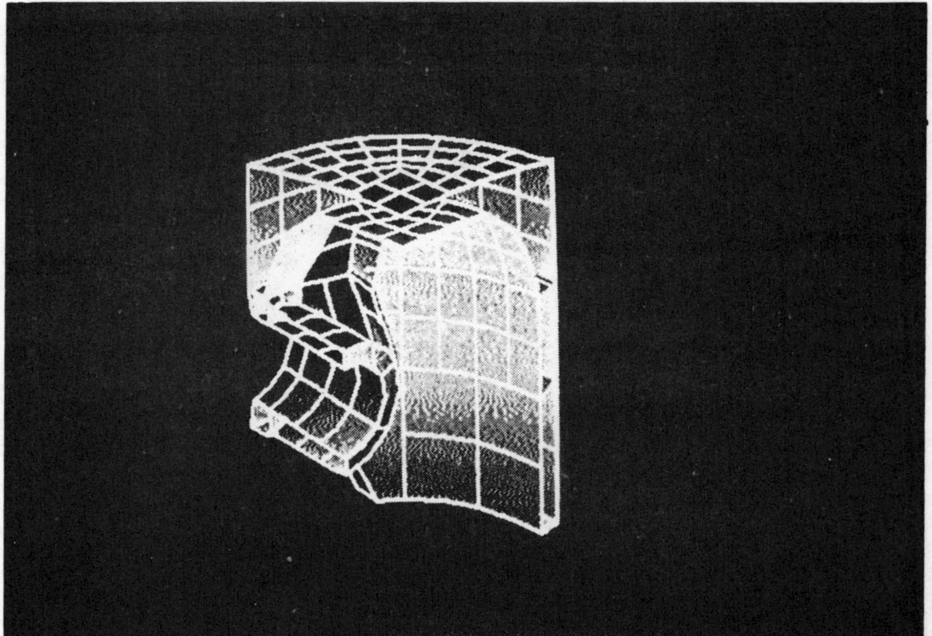

Figure 5. Part of a piston discretized using boundary elements (from a colour original)

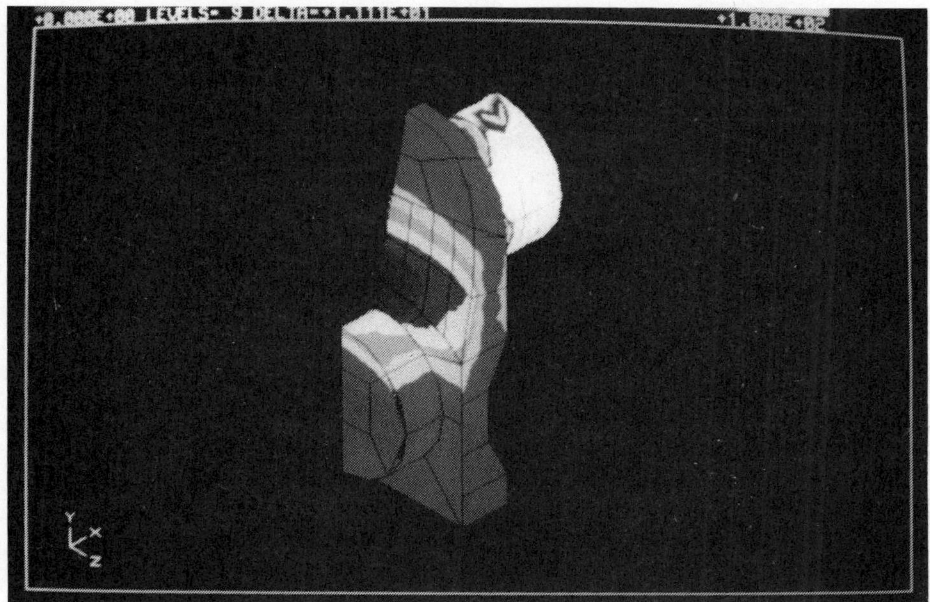

Figure 6. Part of a crankshaft discretized into elements (from a colour original)

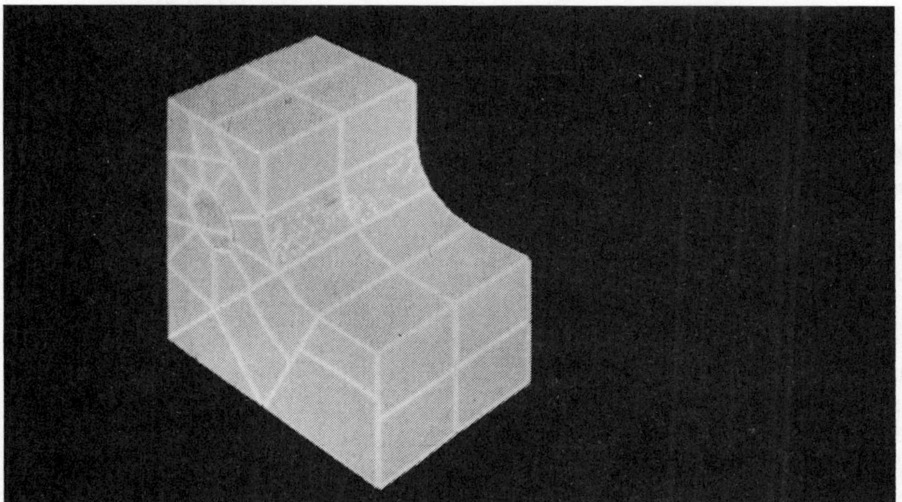

Figure 7. Discretization of a section with hole into elements (from a colour original)

its more complex mathematical foundations. This provides a high degree of versatility and accuracy in well-written codes but can have disastrous consequences in the case of poorly written BEM codes. The BEM is more susceptible to errors when the appropriate numerical techniques are not used and it is then important for developers to understand properly the theory of the method.

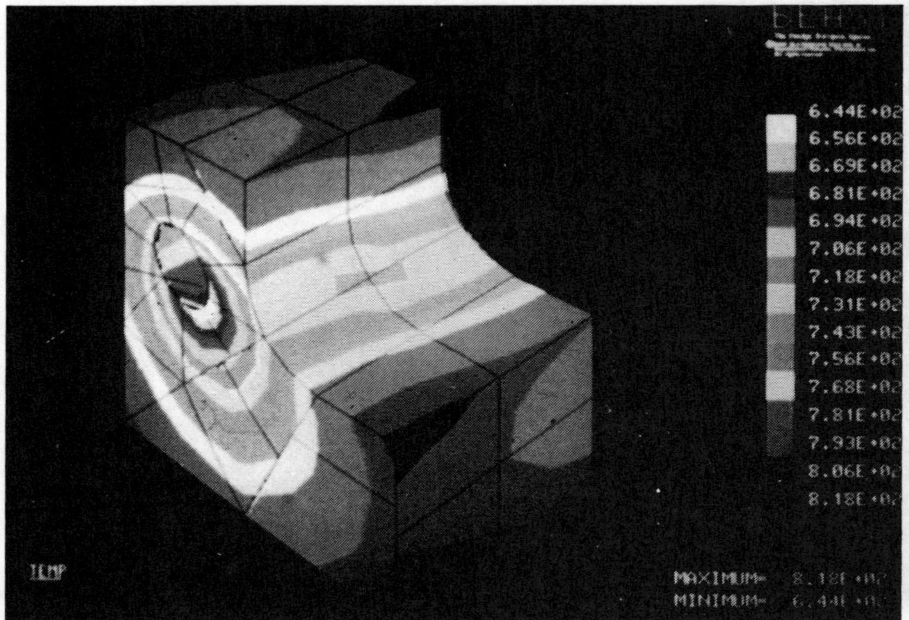

Figure 8. Distribution of temperatures over the surface (from a colour original)

Although better computational performance is important in BEM, particularly for three dimensional problems, improvements in CPU times should not come at the expense of precision and accuracy. For instance, applying coarse numerical integration techniques to BEM codes can result in large savings in computer codes and give reasonable results in many cases. For other cases however the solution may be of very poor accuracy or give non-convergent results. This makes such codes unreliable.

Another important advantage of BEM over FEM is when analysing problems with stress (or other type) concentrations. Many such studies have now been carried out and they tend to demonstrate the high accuracy of boundary elements for problems such as re-entry corners, stress intensity problems and even fracture mechanics applications. It is not our intention in this introduction to review all these studies but rather to point out the difference in results that can be obtained using one or the other numerical method. As an illustration one of the finite element solutions found along a line in the neighbourhood of a re-entry corner (figure 9) of a pressure vessel is shown in figure 10. The problem was also analysed using a photo-elastic model and boundary elements. Results for a finite element mesh consisting of approximately 500 degrees of freedom (69 elements) and using eight nodes elements are compared against BEM solutions obtained using only 20 elements. It is evident from the figures that while the finite element results show lack of equilibrium in the domain as well as on the boundary, reasonably accurate solutions were obtained using boundary elements. It was only when using a very refined finite element mesh that the FE results were in agreement with the boundary

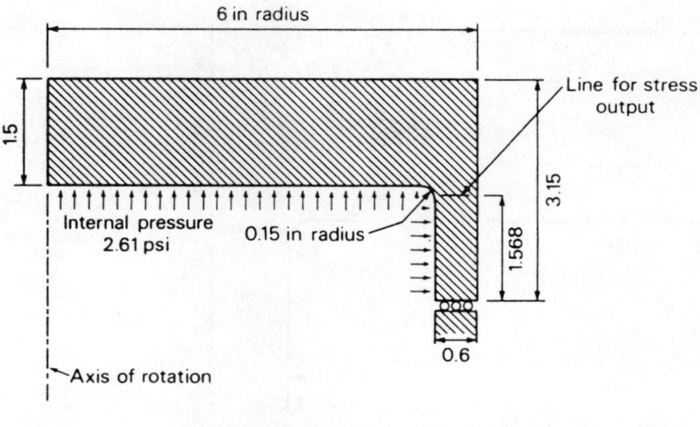

Geometry of the region under consideration

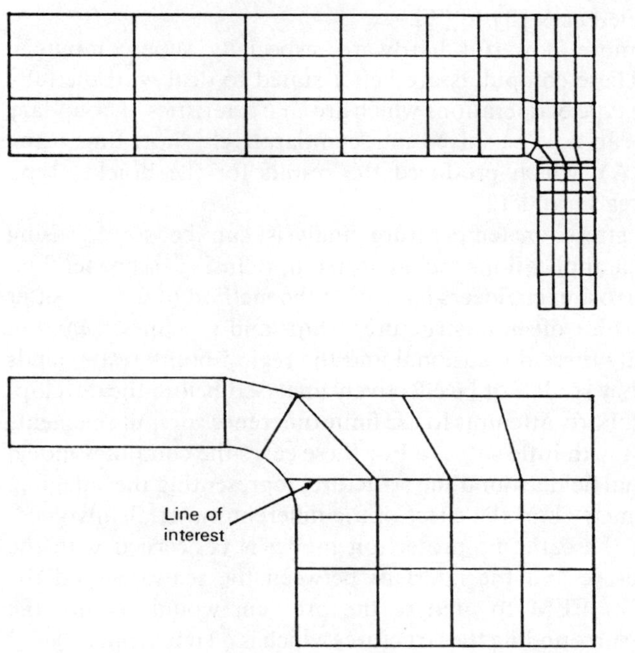

69 finite elements mesh (with mid side nodes) approximately 500 degrees of freedom

Figure 9. Re-entry corner in pressure vessel

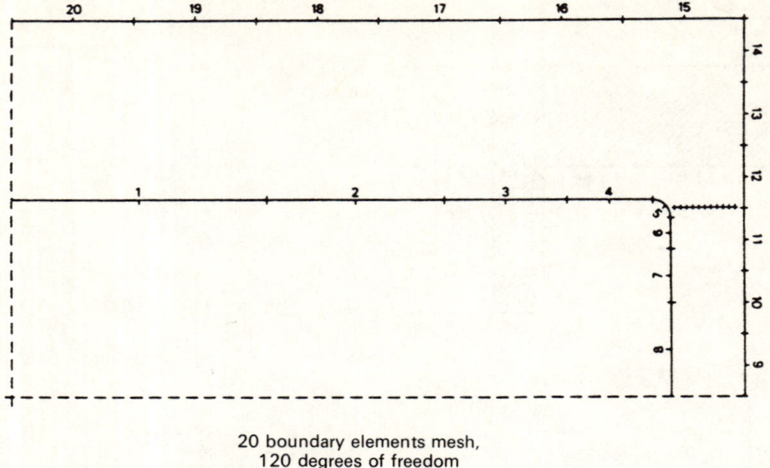

Figure 9 *continued*

elements and photo-elastic model solution. For a full discussion of these results the reader is directed to references [3] to [5].

The development of more powerful hardware, especially supercomputers, favours the use of BEM. These computers are better suited to deal with the fully populated matrices and the type of operations which are characteristics of boundary elements. Large problems can now be solved in a comparatively short time using machines such as the CRAY which produced the results for the bracket type component shown in figures 11 and 12.

Problems other than stress or temperature analysis can be solved using boundary elements. Typical applications include torsion, diffusion, seepage, fluid flow and electrostatics. Corrosion engineers have used the method to design better cathodic protection systems for offshore structures, ships and pipelines. Many of these structures are basically three dimensional and the region of interest extends to infinity. Consequently they could not be effectively analysed before the development of boundary elements. Early attempts to use finite differences or finite elements to solve these problems met with little success. For these cases the computer model has to represent the potential field around the structure, representing the shielding effect of the structural geometry and the effect of the different materials involved. Unlike a structural model the cathodic protection model is concerned with the seawater around the structure and the interface between the seawater and the structure. Hence the use of FEM to analyse the problem would require the subdivision of the seawater surrounding the structures which is a Herculean task.

The use of boundary element method represents the only practical solution for this problem. The advantage of the method is that only the structure needs to be defined as the BEM automatically takes care of the field – i.e. the seawater – extending to infinity. Figure 13 shows the first three dimensional BEM cathodic protection application which was the study of the tension leg platform (TLP) built by CONOCO in the Hutton Field in the North Sea. Figure 14 shows the

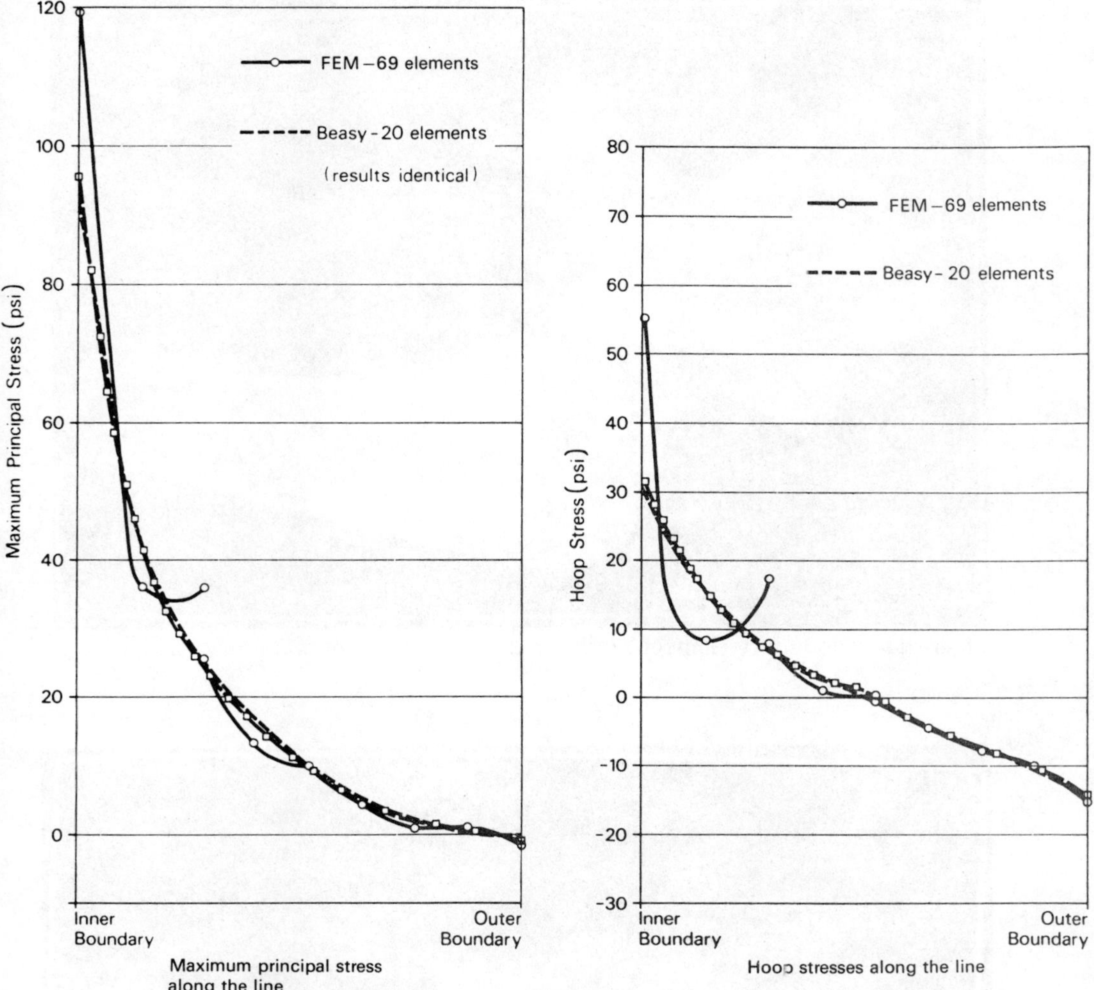

Figure 10. Comparison of FEM and BEM results along the line indicated in figure 9

discretization of a quarter of the structure into the boundary elements used in the analysis and figure 15 the results obtained for the potentials on the surface for a particular configuration of the improved code system used. Since then the boundary element method has become the key to the successful and practical analysis of cathodic protection systems and further work has been carried out in this regard particularly at the Computational Mechanics Institute, Southampton, UK. A system is now available which allows the corrosion engineer to evaluate design options, look at problem areas, interpret experimental observations, optimize the design and predict with accuracy and confidence the degree of protection and life expectancy of a cathodic protection system.

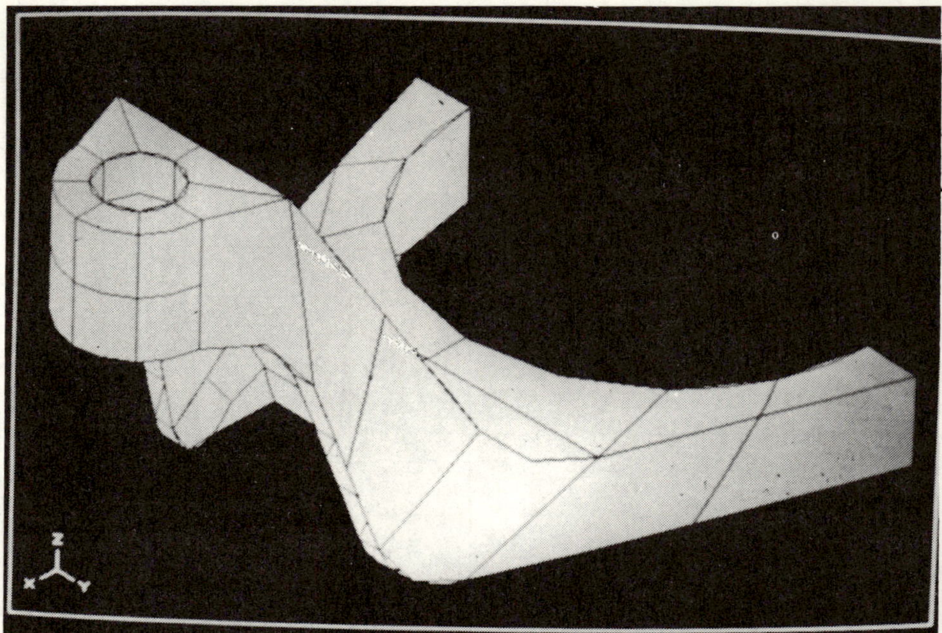

Figure 11. Boundary element mesh for the bracket (from a colour original)

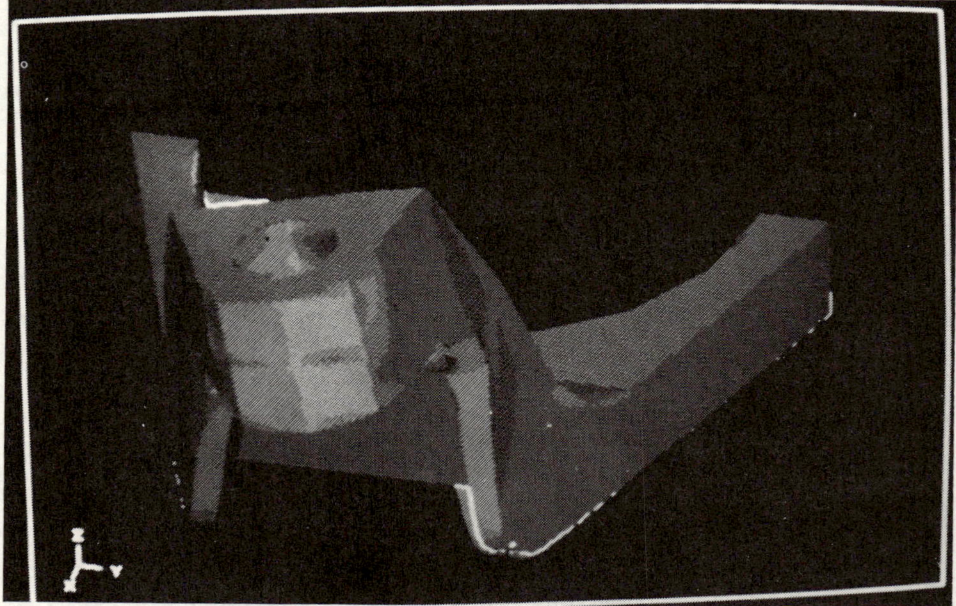

Figure 12. Defined shape for above. Results obtained in a CRAY XMP plotted with SUPERTAB (from a colour original)

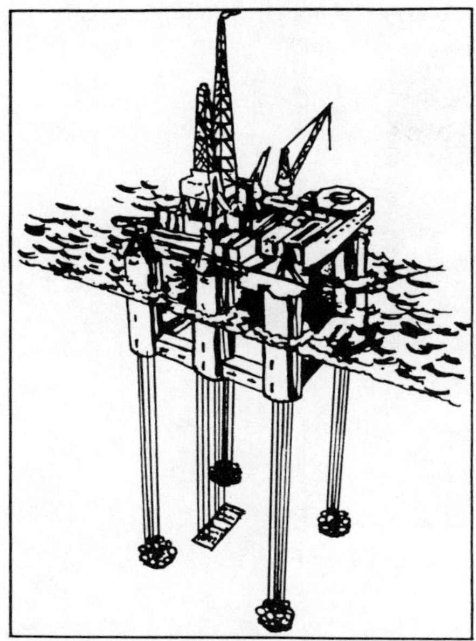

Figure 13. The CONOCO Hutton TLP (Tension Leg Platform)

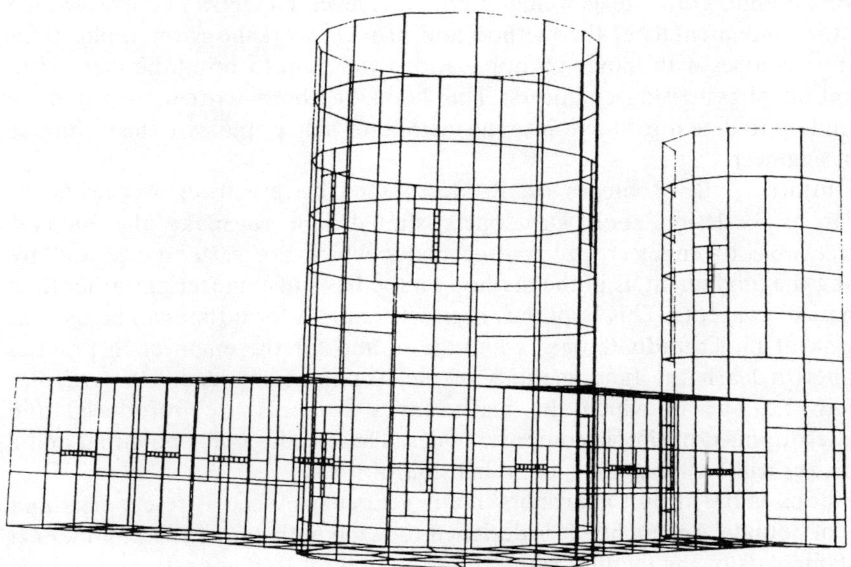

Figure 14. Discretization of a quarter of the platform into boundary elements

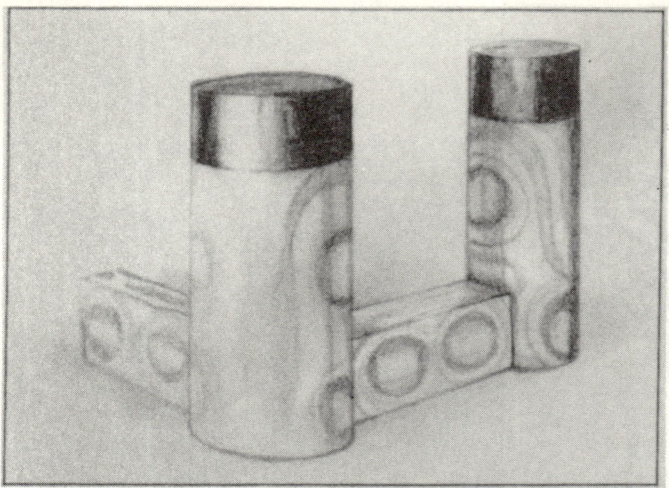

Figure 15. Model of the TLP showing contours of potentials

The advances made in cathodic protection modelling using boundary elements are just one of the applications of the technique for systems extending to infinity. The method is nowadays extensively used in other problems with infinite or semi-infinite domains such as those occurring in geomechanics, ocean engineering, foundations, aerodynamics, flow through porous media and many others.

This brief introduction has attempted to point out the advantages of BEM for a wide variety of engineering problems and the reason why the method should be taught on an undergraduate as well as a graduate level. University courses should include the fundamentals of the method and provide workshops on applications while short courses with hands-on applications will help to bring the method to the attention of practising engineers. This book has been written to provide a simple and up to date introduction to the method to help popularize the technique amongst engineers.

The future of BEM hinges on its acceptance by practising engineers, in particular as a design tool. Developers should aim to make the method more accessible to engineers by writing codes which are easy to use and by explaining the fundamentals of the method on the basis of engineering rather than mathematical concepts. This book has been written in a form that can be used as a textbook at undergraduate or graduate level and for the engineer in practice who wants to learn the fundamentals of the technique unaided. Of particular interest is the way in which the mathematics concepts are introduced and immediately applied in simple computer codes. These codes (4 for potential and 2 for elasticity) will facilitate the comprehension of BEM.

This book is based on the authors' many years experience as researchers and teachers of boundary elements. It is designed to teach in the most effective manner the fundamentals of the method rather than to attempt to demonstrate erudition on the subject. Many topics have been deliberately omitted to avoid confusing the reader. The essentials however are all here. It is now left to the reader to build on this knowledge.

References

[1] Brebbia, C. A., J. Telles and L. Wrobel, *Boundary Element Techniques – Theory and Applications in Engineering*, Springer-Verlag, Berlin and NY, 1984.
[2] Brebbia, C. A., D. Danson and J. Baynham, BEASY. Boundary Element Analysis System, in *Finite Elements Systems Handbook*, Springer-Verlag, Berlin, NY and Computational Mechanics Publications, Southampton and Boston, 1982.
[3] Floyd, C. G. The Determination of Stress using a Combined Theoretical and Experimental Analysis Approach, Computational Methods and Experimental Measurements, *Proc. 2nd Int. Conf. June/July 1984* (C. A. Brebbia, Ed.) Springer-Verlag, Berlin and NY, and Computational Mechanics Publications, Southampton and Boston, 1984.
[4] Sussmans, T. and K. J. Bathe, Studies of Finite Element Procedures – on Mesh Selection, *Computers and Structures*, **21**, 257–264, 1985.
[5] Brebbia, C. A. and J. Trevelyan, On the Accuracy and Convergence of Boundary Element Results for the Floyd Pressure Vessel Problem, *Technical Note*, *Computers and Structures*, **24**, 513–516, 1986.

Further Reading

* Brebbia, C. A. *The Boundary Element Method for Engineers*, Pentech Press, London, Computational Mechanics Publications, Boston, 1978.
* Gipson, G. S. Boundary Element Fundamentals – Basic Concepts and Recent Developments in the Poisson Equation, in *Topics in Engineering Series, Vol. 2* (Ed. C. A. Brebbia and J. J. Connor), Computational Mechanics Publications, Southampton, and Boston, 1987.

Chapter 1

Basic Concepts

1.1 Fundamental concepts

Consider a very simple differential equation applying in a one-dimensional domain x, from $x = 0$ to $x = 1$, i.e.,

$$\frac{d^2u}{dx^2} + \lambda^2 u - b = 0 \qquad \text{in } x \tag{1.1}$$

u is the function which governs the equation and we usually need to find it using a numerical technique which gives an approximate solution. λ^2 is a known positive constant and b is a known function of x.

The solution of equation (1.1) can be found by assuming a variation for u consisting of a series of known shapes (or functions) multiplied by unknown coefficients. These coefficients can then be found by forcing (1.1) to be satisfied at a series of points. This is the basis of the collocation (or point collocation to be precise) method and is essentially what one does when using finite differences. In finite elements instead the solution is found using the concept of distribution of error within the domain. This is somewhat a process of 'smoothing' and it is then not surprising that finite element solutions tend to have less 'noise' than finite difference ones.

The concept of distribution or weighting of a differential equation is not only valid for approximate solutions but it is a fundamental mathematical concept, which can be used in countless engineering applications. Engineers for instance are very familiar with the principle of virtual work which is usually formulated in terms of work done by internal and external forces. They are usually unaware however that the first 'demonstration' of the principle was proposed by Lagrange using the concepts of distributions, applying what are now called the 'Lagrangian' multipliers. These concepts are also essential to study the behaviour of the differential equations, and in particular the type of boundary conditions they require and which are consistent with them.

To understand what these concepts mean before proposing any approximation, one can consider another function w, arbitrary except for being continuous in the domain x and whose derivatives are continuous up to a required degree (the degree of continuity will vary with the problem as will be shown shortly). One can now multiply the whole of equation (1.1) by this w function and integrate on the domain x as follows:

$$\int_0^1 \left(\frac{d^2u}{dx^2} + \lambda^2 u - b \right) w \, dx = 0 \tag{1.2}$$

This operation is called an 'inner' product in mathematics and although does not imply any new concepts, allows us to investigate the properties of the governing equation. This is done by integrating by parts terms with derivatives in the above expression. In this case one can only 'manipulate' in this manner the first term, i.e. d^2u/dx^2, which gives

$$\int_0^1 \left\{ -\frac{du}{dx}\frac{dw}{dx} + (\lambda^2 u - b)w \right\} dx + \left[\frac{du}{dx} w \right]_0^1 = 0 \tag{1.3}$$

Notice that the integration by parts has produced two terms, one in the domain with first derivatives of u and w, and the other on the boundaries (which in this case are simply the two points $x = 0$, $x = 1$).

Furthermore, if the w function has sufficient degree of continuity one can integrate by parts again to obtain

$$\int_0^1 \left\{ u\frac{d^2w}{dx^2} + (\lambda^2 u - b)w \right\} dx + \left[\frac{du}{dx} w \right]_0^1 - \left[u\frac{dw}{dx} \right]_0^1 = 0 \tag{1.4}$$

Expression (1.4) is of course equivalent to (1.3) but here not only has one passed all derivatives to the newly defined w function but the two terms at $x = 0$ and $x = 1$ give us an insight into the boundary conditions required to solve the problem. In this case,

$$u \text{ or } \frac{du}{dx} \qquad \text{needs to be known at } x = 0 \text{ and } x = 1 \tag{1.5}$$

Notice that the w function which in principle was an arbitrary function with a certain degree of continuity can be made to satisfy certain boundary conditions if one wishes to do so. In the principle of virtual displacements for instance, arbitrary functions of this type are defined as virtual displacements but they are assumed to satisfy the homogeneous version of the displacement boundary conditions, i.e. they are set identically to zero at any points where the displacements are prescribed even if those displacements (represented by u) are not set to zero, i.e. $w \equiv 0$ on the parts of the boundary where u is given. This is done in order to eliminate terms of the type $\left[\frac{du}{dx} w \right]$ which give rise to a type of 'work' one does not wish to have. In general however one can assume that w and dw/dx can have values different from zero on the boundaries and this makes expression (1.4) more general.

The concept of an arbitrary function w used as a distribution function is related not only to virtual functions and consequently to virtual work but also to the idea of Lagrangian multipliers. These are functions of the w type defined in order to satisfy certain equations. They will be defined better in what follows.

Although equation (1.4) gives the user an insight into the type of boundary conditions required to solve the problem, these conditions have not yet been explicitly incorporated into the problem. In order to do so let us consider that the boundary conditions are as follows:

$$
\begin{aligned}
u &= \bar{u} && \text{at } x = 0 \\
q &= \frac{du}{dx} = \bar{q} && \text{at } x = 1
\end{aligned}
\tag{1.6}
$$

where the derivatives of u are now defined as q and the terms with bars represent known values of the function and its derivatives. It is usual to call the first type of conditions in (1.6) 'essential' and those like q involving derivatives as 'natural'.

Substituting those values into (1.4) gives

$$
\int_0^1 \left\{ u \frac{d^2 w}{dx^2} + (\lambda^2 u - b) w \right\} dx + \{[\bar{q} w]_{x=1} - [q w]_{x=0}\}
$$
$$
- \left\{ \left[u \frac{dw}{dx} \right]_{x=1} - \left[\bar{u} \frac{dw}{dx} \right]_{x=0} \right\} = 0 \tag{1.7}
$$

It is now interesting to try to return to the original expression (1.2) by integrating by parts again, but this time passing the derivatives for w to u. The first integration gives,

$$
\int_0^1 \left\{ -\frac{du}{dx}\frac{dw}{dx} + (\lambda^2 u - b) w \right\} dx + \left[u \frac{dw}{dx} \right]_{x=1}
$$
$$
- \left[u \frac{dw}{dx} \right]_{x=0} + [\bar{q} w]_{x=1} - [q w]_{x=0} - \left[u \frac{dw}{dx} \right]_{x=1} + \left[\bar{u} \frac{dw}{dx} \right]_{x=0} = 0
\tag{1.8}
$$

Notice that only the term in $[u\, dw/dx]_{x=1}$ disappears.

Furthermore the following expression results after carrying out a second integration,

$$
\int_0^1 \left\{ \frac{d^2 u}{dx} + (\lambda^2 u - b) w \right\} dx - \left[\frac{du}{dx} w \right]_{x=1} + \left[\frac{du}{dx} w \right]_{x=0} - \left[u \frac{dw}{dx} \right]_{x=0}
$$
$$
+ [\bar{q} w]_{x=1} - [q w]_{x=0} + \left[\bar{u} \frac{dw}{dx} \right]_{x=0} = 0 \tag{1.9}
$$

Once again only one term disappears, in this case $[qw]_{x=0}$ – Notice that $q = du/dx$ as defined earlier. – Grouping the terms together one now arrives at an interesting expression, different from the original formula (1.4) i.e.

$$
\int_0^1 \left\{ \frac{d^2 u}{dx^2} w + (\lambda^2 u - b) w \right\} dx - [(q - \bar{q}) w]_{x=1} + \left[(\bar{u} - u) \frac{dw}{dx} \right]_{x=0} = 0 \tag{1.10}
$$

This expression implies that one is trying to enforce not only satisfaction of the differential equation in x but the two boundary conditions. The w and dw/dx functions can be seen as Lagrangian multipliers.

Furthermore nothing has yet been said about approximations; the above expressions are valid for exact solutions as well. In other words the procedure describes a general tool for the investigation of differential equations.

1.2 The Poisson's Equation

An important equation in engineering analysis is the so-called Poisson's equation which for two dimensions can be written as

$$\frac{\partial^2 u}{\partial x_1^2} + \frac{\partial^2 u}{\partial x_2^2} = b \qquad \text{in } \Omega \tag{1.11}$$

or

$$\nabla^2 u = b \qquad \text{in } \Omega \tag{1.12}$$

where $\nabla^2(\ ) = \frac{\partial^2(\ )}{\partial x_1^2} + \frac{\partial^2(\ )}{\partial x_2^2}$, is called the Laplace operator. x_1 and x_2 are the two coordinates and b is a known function of x_1, x_2. Ω is the domain on which the equation applies and is assumed to be bounded by Γ. The outward normal to the boundary is defined as n (figure 1.1).

The Poisson equation or its homogeneous form (i.e. $b = 0$) which is the Laplace equation, governs many types of engineering problems, such as seepage and aquifer analysis, heat conduction, diffusion processes, torsion, fluid motion and others. Consequently it is a very important equation in engineering analysis.

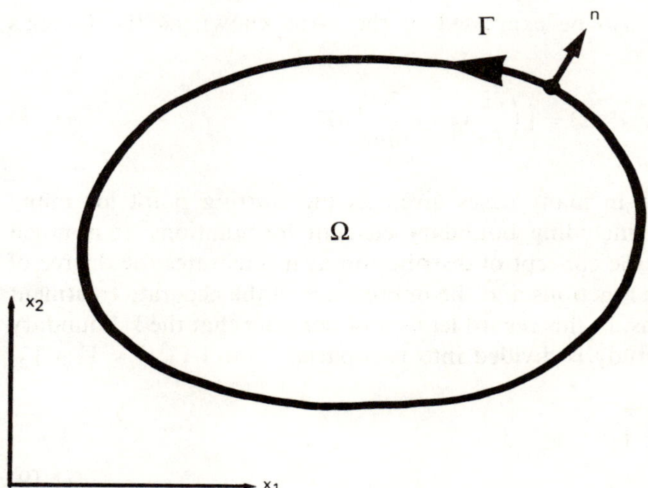

Figure 1.1 Domain under Consideration for Poisson's Equation Basic Definitions

Here one can also introduce the idea of multiplying equation (1.12) by an arbitrary w function, continuous up to the second derivative. This gives,

$$\int_\Omega (\nabla^2 u - b) w \, d\Omega = 0 \tag{1.13}$$

Integrating by parts the terms in x_1 and x_2 gives

$$\int_\Omega \left(-\frac{\partial u}{\partial x_1}\frac{\partial w}{\partial x_1} - \frac{\partial u}{\partial x_2}\frac{\partial w}{\partial x_2} - bw \right) d\Omega + \int_\Gamma \frac{\partial u}{\partial n} w \, d\Gamma = 0 \tag{1.14}$$

In this case the integration by parts of the two terms produces the derivative of u with respect to the normal, i.e. $\partial u/\partial n$ which will later on be called q, i.e. $q = \partial u/\partial n$.

Integrating by parts again, one obtains,

$$\int_\Omega \left(\frac{\partial^2 w}{\partial x_1^2} u + \frac{\partial^2 w}{\partial x_2^2} u - bw \right) d\Omega + \int_\Gamma \frac{\partial u}{\partial n} w \, d\Gamma - \int_\Gamma u \frac{\partial w}{\partial n} \, d\Gamma = 0 \tag{1.15}$$

or

$$\int_\Omega \{(\nabla^2 w) u - bw\} \, d\Omega + \int_\Gamma \frac{\partial u}{\partial n} w \, d\Gamma - \int_\Gamma u \frac{\partial w}{\partial n} \, d\Gamma = 0 \tag{1.16}$$

Expression (1.16) is equal to (1.13) and hence one can write,

$$\int_\Omega (\nabla^2 u) w \, d\Omega = \int_\Omega (\nabla^2 w) u \, d\Omega + \int_\Gamma \frac{\partial u}{\partial n} w \, d\Gamma - \int_\Gamma u \frac{\partial w}{\partial n} \, d\Gamma = 0 \tag{1.17}$$

where the term in b has been eliminated as it appears on the two sides of the equation.

Equation (1.17) can also be expressed in the form known as the Green's theorem, i.e.

$$\int_\Omega \{(\nabla^2 u) w - (\nabla^2 w) u\} \, d\Omega = \int_\Gamma \left(\frac{\partial u}{\partial n} w - u \frac{\partial w}{\partial n} \right) d\Gamma \tag{1.18}$$

Although this theorem is in many cases given as the starting point for many engineering applications, including boundary element formulations, it is much more illuminating to use the concept of distribution as it illustrates the degree of continuity required of the functions and the importance of the accurate treatment of the boundary conditions. In this regard let us now consider that the Γ boundary of the Ω domain under study is divided into two parts, Γ_1 and Γ_2 ($\Gamma = \Gamma_1 + \Gamma_2$) such that,

$$\begin{aligned} u &= \bar{u} && \text{on } \Gamma_1 \\ q &= \frac{\partial u}{\partial n} = \bar{q} && \text{on } \Gamma_2 \end{aligned} \tag{1.19}$$

Hence equation (1.16) can now be written as,

$$\int_{\Omega} \{(\nabla^2 w)u - bw\}\, d\Omega$$

$$+ \int_{\Gamma_1} qw\, d\Gamma + \int_{\Gamma_2} \bar{q}w\, d\Gamma - \int_{\Gamma_1} \bar{u}\frac{\partial w}{\partial n}\, d\Gamma - \int_{\Gamma_2} u\frac{\partial w}{\partial n}\, d\Gamma = 0 \tag{1.20}$$

Once again one can integrate by parts to retrieve the original Laplacian $\nabla^2 u$ in order to see how the importance of the boundary conditions affect the equation. Integrating by parts once we have,

$$\int_{\Omega} \left\{ -\frac{\partial w}{\partial x_1}\frac{\partial u}{\partial x_1} - \frac{\partial w}{\partial x_2}\frac{\partial u}{\partial x_2} - bw \right\} d\Omega$$

$$+ \int_{\Gamma} \frac{\partial w}{\partial n} u\, d\Gamma + \int_{\Gamma_1} qw\, d\Gamma + \int_{\Gamma_2} \bar{q}w\, d\Gamma - \int_{\Gamma_1} \bar{u}\frac{\partial w}{\partial n}\, d\Gamma - \int_{\Gamma_2} u\frac{\partial w}{\partial n}\, d\Gamma = 0 \tag{1.21}$$

One can split the first integral on Γ into two terms (one on Γ_1 and the other on Γ_2), the second of which can be cancelled with the last integral in (1.21). This gives

$$\int_{\Omega} \left\{ -\frac{\partial w}{\partial x_1}\frac{\partial u}{\partial x_1} - \frac{\partial w}{\partial x_2}\frac{\partial u}{\partial x_2} - bw \right\} d\Omega$$

$$+ \int_{\Gamma_1} \frac{\partial w}{\partial n} u\, d\Gamma + \int_{\Gamma_1} qw\, d\Gamma + \int_{\Gamma_2} \bar{q}w\, d\Gamma - \int_{\Gamma_1} \bar{u}\frac{\partial w}{\partial n}\, d\Gamma = 0 \tag{1.22}$$

Integrating again by parts the following expression is obtained

$$\int_{\Omega} \{(\nabla^2 u)w - bw\}\, d\Omega$$

$$- \int_{\Gamma} wq\, d\Gamma + \int_{\Gamma_1} \frac{\partial w}{\partial n} u\, d\Gamma + \int_{\Gamma_1} qw\, d\Gamma + \int_{\Gamma_2} \bar{q}w\, d\Gamma$$

$$- \int_{\Gamma_1} \bar{u}\frac{\partial w}{\partial n}\, d\Gamma = 0 \tag{1.23}$$

The first integral in Γ can again be written as a summation of two integrals, one on Γ_1 and the other on Γ_2. The one on Γ_1 can be cancelled with the integral on Γ_1 of qw in the above equation. This gives

$$\int_{\Omega} \{(\nabla^2 u)w - bw\}\, d\Omega - \int_{\Gamma_2} wq\, d\Gamma + \int_{\Gamma_1} \frac{\partial w}{\partial n} u\, d\Gamma + \int_{\Gamma_2} \bar{q}w\, d\Gamma$$

$$- \int_{\Gamma_1} \bar{u}\frac{\partial w}{\partial n}\, d\Gamma = 0 \tag{1.24}$$

This formula can be written as,

$$\int_{\Omega} \{(\nabla^2 u - b)w\}\, d\Omega - \int_{\Gamma_2} (q - \bar{q})w\, d\Gamma + \int_{\Gamma_1} (u - \bar{u})\frac{\partial w}{\partial n}\, d\Gamma = 0 \tag{1.25}$$

Once again this expression shows that one is trying to satisfy a differential equation in the domain plus two types of boundary conditions, the 'essential' conditions $u = \bar{u}$ on Γ_1 plus the 'natural' conditions $q = \bar{q}$ on Γ_2. This is very much what has been shown in equation (1.10) with the only exception that the sign of the last term is different in both expressions. This is because in (1.10) the derivatives were taken with respect to x rather than with respect to the normal, as they are now.

1.3 Approximate Solutions

Although the previous sections have introduced the concept of distributions, the formulations apply irrespective of the type of solution one finds, i.e. they are valid for exact as well as approximate solutions. This section however will investigate what happens when the concept of an approximate solution is introduced in the formulation. In engineering practice the exact solution can only be known in a few simple cases and it is hence important to see how the solution behaves when one introduces an approximation. Let us consider now that the function u defines an approximate rather than the exact solution. In this case one can write for instance,

$$u = \alpha_1 \phi_1 + \alpha_2 \phi_2 + \ldots \tag{1.26}$$

where α_i are unknown coefficients and the ϕ_i are a set of linearly independent functions which are known. α_i are generalized coefficients although in some cases they can be associated with nodal values of the variable under consideration. In general in engineering problems, one prefers to use nodal values as they have a clear physical meaning and this is done in finite elements, finite differences or the boundary element method. In such cases the approximation for u can be written as

$$\begin{aligned} u &= u_1 \phi_1 + u_2 \phi_2 + \ldots \\ u &= \sum_{j=1}^{N} u_j \phi_j \end{aligned} \tag{1.27}$$

where ϕ_j are a set of linearly independent functions which are sometimes called interpolation functions. u_j are the nodal values of the field variable or its derivative (or more generally the nodal value of any variable with physical meaning directly related to u or its derivatives).

Introducing the approximation for u into the governing differential equation one finds that the equation is no longer identically satisfied except for the case in which (1.26) or (1.27) can represent the exact solution. This produces an error or residual function which will soon be defined.

For instance, introducing an approximate value of u into equation (1.1) one generally finds that

$$\frac{d^2u}{dx^2} + \lambda^2 u - b \neq 0 \qquad \text{in } x \tag{1.28}$$

The same will generally occur with the boundary conditions corresponding to this equation, i.e.

$$\begin{aligned} u - \bar{u} &\neq 0 \qquad \text{at } x = 0 \\ q - \bar{q} &\neq 0 \qquad \text{at } x = 1 \end{aligned} \tag{1.29}$$

One can now introduce the concept of an error function or residual which represents the errors occurring in the domain or on the boundary due to non-satisfaction of the above equations. The error function in the domain is called R and is given by

$$R = \frac{d^2u}{dx^2} + \lambda^2 u - b \tag{1.30}$$

and on the boundary one has,

$$\begin{aligned} R_1 &= u - \bar{u} \\ \text{and} \qquad & \\ R_2 &= q - \bar{q} \end{aligned} \tag{1.31}$$

Although the above case is a particular and relatively simple equation the same occurs for any other problem. If one considers the Poisson's equation (1.12) for instance, the error function in the domain is

$$R = \nabla^2 u - b \qquad \text{in } \Omega \tag{1.32}$$

and the errors for the boundary conditions (equation (1.19)) are defined by

$$\begin{aligned} R_1 &= u - \bar{u} \qquad \text{on } \Gamma_1 \\ R_2 &= q - \bar{q} \qquad \text{on } \Gamma_2 \end{aligned} \tag{1.33}$$

The numerical methods used in engineering try to reduce these errors to a minimum by applying different techniques. This reduction is carried out by forcing the errors to be zero at certain points, regions or in a mean sense. This operation can be generally interpreted as *distributing* these errors. The way in which this distribution is carried out produces different types of error distribution techniques which, in general, force the integrals of the residuals weighted by a certain function to be zero. Because of this they are called weighted residual techniques.

1.4 Weighted Residual Techniques

The solution of the boundary value problem defined by equations (1.28) and (1.29), (1.32) and (1.33) or similar sets for other problems can be attempted by choosing an approximation for the function u. One can then have three types of method:

(i) If the assumed approximate solution identically satisfies all boundary conditions but not the governing equations in Ω, one has a purely 'domain' method.
(ii) If the approximate solution satisfies the field or governing equations but not the boundary conditions one has a 'boundary' method.
(iii) If the assumed solution satisfies neither the field equation nor the boundary conditions, one has a 'mixed' method.

Let us first assume that the functions ϕ_j which are defined to approximate u, satisfy all boundary conditions. One then has a residual R function in the domain as the field equations are generally not identically satisfied. The idea is now to make R as small as possible by setting its weighted residual equal to zero for various values of the weighting functions, ψ_j, such that

$$\int_\Omega R\psi_j \, d\Omega = 0 \qquad \text{in } \Omega \qquad j = 1, 2, \ldots, N \tag{1.34}$$

These functions have to be linearly independent.

Notice that another way of writing (1.34) in a form that is more compact and easy to operate with, is by defining a new function w, such that

$$w = \beta_1\psi_1 + \beta_2\psi_2 + \ldots + \beta_N\psi_N = \sum_{j=1}^{N} \beta_j\psi_j \tag{1.35}$$

where β_j are arbitrary coefficients. Hence equation (1.34) can now be written in a more compact form as,

$$\int_\Omega Rw \, d\Omega = 0 \qquad \text{in } \Omega \tag{1.36}$$

Different types of weighting functions ψ_j (or w) will define different approximate methods. Equation (1.34) or (1.35) will produce a system of algebraic equations from which the unknown values of the α_i or u_i coefficients used in u (equation (1.26) or (1.27)) can be obtained.

The approximation can always be improved by increasing the number of N functions used. (N is the number of terms in the approximate solution equal to the number of weighting functions required.)

Approximate methods based on equation (1.36) are called weighted residual methods and, given an approximate solution, the method will vary in accordance with the functions used as weighting functions. In what follows a few will be reviewed.

(i) Subdomain Collocation

For this method the domain Ω is divided in M subdomains and the integral of the error in each of them is set to zero. The weighted functions are simply chosen as,

$$\psi_j = \begin{cases} 1 & \text{for } x \in \Omega_j \\ 0 & \text{for } x \notin \Omega_j \end{cases} \tag{1.37}$$

($\in$ indicates belonging to and Ω_j is the 'j' subdomain). Equation (1.34) becomes,

$$\int_{\Omega_j} R\, dx = 0; \qquad j = 1, 2, \ldots, N \tag{1.38}$$

(ii) Galerkin Method

In the case of Galerkin's method the weighting functions are the same as the appoximating functions, i.e.

$$\phi_j = \psi_j \tag{1.39}$$

hence equation (1.34) becomes,

$$\int_{\Omega} R\phi_j\, d\Omega = 0 \qquad j = 1, 2, \ldots, N \tag{1.40}$$

Using the same definition as in (1.35) this can be written as,

$$\int_{\Omega} Rw\, d\Omega = 0, \tag{1.41}$$

with,

$$w = \beta_1\phi_1 + \beta_2\phi_2 + \ldots + \beta_N\phi_N \tag{1.42}$$

This method is the starting point of many finite element formulations for which the symmetry of $\phi_j = \psi_j$ coupled to inherently symmetric field equations, lead to symmetric algebraic matrices.

(iii) Point Collocation Method

In this case N points $x_1, x_2, \ldots, x_N$ are chosen in the domain and the residual is set to zero at these points This operation can be interpreted as defining weighting functions in terms of Dirac deltas, i.e.

$$\psi_j = \Delta(x - x_j); \qquad j = 1, 2, \ldots, N \tag{1.43}$$

$\Delta(x-x_j)$ at point $x-x_j$ has an infinite value but is such that its integral gives unity, i.e.

$$\int_\Omega \Delta(x-x_j)\,d\Omega = 1; \qquad j=1,2,\ldots,N \tag{1.44}$$

The Dirac function can be interpreted as the limit of a regular function when its base tends to zero.

Hence equation (1.34) can now be written as,

$$\int_\Omega R\Delta(x-x_j)\,d\Omega = 0; \qquad j=1,2,\ldots,N \tag{1.45}$$

which simply says that the error function is zero at a series of points, i.e.

$$R\big|_{x=x_j} = 0; \qquad j=1,2,\ldots,N \tag{1.46}$$

The method consists of setting the residual or error function equal to zero at as many points as there are unknown coefficients in the approximate solution. The distribution of the collocation points is in principle arbitrary, but in practice better results are obtained if they are uniformly distributed.

Example 1.1

As an illustration of how to use weighted residuals, consider the following differential or field equation in the one dimensional domain x (where x varies from $x=0$ to $x=1$), i.e.

$$\frac{d^2u}{dx^2} + x = 0 \tag{a}$$

with homogeneous boundary conditions, i.e.

$$u=0 \text{ at } x=0 \text{ and } x=1 \tag{b}$$

(Notice that equation (a) is a particular case of equation (1.1) when $\lambda=0$ and $b=-x$.)

The exact solution of (a) can be found by integration and gives,

$$u_{\text{exact}} = \frac{x}{6} - \frac{x^3}{6} \tag{c}$$

Let us now attempt to solve (a) using the weighted residual techniques described above, starting by defining an approximate solution which satisfies the boundary conditions and can be written as

$$u = \alpha_1\phi_1 + \alpha_2\phi_2 + \ldots \tag{d}$$

One can use Hermitian polynomials for ϕ_j but since only two of them satisfy the homogeneous boundary conditions, only these two will be used, i.e.

$$u = \alpha_1 \phi_1 + \alpha_2 \phi_2 \qquad \text{(e)}$$

where

$$\begin{aligned} \phi_1 &= x - 2x^2 + x^3 \\ \phi_2 &= x^3 - x^2 \end{aligned} \qquad \text{(f)}$$

The residual or error function in this case is obtained by substituting (e) into equation (a) which gives,

$$\begin{aligned} R(x_1 \alpha_1 \alpha_2) &= \frac{d^2 u}{dx^2} + x \\ &= \alpha_1 \frac{d^2 \phi_1}{dx^2} + \alpha_2 \frac{d^2 \phi_2}{dx^2} + x \\ &= \alpha_1 (6x - 4) + \alpha_2 (6x - 2) + x \end{aligned} \qquad \text{(g)}$$

Let us now reduce (g) using the various techniques previously described.

(i) Subdomain Collocation

Consider the domain divided into 2 equal parts, one from 0 to $\frac{1}{2}$ and the other from $\frac{1}{2}$ to 1. In this case one can write,

$$\int_0^{1/2} R\, dx = \int_0^{1/2} [\alpha_1 (6x - 4) + \alpha_2 (6x - 2) + x]\, dx = 0$$

and

$$\int_{1/2}^{1} R\, dx = \int_{1/2}^{1} [\alpha_1 (6x - 4) + \alpha_2 (6x - 2) + x]\, dx = 0 \qquad \text{(h)}$$

which produce the following system of equations

$$\begin{aligned} -1.2\alpha_1 - 0.25\alpha_2 + 0.125 &= 0 \\ 0.25\alpha_1 + 1.2\alpha_2 + 0.375 &= 0 \end{aligned} \qquad \text{(i)}$$

from which one can obtain,

$$\alpha_1 = \tfrac{1}{6}; \qquad \alpha_2 = -\tfrac{1}{3} \qquad \text{(j)}$$

Substituting (j) into (e) gives the following result

$$u = \frac{x}{6} - \frac{x^3}{6} \qquad \text{(k)}$$

Notice that the exact solution (c) has been obtained since the assumed shapes of u are able to represent it.

(ii) Galerkin

In this case the weighting functions are,

$$\psi_1 = \phi_1$$
$$\psi_2 = \phi_2$$

and the weighted residual expressions are

$$\int_0^1 [\alpha_1(6x-4)+\alpha_2(6x-2)+x](x-2x^2+x^3)\,dx=0$$
$$\int_0^1 [\alpha_1(6x-4)+\alpha_2(6x-2)+x](x^3-x^2)\,dx=0 \tag{l}$$

which produces the following algebraic equations in α_1 and α_2.

$$-4\alpha_1+\alpha_2+1=0$$
$$\alpha_1-4\alpha_2-1.5=0 \tag{m}$$

This also results in

$$\alpha_1=\tfrac{1}{6}, \qquad \alpha_2=-\tfrac{1}{3}$$

(iii) Point Collocation

Here one forces the residual to be zero at a series of points. Consider in this case that R is zero at the two points $x=0.25$ and $x=0.75$. This gives

$$R|_{x=0.25}=-10\alpha_1-2\alpha_2+1=0$$
$$R|_{x=0.25}=2\alpha_1+10\alpha_2+3=0 \tag{n}$$

with the same results for α_1 and α_2, i.e.

$$\alpha_1=\tfrac{1}{6}; \qquad \alpha_2=-\tfrac{1}{3}$$

Notice that this case is rather trivial and the same results have been obtained for all the methods. In general this will not be true when the exact solution cannot be reproduced by the proposed value of u and one will find different results depending on the method used.

Example 1.2

Let us now study another equation using point collocation such that in this case we will obtain an approximate rather than the exact solution.

Consider the equation (1.1), with $\lambda^2 = 1$ and $x = -b$, i.e.

$$\frac{d^2u}{dx^2} + u + x = 0 \tag{a}$$

and the homogeneous boundary conditions, $u \equiv 0$ at $x = 0$ and $x = 1$.

The exact solution of (a) can be easily obtained by integration and gives

$$u = \frac{\sin x}{\sin 1} - x \tag{b}$$

Instead of using (b) we will try to approximate it defining a solution

$$u = a_1\phi_1 + a_2\phi_2 + a_3\phi_3 + \dots \tag{c}$$

where the ϕ_i are terms of a polynomial in x, i.e.

$$\phi_1 = 1, \qquad \phi_2 = x, \qquad \phi_3 = x^2 \dots \tag{d}$$

In order to satisfy the boundary conditions exactly, equation (c) has to give,

$$u \equiv 0 \text{ at } x = 0 \text{ and } x = 1 \tag{e}$$

which implies that,

$$\begin{aligned} &\text{at } x = 0 \rightarrow u = a_1 = 0 \\ &\text{at } x = 1 \rightarrow u = a_1 + a_2 + a_3 + \dots = 0 \end{aligned} \tag{f}$$

Hence $a_1 \equiv 0$ and a_2 can be expressed in function of the other a_i parameter, i.e.

$$a_2 \equiv -(a_3 + a_4 + \dots) \tag{g}$$

Substituting $a_1 \equiv 0$ and (g) into (c) one can write,

$$\begin{aligned} u &= a_3(x^2 - x) + a_4(x^3 - x) + a_5(x^4 - x) + \dots \\ &= x(1-x)(-a_3 - a_4) + x(1-x)(-a_4)x + \dots \end{aligned} \tag{h}$$

Defining now a new set of unknown parameters α_i such that,

$$\alpha_1 = -a_3 - a_4; \qquad \alpha_2 = -a_4 \dots \tag{i}$$

one can write,

$$u = x(1-x)(\alpha_1 + \alpha_2 x + \dots) \tag{j}$$

This function satisfies the boundary conditions in u and has the degree of continuity required by the derivatives in equation (a), hence it is said to be 'admissible'. We will also see that the 'distance' between the approximate and exact solution decreases when the number of terms in (j) increases and this implies that the approximate formulation u is 'complete', i.e. tends to represent the exact solution better and better when the number of terms increases.

In order to apply the point collocation technique we will restrict ourselves to two terms in the (j) expression, i.e.

$$u = x(1-x)(\alpha_1 + \alpha_2 x) \tag{k}$$

Substituting this function into the governing equation (a) one finds the following residual, i.e.

$$R = \frac{d^2u}{dx^2} + u + x = (-2 + x - x^2)\alpha_1 + (2 - 6x + x^2 - x^3)\alpha_2 + x \tag{l}$$

Collocation can now be interpreted as setting $R \equiv 0$ at two points, say $x = \frac{1}{4}$ and $x = \frac{1}{2}$. This can also be expressed in terms of Dirac delta functions applied at these two points, i.e. the weighting function is,

$$w = \beta_1 \Delta_1(x - \tfrac{1}{4}) + \beta_2 \Delta_2(x - \tfrac{1}{2}) \tag{m}$$

The weighted residual integrals are represented by

$$\int_0^1 Rw \, dx = 0 \tag{n}$$

or simply,

$$R \equiv 0 \qquad \text{at } x = \tfrac{1}{4} \text{ and } x = \tfrac{1}{2} \tag{o}$$

Substituting these values of x into (l) one obtains two equations in α_1 and α_2. They can be written in matrix form as follows,

$$\begin{bmatrix} \frac{29}{16} & -\frac{35}{64} \\ \frac{7}{4} & \frac{7}{8} \end{bmatrix} \begin{Bmatrix} \alpha_1 \\ \alpha_2 \end{Bmatrix} = \begin{Bmatrix} \frac{1}{4} \\ \frac{1}{2} \end{Bmatrix} \tag{p}$$

The solution of this system gives

$$\alpha_1 = \tfrac{6}{31}, \qquad \alpha_2 = \tfrac{40}{217} \tag{q}$$

The approximate value of u – equation (k) – can now be written as,

$$u = \frac{x(1-x)}{217}(42 + 40x) \tag{r}$$

Table 1.1 Results for Point Collocation

x	u (exact)	u (approximate)	R
0.10	0.018641	0.019078	−0.009953
0.30	0.051194	0.052258	+0.002027
0.50	0.069746	0.071428	+0.00000
0.70	0.065585	0.065806	−0.024884
0.90	0.030901	0.032350	−0.081474

Notice that the error function can now also be fully defined in terms of x, by substituting α_1 and α_2 into (l). This gives,

$$R = \frac{1}{217}(-4 + 19x - 2x^2 - 40x^3) \tag{s}$$

These results can be tabulated in table 1.1 where they are compared against the exact solution for u. Notice that the values of R are identically zero at $x = \frac{1}{4}$ and $x = \frac{1}{2}$ but that this does not mean that the solution for u is exact at those points.

Example 1.3

Let us apply Galerkin's technique to equation (1.1) for which $\lambda^2 = 1$ and $b = -x$ with homogeneous boundary conditions $u \equiv 0$ at $x = 0$ and $x = 1$. The approximate solution will be the same as in example 1.2., i.e.

$$u = \alpha_1 x(1 - x) + \alpha_2 x^2(1 - x) \tag{a}$$

which can be written as

$$u = \alpha_1 \phi_1 + \alpha_2 \phi_2 \tag{b}$$

where ϕ_1 and ϕ_2 are the shape functions ($\phi_1 = x(1 - x)$; $\phi_2 = x^2(1 - x)$). The residual is the same as previously, i.e.

$$\begin{aligned} R &= \frac{d^2u}{dx^2} + u + x \\ &= (-2 + x - x^2)\alpha_1 + (2 - 6x + x^2 - x^3)\alpha_2 + x \end{aligned} \tag{c}$$

The weighting function w in Galerkin is assumed to have the same shape function as the approximate solution (b), i.e.

$$w = \beta_1 \phi_1 + \beta_2 \phi_2 \tag{d}$$

The coefficients β_1 and β_2 are arbitrary.

The weighted residual statement is,

$$\int_0^1 Rw\,dx = 0 \tag{e}$$

which produces two integral expressions as β_1 and β_2 are arbitrary, i.e.

$$\int_0^1 R(\beta_1\phi_1 + \beta_2\phi_2)\,dx = 0 \tag{f}$$

or simply,

$$\int_0^1 R\phi_1\,dx = 0 \qquad \text{and} \qquad \int_0^1 R\phi_2\,dx = 0 \tag{g}$$

Substituting (c) and the functions ϕ_1 and ϕ_2 into (g) gives

$$\begin{aligned} &\int_0^1 [(-2 + x - x^2)\alpha_1 + (2 - 6x + x^2 - x^3)\alpha_2 + x][x(1-x)]\,dx = 0 \\ &\int_0^1 [(-2 + x - x^2)\alpha_1 + (2 - 6x + x^2 - x^3)\alpha_2 + x][x^2(1-x)]\,dx = 0 \end{aligned} \tag{h}$$

After integration this gives the following system

$$\begin{bmatrix} \frac{3}{10} & \frac{3}{10} \\ \frac{3}{10} & \frac{13}{105} \end{bmatrix} \begin{Bmatrix} \alpha_1 \\ \alpha_2 \end{Bmatrix} = \begin{Bmatrix} \frac{1}{12} \\ \frac{1}{20} \end{Bmatrix} \tag{i}$$

Notice that the matrix is symmetric because the equation is of an even order and the approximate and weighting functions are the same. Solving (i) gives

$$\alpha_1 = \tfrac{71}{369}; \qquad \alpha_2 = \tfrac{7}{41} \tag{j}$$

Substituting these values into (a) produces the approximate solution for u, i.e.

$$u = x(1-x)(\tfrac{71}{369} + \tfrac{7}{41}x) \tag{k}$$

One can also find the residual function R (equation c) which is now

$$R = \tfrac{1}{369}(-16 + 62x - 8x^2 - 63x^3)$$

The results for u and R are given in table 1.2 where they are compared against the exact solution of u. Notice that although the solution is overall more accurate than in the case of using the collocation technique, one now needs to carry out some integrations as shown in formula (h). This operation was not required for the case of point collocation.

Table 1.2 Results for Galerkin

x	u (exact)	u (approximate)	R
0.1	0.018641	0.018853	−0.026945
0.3	0.051194	0.051162	+0.000485
0.5	0.069746	0.069444	+0.013888
0.7	0.065582	0.065505	+0.005070
0.9	0.030901	0.031146	−0.034165

1.5 Weak Formulations

The fundamental integral statements of the boundary element and the finite element methods can be interpreted as a combination of a weighted residual statement and a process of integration by parts that reduces or 'weakens' the order of the continuity required for the u function.

If one returns to equation (1.12) with $b \equiv 0$ for simplicity, i.e.

$$\nabla^2 u = 0 \qquad \text{in } \Omega \tag{1.47}$$

one can write formula (1.25) as,

$$\int_\Omega (\nabla^2 u) w \, d\Omega - \int_{\Gamma_2} (q - \bar{q}) w \, d\Gamma + \int_{\Gamma_1} (u - \bar{u}) \frac{\partial w}{\partial n} \, d\Gamma = 0 \tag{1.48}$$

or in terms of residual functions,

$$\int_\Omega R w \, d\Omega - \int_{\Gamma_2} R_2 w \, d\Gamma + \int_{\Gamma_1} R_1 \frac{\partial w}{\partial n} \, d\Gamma = 0 \tag{1.49}$$

A special case of this equation is the case for which the function u exactly satisfies the 'essential' boundary conditions, $u = \bar{u}$ on Γ_1, which results in $R_1 \equiv 0$. In this case equation (1.49) becomes

$$\int_\Omega R w \, d\Omega = \int_{\Gamma_2} R_2 w \, d\Gamma \tag{1.50}$$

or,

$$\int_\Omega (\nabla^2 u) w \, d\Omega = \int_{\Gamma_2} (q - \bar{q}) w \, d\Gamma \tag{1.51}$$

A more usual form of this expression can be obtained by integrating by parts once which gives

$$-\int_\Omega \left(\frac{\partial u}{\partial x_1} \frac{\partial w}{\partial x_1} + \frac{\partial u}{\partial x_2} \frac{\partial w}{\partial x_2} \right) d\Omega = -\int_{\Gamma_2} \bar{q} w \, d\Gamma - \int_{\Gamma_1} q w \, d\Gamma \tag{1.52}$$

It should be pointed out that equation (1.52) could also be obtained by integrating by parts over the domain the weighted residual statement for $\nabla^2 u$ and then introducing the boundary conditions, i.e. starting with

$$\int_\Omega (\nabla^2 u) w \, d\Omega = 0 \tag{1.53}$$

one can integrate by parts once to produce the following expression,

$$-\int_\Omega \left(\frac{\partial u}{\partial x_1} \frac{\partial w}{\partial x_1} + \frac{\partial u}{\partial x_2} \frac{\partial w}{\partial x_2} \right) d\Omega + \int_\Gamma \frac{\partial u}{\partial n} w \, d\Gamma = 0 \tag{1.54}$$

Introducing then the corresponding boundary conditions in Γ ($\Gamma = \Gamma_1 + \Gamma_2$) results in equation (1.52).

The last term in equation (1.52) is usually forced to be identically equal to zero by the requirement that the w functions have to satisfy the Lagrangian version of the essential boundary conditions, or condition on Γ_1, i.e. $w \equiv 0$ on Γ_1. This gives a relationship well known in finite elements, i.e.

$$\int_\Omega \left(\frac{\partial u}{\partial x_1} \frac{\partial w}{\partial x_1} + \frac{\partial u}{\partial x_2} \frac{\partial w}{\partial x_2} \right) d\Omega = \int_{\Gamma_2} \bar{q} w \, d\Gamma \tag{1.55}$$

Equation (1.55) is usually interpreted in terms of virtual work or virtual power, by associating w with a virtual function. Notice that the integral on the left hand side is a measure of the internal virtual work and the one on the right the virtual work done by the external forces $\bar{q}$. Equation (1.55) is the starting point of most finite element schemes for Laplacian problems and is usually called a 'weak' variational formulation. The 'weakness' can be interpreted as due to two reasons, (i) the order of u function continuity has been reduced as its derivatives are now of a lower order (i.e. first rather than second order); (ii) satisfaction of the natural boundary conditions is done in an approximate rather than exact manner, which reduces the accuracy of boundary values of this variable. (Notice that R_2 is generally different from zero.)

The boundary element formulation can be interpreted as introducing a further formal step in the process of integration by parts on the derivatives of u, and consequently weakening the continuity requirements for u.

If one starts again from equation (1.48) and integrates by parts as before, the more complete expression obtained is as follows:

$$-\int_\Omega \left(\frac{\partial u}{\partial x_1} \frac{\partial w}{\partial x_1} + \frac{\partial u}{\partial x_2} \frac{\partial w}{\partial x_2} \right) d\Omega = -\int_{\Gamma_2} \bar{q} w \, d\Gamma - \int_{\Gamma_1} q w \, d\Gamma - \int_{\Gamma_1} (u - \bar{u}) \frac{\partial w}{\partial n} d\Gamma \tag{1.56}$$

Integrating again in order to eliminate all derivatives in u on the left hand side integral, one finds,

$$\int_\Omega (\nabla^2 w) u \, d\Omega = -\int_{\Gamma_2} \bar{q} w \, d\Gamma - \int_{\Gamma_1} q w \, d\Gamma + \int_{\Gamma_1} \bar{u} \frac{\partial w}{\partial n} d\Gamma + \int_{\Gamma_2} u \frac{\partial w}{\partial n} d\Gamma \tag{1.57}$$

This is the starting statement for the Boundary Element formulation of the Laplace equation. The same equation can be obtained starting from the integral of the weighted residual over the domain Ω (equation (1.53)), integrating by parts twice and *then* introducing the boundary conditions. The processes have already been shown from another field equation in formulae (1.13) to (1.16) and then (1.19) and (1.20), the only difference now being that b is zero.

Consider now equation (1.1) again to illustrate how a weak formulation can be used and the domain and boundary element statements are obtained. Let us start with equation (1.10) which was deduced from (1.1) by a process of integrations by parts and application of boundary conditions, i.e.

$$\int_0^1 \left\{\frac{d^2u}{dx^2} w + (\lambda^2 u - b)w\right\} dx - [(q-\bar{q})w]_{x=1} + \left[(u-\bar{u})\frac{dw}{dx}\right]_{x=0} = 0 \quad (1.58)$$

which can also be expressed in a more compact form in function of residuals, i.e.

$$\int_0^1 Rw\, dx - [R_2 w]_{x=1} + \left[R_1 \frac{dw}{dx}\right]_{x=0} = 0 \quad (1.59)$$

The function u will now be assumed to satisfy exactly the 'essential' boundary conditions $u = \bar{u}$ at $x = 0$. In this case (1.58) becomes,

$$\int_0^1 \left[\frac{d^2u}{dx^2} w + (\lambda^2 u - b)w\right] dx = [(q-\bar{q})w]_{x=1} \quad (1.60)$$

or in terms of (1.59), simply

$$\int_0^1 Rw\, dx = [R_2 w]_{x=1} \quad (1.61)$$

Integrating by parts equation (1.60) one can write

$$\int_0^1 \left\{-\frac{du}{dx}\frac{dw}{dx} + (\lambda^2 u - b)w\right\} dx = [qw]_{x=0} - [qw]_{x=1} \quad (1.62)$$

If the weighting function w is forced to satisfy the homogeneous version of the essential boundary conditions at $x = 0$, equation (1.62) becomes,

$$\int_0^1 \left\{-\frac{du}{dx}\frac{dw}{dx} + (\lambda^2 u - b)u\right\} dx = -[\bar{q}w]_{x=1} \quad (1.63)$$

which is analogous to equation (1.55) obtained for the Laplace field equation.

Notice that equation (1.59) can also be obtained by applying the boundary conditions into statement (1.3) and that this statement was simply obtained by integrating by parts weighted residual expression (1.2).

The Boundary Element type governing statement for the example under discussion is found by carrying out two consecutive integrations by parts of (1.10) and this gives the previously obtained formula (1.7), i.e.

$$\int_0^1 \left\{ u\frac{d^2w}{dx^2} + (\lambda^2 u - b)w \right\} dx + \{[\bar{q}w]_{x=1} - [qw]_{x=0}\}$$
$$-\left\{\left[u\frac{dw}{dx}\right]_{x=1} - \left[\bar{u}\frac{dw}{dx}\right]_{x=0}\right\} = 0 \qquad (1.64)$$

This expression could also have been obtained by carrying out a double integration by parts of the weighted residual equation (1.2) and applying afterwards the boundary conditions.

It is worth noting that both in this one dimensional example and the two dimensional Laplace equations, a Finite Element type statement has been obtained after the first integration by parts (equations (1.3) and (1.14)), and Boundary Element type integral equation after the second integration (equations (1.4) and (1.15)).

Example 1.4

In order to understand the effect of weak formulations on the satisfaction of boundary conditions, we will now consider again equation (1.1) but assume that the boundary conditions are of two types, i.e.

$$\begin{aligned} &\text{at } x = 0 \rightarrow u = 0 && \text{('essential' condition)} \\ &\text{at } x = 1 \rightarrow q = \frac{du}{dx} = \bar{q} && \text{('natural' condition)} \end{aligned} \qquad (a)$$

The expression previously used for the approximate values of u can not now be applied as the boundary conditions are different. Let us consider again the starting expression,

$$u = a_1 + a_2 x + a_3 x^2 + \ldots \qquad (b)$$

and satisfy exactly the essential condition, at $x = 0$, i.e.

$$\text{at } x = 0; \qquad u = a_1 \equiv 0 \qquad (c)$$

but not the natural condition.

Hence the approximate solution is now,

$$u = \alpha_1 x + \alpha_2 x^2 + \ldots \qquad (d)$$

where $\alpha_1 = a_2$, $\alpha_2 = a_3 \ldots$

The residual will now be different from the one in the previous examples, i.e.

$$R = \frac{d^2u}{dx^2} + u + x = \alpha_1 x + \alpha_2(2 + x^2) + x \tag{e}$$

The weighted residual statement has to include now the natural boundary condition R_2 residual which is not identically satisfied, i.e.

$$\int_0^1 Rw\, dx = [R_2 w]_{x=1} \tag{f}$$

or in expanded form,

$$\int_0^1 \left(\frac{d^2u}{dx^2} + u + x\right) w\, dx = [(q - \bar{q})w]_{x=1} \tag{g}$$

One can now solve equation (g) in its present form or reduce the order of derivatives in the domain and the number of terms on the right hand side by integrating by parts the d^2u/dx^2 term. This gives

$$\int_0^1 \left\{\frac{du}{dx}\frac{dw}{dx} - (u + x)w\right\} dx = [\bar{q}w]_{x=1} - [qw]_{x=0} \tag{h}$$

Notice that in Galerkin the weighting function w has the same shapes as the approximation for u (equation (d)). Hence for two terms,

$$u = \alpha_1\phi_1 + \alpha_2\phi_2 \tag{i}$$

and

$$w = \beta_1\phi_1 + \beta_2\phi_2$$

where

$$\phi_1 = x \quad \text{and} \quad \phi_2 = x^2.$$

Substituting these values into (h) one finds,

$$\int_0^1 \left\{\left(\alpha_1 \frac{d\phi_1}{dx} + \alpha_2 \frac{d\phi_2}{dx}\right)\left(\beta_1 \frac{d\phi_1}{dx} + \beta_2 \frac{d\phi_2}{dx}\right) - (\alpha_1\phi_1 + \alpha_2\phi_2 + x)(\beta_1\phi_1 + \beta_2\phi_2)\right\} dx$$

$$= [\bar{q}(\beta_1\phi_1 + \beta_2\phi_2)]_{x=1} \tag{j}$$

or

$$\int_0^1 \{(\alpha_1 + 2\alpha_2 x)(\beta_1 + 2\beta_2 x) - (\alpha_1 x + \alpha_2 x^2 + x)(\beta_1 x + \beta_2 x^2)\}\, dx$$

$$= [\bar{q}(\beta_1 + \beta_2)] \tag{k}$$

As the β_1 and β_2 terms are arbitrary this implies satisfaction of the following two equations,

$$\int_0^1 \{(\alpha_1 + 2\alpha_2 x) - (\alpha_1 x^2 + \alpha_2 x^3 + x^2)\}\, dx = \bar{q}$$
$$\int_0^1 \{2(\alpha_1 x + 2\alpha_2 x^2) - (\alpha_1 x^3 + \alpha_2 x^4 + x^3)\}\, dx = \bar{q} \tag{l}$$

Integrating the above equations and writing the results in matrix form one finds,

$$\begin{bmatrix} \frac{2}{3} & \frac{3}{4} \\ \frac{3}{4} & \frac{17}{15} \end{bmatrix} \begin{Bmatrix} \alpha_1 \\ \alpha_2 \end{Bmatrix} = \begin{Bmatrix} \bar{q} + \frac{1}{3} \\ \bar{q} + \frac{1}{4} \end{Bmatrix} \tag{m}$$

The values of α_1 and α_2 are,

$$\alpha_1 = 0.9859 + 1.9864\bar{q}$$
$$\alpha_2 = -0.4319 - 0.4322\bar{q} \tag{n}$$

Notice that an error will now appear when we try to compute the value of q at $x = 1$, i.e.

$$q = \left[\frac{du}{dx}\right]_{x=1} = \alpha_1 + 2\alpha_2 = 0.1221 + 1.122\bar{q} \tag{o}$$

and hence this value will never be equal to the applied $\bar{q}$, i.e.

$$q = 0.1221 + 1.122\bar{q} \neq \bar{q} \tag{p}$$

This peculiar result is characteristic of weak formulations such as those used in finite elements. Because of this approximate satisfaction of the natural boundary conditions, f.e. solutions used in engineering practice tend to give poor results for surface fluxes or tractions. The resulting errors in many cases 'pollute' the results to such an extent that the finite element solutions are unreliable for many cases of stress or flux concentration except when using very fine meshes.

Results for u and R are given in table 1.3 for the case in which $\bar{q} = 0$. The exact solution is

$$u = \frac{\sin x}{\cos 1} - x \tag{q}$$

Table 1.3 Results for Weak Formulation and Galerkin Method

x	u (exact)	u (approximate	R
0.1	0.084773	0.094271	−0.669519
0.3	0.246953	0.256899	−0.306901
0.5	0.387328	0.384975	0.021175
0.7	0.492328	0.478499	0.314699
0.9	0.549794	0.537471	0.573671

1.6 Boundary and Domain Solutions

In section 4 the weighted residual technique was classified into boundary, domain and mixed methods. Boundary methods were defined as those for which the assumed approximate solution satisfies the governing or field equation in such a way that the only unknowns of the problem remain on the boundary. The satisfaction of the field equation may be of its homogeneous form or a special form with a singular right hand side.

In the process of double integration described earlier one had transferred the derivatives of the approximate solution u to the weighting function w and so the conditions previously imposed on the former apply now to the latter. A boundary method can be obtained by choosing a weighting function w in either of the following two ways, i.e.

(i) By selecting a function w which satisfies the governing equation in its homogeneous form, or
(ii) By using special types of functions which satisfy those equations in a way that it is still possible to reduce the problems to the boundary only. The best known of the functions applied as right hand side of the equation in the second method are the Dirac delta functions which give simply a value at a point when integrated over the domain.

It is important however to realize that other functions could also be proposed and may be very appropriate for other cases, provided that they can be reduced to the boundary.

We will now apply both techniques to our simple equation (1.1), i.e.

$$\frac{d^2u}{dx^2} + \lambda^2 u - b(x) = 0 \tag{1.65}$$

or its weighted residual statement,

$$\int_0^1 \left[u\left(\frac{d^2w}{dx^2} + \lambda^2 w\right) - bw \right] dx + \left[\frac{du}{dx} w\right]_0^1 - \left[u \frac{dw}{dx}\right]_0^1 = 0 \tag{1.66}$$

The *first approach* implies that a solution is known such that

$$\frac{d^2w}{dx^2} + \lambda^2 w \equiv 0 \tag{1.67}$$

without taking into account the actual boundary conditions of the problem.

Hence statement (1.66) reduces to

$$-\int_0^1 bw\, dx + \left[\frac{du}{dx} w\right]_0^1 - \left[u \frac{dw}{dx}\right]_0^1 = 0 \tag{1.68}$$

This approach is associated with the method called Trefftz.

The *second approach* is based on a function w such that

$$\frac{d^2w}{dx^2} + \lambda^2 w = -\Delta_i \tag{1.69}$$

where Δ_i indicates the Dirac function such that

$$\Delta_i \begin{cases} \text{singular at the } x_i \text{ point with } \int\limits_{x_i-\varepsilon}^{x_i+\varepsilon} \Delta_i \, dx = 1 \\ = 0 \quad \text{at any other point} \end{cases} \tag{1.70}$$

Notice that in this case

$$\int\limits_0^1 \left[u\left(\frac{d^2w}{dx^2} + \lambda^2 w\right)\right] dx = -\int\limits_0^1 u\Delta_i \, dx = -u_i \tag{1.71}$$

where u_i represents the value of the function u at the point x_i. In this case equation (1.66) becomes,

$$-u_i - \int\limits_0^1 bw \, dx + \left[\frac{du}{dx} w\right]_0^1 - \left[u \frac{dw}{dx}\right]_0^1 = 0 \tag{1.72}$$

When the x_i point is chosen on the boundary, then equation (1.72) gives a relationship between boundary variables.

The *second* approach is the one usually applied in boundary elements where the function w is called the 'fundamental' solution of the governing equation, or solution of (1.69). Notice that this solution is obtained without taking into consideration the boundary conditions of the problem.

Domain solutions are obtained from weighted residual statements when the assumed approximate solutions do not satisfy the governing equations One can return to equation (1.1) which after integration by parts gives the following statement,

$$\int\limits_0^1 \left\{ -\frac{du}{dx}\frac{dw}{dx} + (\lambda^2 u - b)w \right\} dx + \left[\frac{du}{dx} w\right]_0^1 = 0 \tag{1.73}$$

This is a finite element type equation for which the last term can be found to be zero at the boundary points where $q = du/dx$ is unknown, by the requirement that $w \equiv 0$ there. Substituting an approximate solution u in terms of unknown coefficients and known weighting functions leads to a system of equations to solve the problem. Notice that in the case of finite elements the unknown function u is explicitly defined over all the domain.

Although the above remarks refer to the starting one dimensional equation (1.1) they also apply for the case of the Poisson equation (1.12) and the associated weighted residual statements (equations (1.14) and (1.16)). Similar considerations can be made for many other types of field equations.

Example 1.5

Let us now return to the same equation as defined in Example 1.1 and try to solve it using a weak formulation and considering boundary as well as domain techniques. The field equation is

$$\frac{d^2u}{dx^2}+x=0 \tag{a}$$

with boundary conditions,

$$u=0 \qquad \text{at } x=0 \text{ and } x=1 \tag{b}$$

(i) Boundary Solution. Homogeneous Approach

A weighting function which will satisfy the homogeneous version of equation (a), i.e.

$$\frac{d^2w}{dx^2}=0 \tag{c}$$

is the simple function

$$w=a_1x+a_2 \qquad \text{with } dw/dx=a_1 \tag{d}$$

Equation (1.68) can be written for the case $\lambda\equiv 0$ and $b=-x$ as,

$$\int_0^1 xw\,dx+[qw]_0^1-\left[u\frac{dw}{dx}\right]_0^1=0 \tag{e}$$

After substituting above the boundary conditions (b) and the expressions for w and dw/dx as given by (d) equation (e) becomes

$$\int_0^1 x(a_1x+a_2)\,dx+q_1(a_1+a_2)-q_0a_2=0 \tag{f}$$

As the above equation has to be satisfied for any arbitrary values of a_1 and a_2, it gives the following two expressions

$$\begin{aligned} q_1&=-\int_0^1 x^2\,dx=-\tfrac{1}{3} \\ q_1-q_0&=-\int_0^1 x\,dx=-\tfrac{1}{2} \end{aligned} \tag{g}$$

and hence,

$$q_0=\tfrac{1}{6} \tag{h}$$

These values of q at $x=0$ and $x=1$ which are now the problem unknowns, are in this case the exact values.

(ii) Boundary Solution. Singular Approach

The weighting function in this case is chosen such that

$$\frac{d^2w}{dx^2} + \Delta_i = 0 \tag{i}$$

A solution of equation (i) regardless of boundary conditions is

$$w = \begin{cases} x, & x \leqslant x_i \\ x_i, & x > x_i \end{cases} \tag{j}$$

Once the boundary conditions are applied, equation (1.72) becomes

$$-u_i + \int_0^1 xw\,dx + q_1 w_1 - q_0 w_0 = 0 \tag{k}$$

Taking into consideration that $w_0 \equiv 0$ and substituting the other values of w as given by (j) one finds,

$$u_i = \int_0^{x_i} x^2\,dx + \int_{x_i}^1 x_i x\,dx + q_1 x_i \tag{l}$$

or,

$$u_i = \frac{x_i}{2} - \frac{x_i^2}{6} + q_1 x_i \tag{m}$$

Notice that only one unknown (q_1) remains, since one of the boundary stresses disappeared because of the variation of the weighting function w. The value of q_1 can be determined by taking the coordinate $x_i = 1$, i.e.

$$u_1 = 0 = \tfrac{1}{2} - \tfrac{1}{6} + q_1 \tag{n}$$
$$\therefore\ q_1 = -\tfrac{1}{3}$$

which is the exact value in this case.

Any value of u inside the domain can be computed from (l), i.e.

$$u_{1/2} = \tfrac{3}{48} \tag{o}$$

which is also the exact solution.

If instead of the fundamental solution given by (j) one had chosen a fundamental solution that also satisfies the boundary conditions, then no unknowns would exist either in the domain or at the boundaries and the value of u at any point would be obtained by a single integration. Consider for instance the solution,

$$w = \begin{cases} (1-x_i)x & x \leqslant x_i \\ (1-x)x_i & x > x_i \end{cases} \tag{p}$$

This function satisfies (i) and the boundary conditions $w=0$ at $x=0$ and $x=1$ ($w_0=w_1=0$). Thus equation (k) gives

$$u_i=\int_0^1 xw\,dx=\int_0^{x_i}(1-x_i)x^2\,dx+\int_{x_i}^1 x_i(1-x)x\,dx \tag{q}$$

resulting in

$$u_i=\frac{x_i}{6}-\frac{x_i^2}{6} \tag{r}$$

which is the exact solution.

Fundamental solutions that satisfy the boundary conditions as well as the governing equations are called Green's functions.

(iii) Domain Solution

The weighted residual statement used here is the one resulting after one integration by parts has been carried out (equation (1.73)), i.e.

$$\int_0^1\left\{-\frac{du}{dx}\frac{dw}{dx}+xw\right\}dx+\left[\frac{du}{dx}w\right]_0^1=0 \tag{s}$$

The proposed approximate solution is the one in Example 1.1 which satisfies the boundary conditions, i.e.

$$u=\alpha_1\phi_1+\alpha_2\phi_2 \tag{t}$$

with

$$\phi_1=x-2x^2+x^3$$

$$\phi_2=x^3-x^2$$

$$\frac{d\phi_1}{dx}=3x^2-4x+1$$

$$\frac{d\phi_2}{dx}=3x^2-2x$$

where ϕ_1 and ϕ_2 are the Hermitian cubic polynomials.

Substituting these values into (s) and using Galerkin, i.e.

$$\psi_1=\phi_1;\qquad \psi_2=\phi_2 \tag{u}$$

one can write,

$$\int_0^1 [\alpha_1(3x^2 - 4x + 1) + \alpha_2(3x^2 - 2x)](3x^2 - 4x + 1)\,dx$$

$$= \int_0^1 x(x - 2x^2 + x^3)\,dx \tag{w}$$

$$\int_0^1 [\alpha_1(3x^2 - 4x + 1) + \alpha_2(3x^2 - 2x)](3x^2 - 2x)\,dx$$

$$= \int_0^1 x(x^3 - x^2)\,dx$$

which after integrating and solving also gives the exact solution, i.e.

$$\alpha_1 = \tfrac{1}{6}, \qquad \alpha_2 = -\tfrac{1}{3} \tag{x}$$

It is worth noticing that if the approximation used for the 'weak' formulation is the same as the one used for the original domain weighted residual equation the result will be the same in all cases. The advantage of the 'weak' formulation is that the order of the derivatives of u is in this case reduced and hence the order of derivability required by the assumed approximate solution is also reduced.

1.7 Concluding Remarks

This chapter has presented the Boundary Element Method as a weighted residual technique. This approach permits relation of the method to other numerical techniques and gives an easy way of introducing boundary elements.

For simplicity one dimensional problems have been discussed throughout the chapter to present the relationship between different integral statements and also between approximate techniques. The presentation was then extended to potential problems governed by the Laplace or Poisson's equations, which will be used in the next chapter. Some authors prefer to deduce the boundary integral equations from Green's theorem instead. Notice that this theorem has also been presented here (equation (1.18)) where it was shown that it can be derived from Lagrangian multipliers or basic residual type statements.

Later on a similar approach will be discussed for elasticity problems as shown in Chapter 3. The beauty of weighted residuals is that they are simple to use and can be applied for a wide range of problems, including some very complex non-linear and time dependent cases which are not discussed in this book.

Exercises

1.1. Solve $\frac{d^2u}{dx^2}+u+x=0$ with boundary conditions $u(0)=u(1)=0$ using a trial function of the form $u=a_0+a_1x+a_2x^2$ and point collocation for $x=1/2$. Plot the solution and compare it with that of example 1.2 of the text and the exact solution given by equation (b) of that example.

1.2. Solve $\frac{d^2u}{dx^2}=\exp(u)$ from $x=0$ to $x=1$with boundary conditions $u(0)=u(1)=0$ using the same trial function and collocation point of exercise 1.1.

1.3. Solve $\nabla^2u=0$ in the plane domain $0\leqslant x\leqslant 1$, $0\leqslant y<\infty$ with boundary conditions

$$u(0, y)=u(1, y)=0$$

$$u(x, \infty)=0$$

$$u(x, 0)=x(1-x)$$

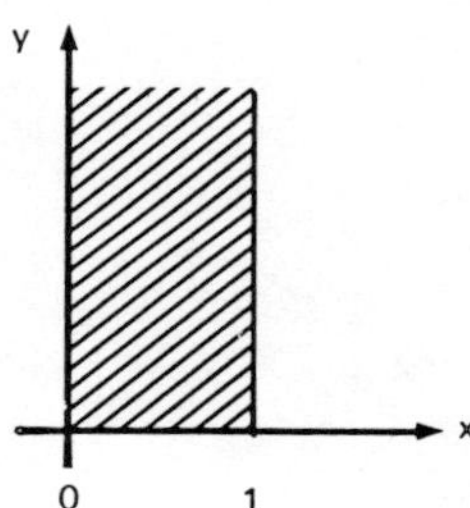

using a trial function of the form $u=A(y)x(x-1)$ and point collocation for $x=1/2$, $0\leqslant y<\infty$ as collocation point.

1.4. Solve exercise 1.3 using the Galerkin method with the integral and weighting function only along the x axis.

1.5. Solve the equation $\frac{d}{dx}\left[(1+u)\frac{du}{dx}\right]=0$ from $x=0$ to $x=1$ with boundary conditions $u(0)=0$ and $u(1)=1$ using $u=a_0+a_1x+a_2x^2$ as trial function and subdomain collocation with only one subdomain (x from 0 to 1).

1.6. The equation of the vertical displacement of a cable suspended between two points is $\frac{d^2u}{dx^2}+p(x)=0$ where $p(x)$ is the ratio between the distributed load and the horizontal force at the extremes.

Use the weak formulation and the homogeneous approach of boundary solution to compute the slope at the extremes for a cable that extends from $x=0$ to $x=1$ with boundary conditions $u(0)=u(1)=0$. The function $p(x)$ is given by

$$p(x)=0 \qquad 0\leqslant x<1/4$$

$$p(x)=1 \qquad 1/4\leqslant x<3/4$$

$$p(x)=0 \qquad 3/4\leqslant x<1$$

1.7. Using the same equation of exercise 1.6 and the singular approach of boundary solution compute the value of u at the mid-point.

1.8. The same as exercise 1.7 using a fundamental solution that also satisfies the boundary conditions (Green's function).

1.9. Solve the same equation of exercise 1.6 by means of a domain solution procedure and Galerkin. Use the following approximate solution

$$u = a_1 \sin \pi x + a_2 \sin 3\pi x$$

Chapter 2

Potential Problems

2.1 Introduction

Since the publication of the first book on Boundary Elements in 1978 [1] many such works have appeared in the literature, some dealing with potential, others with elastostatics problems as described in Chapter 3 and many other engineering applications. The importance of 1978 is that on that date publication of the first book on boundary elements coincided with holding the first conference in which the method was established, although the name in the context of the now classical B.E.M. appears to have been used for the first time in two papers by Brebbia and Dominguez and dated 1977 [2], [3]. Up to that time boundary integral equations solutions were almost exclusively the domain of mathematicians and physicists, with very little work being done to apply them to realistic engineering problems. Efforts such as the pioneering work by Hess and Smith [4] remained as special cases rather than being interpreted as a way of generating a whole new method of solutions for general engineering problems. Nevertheless Hess and Smith developed many powerful programs for the solution of Laplace type boundary-value problems which were applied to potential flow and arbitrary bodies, using what is now called the indirect boundary element technique. They extended their formulation to analyse three dimensional objects as well as two dimensional and their codes are still popular in aerodynamics. Equally important although in a different field was the work of Harrington and his collaborators [5], [6] who applied the technique to solve electrical engineering problems using more general impedance boundary conditions, of the type now called Robin or mixed conditions. They were still using the indirect formulation.

Boundary elements are usually associated with the direct formulation which is the one discussed most of the time in this book. This formulation in potential problems can be traced to Jaswon. As early as 1963 Jaswon [7] and Symm [8] presented a numerical technique to solve Fredholm boundary integral equations, which consisted of discretizing the boundary into a series of small segments (elements) and assuming a constant source density within each segment. They employed collocation to obtain the governing system of equations and compute the influence coefficients using numerical – Simpson rule – techniques, with the exception of the singular coefficients which were computed either analytically or by the summation of off-diagonal terms. They even proposed a more general formulation through the application of Green's third identity with potentials and their derivatives as boundary unknowns and results for this formulation were represented in [7] and [9]. All the bases of boundary elements were there but somehow their work failed to attract the attention it deserved, probably due to the simultaneous emergence of the finite element method.

Since 1978 the boundary element method is seen as related to other numerical techniques such as finite elements and finite differences, mainly through the work of Brebbia and his collaborators. This relationship is sometimes highlighted by using weighted residual or variational type techniques.

In this chapter the formulation of boundary elements for potential problems will be discussed. The method is presented using the same consideration of weighted residuals as used in Chapter 1. The numerical implementation of the method will be described in detail, and used in simple computer codes which can be run in the type of PC used by engineers. The chapter presents formulations using different types of elements, i.e. those with constant and linear variations. Special consideration is given to the proper treatment of the corner points. Quadratic elements are also discussed and implemented in a computer code.

Other sections in this chapter deal with the Poisson equation and the treatment of distributed sources, problems with more than one surface, multizones applications, anisotropy and the Helmholtz equations. Computer codes are not provided for these cases but the existing programs can be easily extended and the boundary element student can do this as a computational exercise.

2.2 Basic Integral Equation

The starting boundary integral equation required by the method can be deduced in a simple way based on considerations of weighted residuals, Betti's reciprocal theorem, Green's third identity or fundamental principles such as virtual work. The advantage of using weighted residuals is its generality; it permits the extension of the method to solve more complex partial differential equations. It can also be used to relate boundary elements to other numerical techniques and can be easily understood by engineers.

Consider that we are seeking to find the solution of a Laplace equation in a Ω (two or three dimensional) domain, (figure 2.1)

$$\nabla^2 u = 0 \quad \text{in } \Omega \tag{2.1}$$

with the following conditions on the Γ boundary

$$\begin{aligned} &\text{(i) '}\textit{Essential' Conditions}\text{ of the type } u = \bar{u} \text{ on } \Gamma_1 \\ &\text{(ii) '}\textit{Natural' Conditions}\text{ such as } q = \partial u/\partial n = \bar{q} \text{ on } \Gamma_2 \end{aligned} \tag{2.2}$$

where n is the normal to the boundary, $\Gamma = \Gamma_1 + \Gamma_2$ and the dashes indicate that those values are known. More complex boundary conditions such as combination of the above two, i.e.

$$\alpha u + \beta q = \gamma \tag{2.3}$$

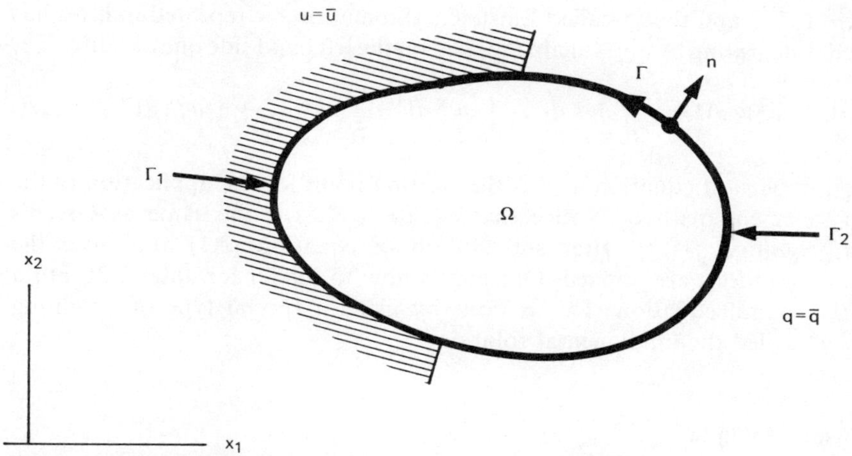

Figure 2.1 Geometrical definitions for Laplace equation

where α, β and γ are known parameters, can be easily included but they will not be considered now for simplicity's sake.

In principle the error introduced in the above equation if the exact (but unknown) values of u and q are replaced by an approximate solution can be minimized by orthogonalizing them with respect to a weighted function u^*, with derivatives on the boundary $q^* = \partial u^*/\partial n$.

In other words if R are the residuals, one can write in general that

$$\begin{aligned} R &= \nabla^2 u \neq 0 \\ R_1 &= u - \bar{u} \neq 0 \\ R_2 &= q - \bar{q} \neq 0 \end{aligned} \tag{2.4}$$

where u and q are approximate values. (The fact that one or more of the residuals may be identically zero does not detract from the generality of the argument.)

The weighting can now be carried out as shown in Chapter 1, i.e.

$$\int_{\Omega} R u^* \, d\Omega = \int_{\Gamma_2} R_2 u^* \, d\Gamma - \int_{\Gamma_1} R_1 q^* \, d\Gamma \tag{2.5}$$

or,

$$\int_{\Omega} (\nabla^2 u) u^* \, d\Omega = \int_{\Gamma_2} (q - \bar{q}) u^* \, d\Gamma - \int_{\Gamma_1} (u - \bar{u}) q^* \, d\Gamma \tag{2.6}$$

Integrating by parts the left hand side of this equation gives,

$$-\int_{\Omega} \left\{ \frac{\partial u}{\partial x_k} \frac{\partial u^*}{\partial x_k} \right\} d\Omega = -\int_{\Gamma_2} \bar{q} u^* \, d\Gamma - \int_{\Gamma_1} q u^* \, d\Gamma - \int_{\Gamma_1} u q^* \, d\Gamma + \int_{\Gamma_1} \bar{u} q^* \, d\Gamma \tag{2.7}$$

where $k = 1, 2, 3$ and the so-called Einstein's summation for repeated indexes has been used. Integrating by parts again the term on the left hand side one obtains,

$$\int_{\Omega} (\nabla^2 u^*) u \, d\Omega = -\int_{\Gamma_2} \bar{q} u^* \, d\Gamma - \int_{\Gamma_1} q u^* \, d\Gamma + \int_{\Gamma_2} u q^* \, d\Gamma + \int_{\Gamma_1} \bar{u} q^* \, d\Gamma \tag{2.8}$$

This is an important equation as it is the starting point for the application of the boundary element method. Notice that equation (2.8) is the same as Green's theorem (equation (1.18)) after substitution of equation (2.1) and once the boundary conditions are applied. Our aim is now to render formula (2.8) into a boundary integral equation. This is done by using a special type of weighting function u^* called the fundamental solution.

Fundamental Solution

The fundamental solution u^* satisfies Laplace's equation and represents the field generated by a concentrated unit charge acting at a point 'i'. The effect of this charge is propagated from i to infinity without any consideration of boundary conditions. Because of this the solution can be written

$$\nabla^2 u^* + \Delta^i = 0 \tag{2.9}$$

where Δ^i represents a Dirac Delta function which tends to infinity at the point $\mathbf{x} = \mathbf{x}^i$ and is equal to zero anywhere else. The integral of Δ^i however is equal to one. The use of Dirac delta function is an elegant way of representing unit concentrated charges as forces when dealing with differential equations.

The integral of a Dirac delta function multiplied by any other function is equal to the value of the latter at the point x^i. Hence

$$\int_{\Omega} u(\nabla^2 u^*) \, d\Omega = \int_{\Omega} u(-\Delta^i) \, d\Omega = -u^i \tag{2.10}$$

Equation (2.8) can now be written as,

$$u^i + \int_{\Gamma_2} u q^* \, d\Gamma + \int_{\Gamma_1} \bar{u} q^* \, d\Gamma = \int_{\Gamma_2} \bar{q} u^* \, d\Gamma + \int_{\Gamma_1} q u^* \, d\Gamma \tag{2.11}$$

It needs to be remembered that equation (2.11) applies for a concentrated charge at 'i' and consequently the values of u^* and q^* are those corresponding to that particular position of the charge. For each other $\mathbf{x}^i$ position one will find a new integral equation.

For an isotropic three dimensional medium the fundamental solution of equation (2.9) is

$$u^* = \frac{1}{4\pi r} \tag{2.12}$$

and for a two dimensional isotropic domain, it is

$$u^* = \frac{1}{2\pi} \ln\left(\frac{1}{r}\right) \tag{2.13}$$

where r is the distance from the point $\mathbf{x}^i$ of application of the delta function to any point under consideration.

It is easy to check that solution (2.12) and (2.13) satisfy the three and two dimensional Laplace equations. Consider for instance the three dimensional equation in terms of polar coordinates after neglecting terms which are zero due to symmetry of the solution, i.e.

$$\nabla^2 u^* \rightarrow \frac{\partial^2 u^*}{\partial r^2} + \frac{2}{r}\frac{\partial u^*}{\partial r} = -\Delta^i \tag{2.14}$$

Simply by substituting solution (2.12) into (2.14) we can check that the equation is satisfied for any value of r different from zero. For the case where $r \equiv 0$ we need to carry out the integration around a sphere of radius ε and then take ε to zero. Consider that the sphere has an Ω_ε domain, and integrate by parts to express the Laplacian in terms of boundary fluxes $\partial u^*/\partial n$, i.e.

$$\int_{\Omega_\varepsilon} (\nabla^2 u^*)\, d\Omega = \int_{\Gamma_\varepsilon} \left(\frac{\partial u^*}{\partial n}\right) d\Gamma = \int_{\Gamma_\varepsilon} \frac{\partial u^*}{\partial r}\, d\Gamma \tag{2.15}$$

Notice that $n \equiv r$ on the surface of the sphere.

Substituting now the fundamental solution (2.12) into (2.15) and making r (or ε) tend to zero gives

$$\begin{aligned}\lim_{\varepsilon\to 0}\left\{\int_{\Gamma_\varepsilon} \frac{\partial u^*}{\partial r}\, d\Gamma\right\} &= \lim_{\varepsilon\to 0}\left\{\int_{\Gamma_\varepsilon} -\frac{1}{4\pi\varepsilon^2}\, d\Gamma\right\} \\ &= \lim_{\varepsilon\to 0}\left\{-\frac{4\pi\varepsilon^2}{4\pi\varepsilon^2}\right\} = -1\end{aligned} \tag{2.16}$$

Notice that the surface of the sphere is $\Gamma_\varepsilon = 4\pi\varepsilon^2$. Similarly for the two dimensional case one can define a small circle of radius ε and then take the limit when $\varepsilon \rightarrow 0$, i.e.

$$\begin{aligned}\lim_{\varepsilon\to 0}\left\{\int_{\Gamma_\varepsilon} \frac{\partial u^*}{\partial n}\, d\Gamma\right\} &= \lim_{\varepsilon\to 0}\left\{\int_{\Gamma} -\frac{1}{2\pi r}\, d\Gamma\right\} \\ &= \lim_{\varepsilon\to 0}\left\{-\frac{2\pi\varepsilon}{2\pi\varepsilon}\right\} = -1\end{aligned} \tag{2.17}$$

Here the perimeter of the small circle is $\Gamma_\varepsilon = 2\pi\varepsilon$.

Boundary Integral Equation

We have now deduced an equation (2.11) which is valid for any point within the Ω domain. In boundary elements it is usually preferable for computational reasons to apply equation (2.11) on the boundary and hence we need to find out what happens when the point $\mathbf{x}^i$ is on Γ. A simple way to do this is to consider that the point i is on the boundary but the domain itself is augmented by a hemisphere of radius ε (in 3D) as shown in figure 2.2 (for 2D the same applies but we will consider a semicircle instead). The point $\mathbf{x}^i$ is considered to be at the centre and then the radius ε is taken to zero. The point will then become a boundary point and the resulting expression the specialization of (2.11) for a point on Γ. At present we will only consider smooth surfaces as represented in figure 2.2 and discuss the case of corners in other sections.

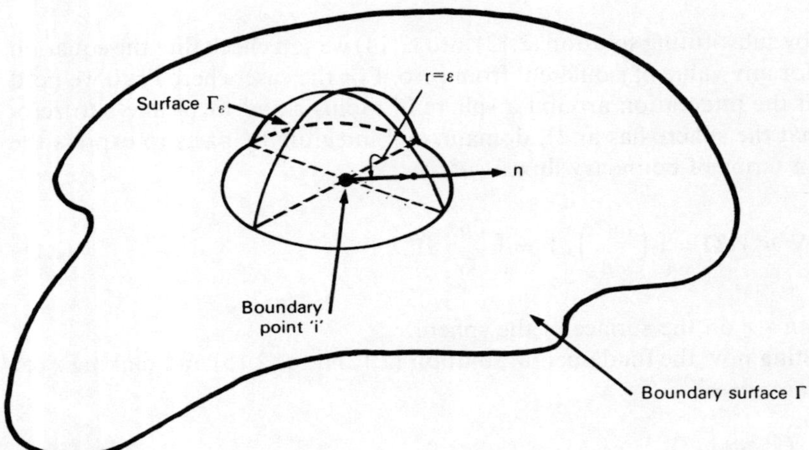

(i) Three Dimensional Case. Hemisphere around i

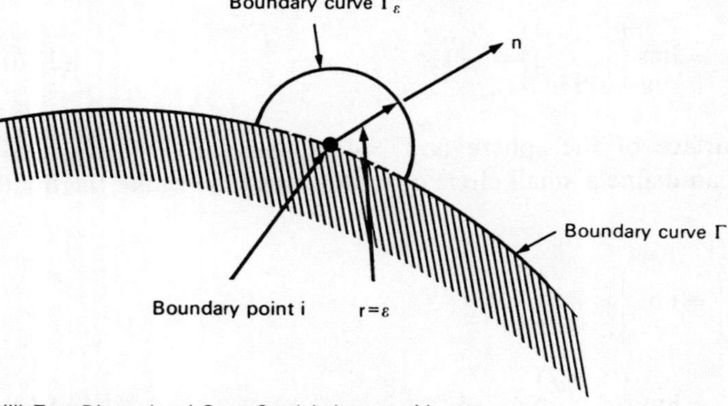

(ii) Two Dimensional Case. Semicircle around i

Figure 2.2 Boundary points for two and three dimensional case, augmented by a small hemisphere or semicircle

It is important at this stage to differentiate between two types of boundary integrals in (2.11) as the fundamental solution and its derivative behave differently. Consider for the sake of simplicity equation (2.11) before any boundary conditions have been applied, i.e.

$$u^i + \int_\Gamma uq^* \, d\Gamma = \int_\Gamma u^* q \, d\Gamma \tag{2.18}$$

Here $\Gamma = \Gamma_1 + \Gamma_2$ and satisfaction of the boundary conditions will be left for later on.

Integrals of the type shown on the right hand side of (2.18) are easy to deal with as they present a lower order singularity, i.e. for three dimensional cases the integral around Γ_ε gives:

$$\begin{aligned}\lim_{\varepsilon\to 0}\left\{\int_{\Gamma_\varepsilon} qu^* \, d\Gamma\right\} &= \lim_{\varepsilon\to 0}\left\{\int_{\Gamma_\varepsilon} q \frac{1}{4\pi\varepsilon} \, d\Gamma\right\} \\ &= \lim_{\varepsilon\to 0}\left\{q \frac{2\pi\varepsilon^2}{4\pi\varepsilon}\right\} \equiv 0\end{aligned} \tag{2.19}$$

In other words nothing occurs to the right hand side integral when (2.11) or (2.18) are taken to the boundary. The left hand side integral however behaves in a different manner. Here we have around Γ_ε the following result,

$$\begin{aligned}\lim_{\varepsilon\to 0}\left\{\int_{\Gamma_\varepsilon} uq^* \, d\Gamma\right\} &= \lim_{\varepsilon\to 0}\left\{-\int_{\Gamma_\varepsilon} u \frac{1}{4\pi\varepsilon^2} \, d\Gamma\right\} \\ &= \lim_{\varepsilon\to 0}\left\{-u \frac{2\pi\varepsilon^2}{4\pi\varepsilon^2}\right\} = -\tfrac{1}{2}u^i\end{aligned} \tag{2.20}$$

They produce what is called a free term. It is easy to check that the same will occur for two dimensional problems in which case the right hand side integral around Γ_ε is also identically equal to zero and the left hand side integral becomes

$$\begin{aligned}\lim_{\varepsilon\to 0}\left\{\int_{\Gamma_\varepsilon} uq^* \, d\Gamma\right\} &= \lim_{\varepsilon\to 0}\left\{-\int_{\Gamma_\varepsilon} u \frac{1}{2\pi\varepsilon} \, d\Gamma\right\} \\ &= \lim_{\varepsilon\to 0}\left\{-u \frac{\pi\varepsilon}{2\pi\varepsilon}\right\} = -\tfrac{1}{2}u^i\end{aligned} \tag{2.21}$$

From (2.19) to (2.21) one can write the following expression for two or three dimensional problems

$$\tfrac{1}{2}u^i + \int_\Gamma uq^* \, d\Gamma = \int_\Gamma qu^* \, d\Gamma \tag{2.22}$$

where the integrals are in the sense of Cauchy Principal Value.

This is the boundary integral equation generally used as a starting point for boundary elements.

2.3 The Boundary Element Method

Let us now consider how expression (2.22) can be discretized to find the system of equations from which the boundary values can be found. Assume for simplicity that the body is two dimensional and its boundary is divided into N segments or elements as shown in figure 2.3. The points where the unknown values are considered are called 'nodes' and taken to be in the middle of the element for the so-called 'constant' elements (figure 2.3(a)). These are going to be the elements considered in this section, but later on we will also discuss the case of linear elements, i.e those elements for which the nodes are at the extremes or ends (figure 2.3(b)) and curved elements such as the quadratic ones shown in figure 2.3(c) and for which a further mid-element node is required.

For the constant elements considered here the boundary is assumed to be divided into N elements. The values of u and q are assumed to be constant over each element and equal to the value at the mid-element node. Equation (2.22) can be discretized for a given point 'i' before applying any boundary conditions, as follows,

$$\tfrac{1}{2}u^i + \sum_{j=1}^{N} \int_{\Gamma_j} uq^* \, d\Gamma = \sum_{j=1}^{N} \int_{\Gamma_j} qu^* \, d\Gamma \tag{2.23}$$

The point i is one of the boundary nodes. Note that for this type of element (i.e. constant) the boundary is always 'smooth' as the node is at the centre of the element, hence the multiplier of u^i is $\frac{1}{2}$. Γ_j is the boundary of the 'j' element.

The u and q values can be taken out of the integrals as they are constant over each element. They will be called u^j and q^j for element 'j'. Hence

$$\tfrac{1}{2}u^i + \sum_{j=1}^{N} \left(\int_{\Gamma_j} q^* \, d\Gamma \right) u^j = \sum_{j=1}^{N} \left(\int_{\Gamma_j} u^* \, d\Gamma \right) q^j \tag{2.24}$$

Notice that there are now two types of integrals to be carried out over the elements, i.e. those of the following types

$$\int_{\Gamma_j} q^* \, d\Gamma \quad \text{and} \quad \int_{\Gamma_j} u^* \, d\Gamma$$

These integrals relate the 'i' node where the fundamental solution is acting to any other 'j' node. Because of this their resulting values are sometimes called influence coefficients. We will call them $\hat{H}^{ij}$ and G^{ij}, i.e.

$$\hat{H}^{ij} = \int_{\Gamma_j} q^* \, d\Gamma; \qquad G^{ij} = \int_{\Gamma_j} u^* \, d\Gamma \tag{2.25}$$

Notice that we are assuming throughout that the fundamental solution is applied at a particular 'i' node, although this is not explicitly indicated in u^*, q^* notation to avoid proliferation of indexes. Hence for a particular 'i' point one can write,

$$\tfrac{1}{2}u^i + \sum_{j=1}^{N} \hat{H}^{ij} u^j = \sum_{j=1}^{N} G^{ij} q^j \tag{2.26}$$

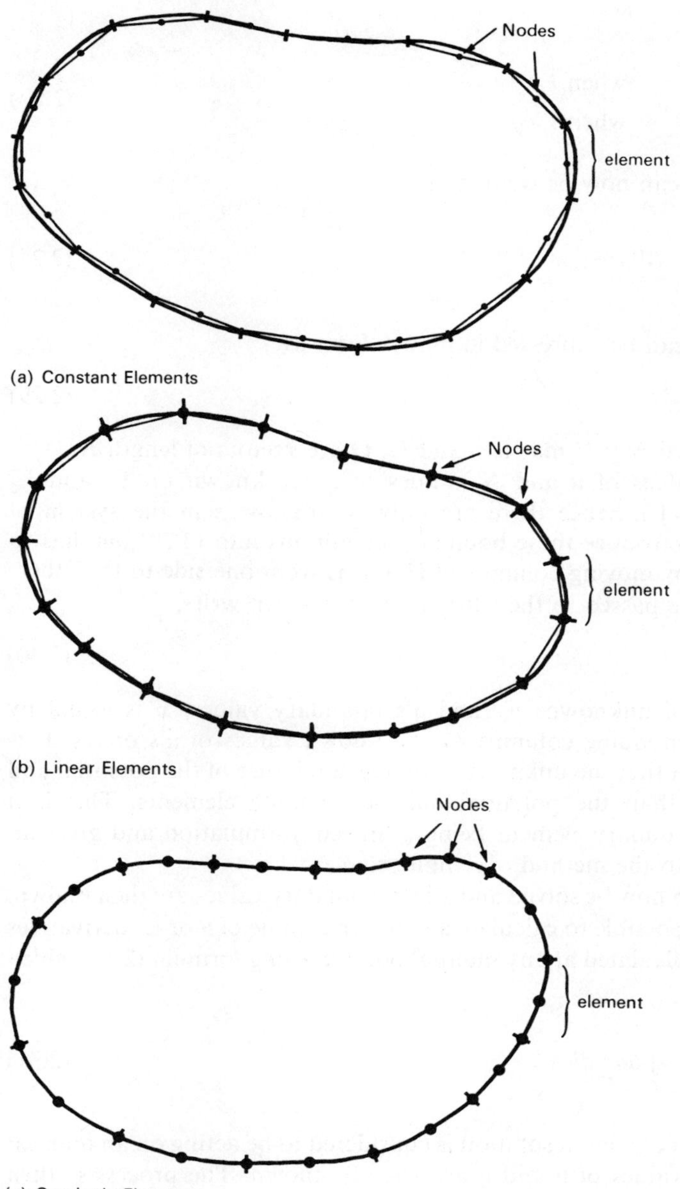

Figure 2.3 Different types of boundary elements

If we now assume that the position of i can also vary from 1 to N, i.e. we assume that the fundamental solution is applied at each node successively one obtains a system of equations resulting from applying (2.26) to each boundary point in turn.

Let us now call

$$H^{ij} = \begin{cases} \hat{H}^{ij} & \text{when } i \neq j \\ \hat{H}^{ij} + \frac{1}{2} & \text{when } i = j \end{cases} \tag{2.27}$$

hence equation (2.26) can now be written as

$$\sum_{j=1}^{N} H^{ij} u^j = \sum_{j=1}^{N} G^{ij} q^j \tag{2.28}$$

This set of equations can be expressed in matrix form as

$$\mathbf{HU} = \mathbf{GQ} \tag{2.29}$$

where **H** and **G** are two $N \times N$ matrices and **U**, **Q** are vectors of length N.

Notice that N_1 values of u and N_2 values of q are known on Γ_1 and Γ_2 respectively ($\Gamma_1 + \Gamma_2 = \Gamma$), hence there are only N unknowns in the system of equations (2.29). To introduce these boundary conditions into (2.29) one has to rearrange the system by moving columns of **H** and **G** from one side to the other. Once all unknowns are passed to the left-hand side one can write,

$$\mathbf{AX} = \mathbf{F} \tag{2.30}$$

where **X** is a vector of unknowns u's and q's boundary values. **F** is found by multiplying the corresponding columns by the known values of u's or q's. It is interesting to point out that the unknowns are now a mixture of the potential and its derivative, rather than the potential only as in finite elements. This is a consequence of the boundary element being a 'mixed' formulation and gives an important advantage to the method over finite elements.

Equation (2.30) can now be solved and all the boundary values are then known. Once this is done it is possible to calculate any internal value of u or its derivatives The values of u's are calculated at any internal point 'i' using formula (2.11) which can be written as,

$$u^i = \int_\Gamma q u^* \, d\Gamma - \int_\Gamma u q^* \, d\Gamma \tag{2.31}$$

Notice that now the fundamental solution is considered to be acting on an internal point 'i' and that all values of u and q are already known. The process is then one of integration (usually numerically). The same discretization is used for the boundary integrals, i.e.

$$u^i = \sum_{j=1}^{N} G^{ij} q^j - \sum_{j=1}^{N} \hat{H}^{ij} u^j \tag{2.32}$$

The coefficients G^{ij} and $\hat{H}^{ij}$ have been calculated anew for each different internal point.

The values of internal fluxes in the two directions, say x_1 and x_2, $q_{x_1} = \partial u/\partial x_1$ and $q_{x_2} = \partial u/\partial x_2$, are calculated by carrying out derivatives on (2.31), i.e.

$$\begin{aligned}(q_{x_1})^i &= \left(\frac{\partial u}{\partial x_1}\right)^i = \int_\Gamma q\left(\frac{\partial u^*}{\partial x_1}\right)^i d\Gamma - \int_\Gamma u\left(\frac{\partial q^*}{\partial x_1}\right)^i d\Gamma \\ (q_{x_2})^i &= \left(\frac{\partial u}{\partial x_2}\right)^i = \int_\Gamma q\left(\frac{\partial u^*}{\partial x_2}\right)^i d\Gamma - \int_\Gamma u\left(\frac{\partial q^*}{\partial x_2}\right)^i d\Gamma\end{aligned} \tag{2.33}$$

Notice that the derivatives are carried out only on the fundamental solution functions u^* and q^* as we are computing the variations of flux around the 'i' point.

Computation of integrals for internal points in (2.32) and (2.33) are usually carried out numerically.

Evaluation of Integrals

Integrals like G^{ij} and $\hat{H}^{ij}$ in the above expressions can be calculated using numerical integration formulae (such as Gauss quadrature rules) for the case $i \neq j$. For the element $i = j$ however the presence on that element of the singularity due to the fundamental solution requires a more accurate integration. For these integrals it is recommended to use higher-order integration rules or a special formula (such as logarithmic and other transformations which will be discussed later on).

For the particular case of constant elements however the $\hat{H}^{ij}$ and G^{ij} integrals can be computed analytically. The $\hat{H}^{ii}$ terms for instance are identically zero, as the normal n and the element coordinate are always perpendicular to each other, i.e.

$$\hat{H}^{ii} = \int_{\Gamma_i} q^* \, d\Gamma = \int_{\Gamma_i} \frac{\partial u^*}{\partial r}\frac{\partial r}{\partial n} \, d\Gamma \equiv 0 \tag{2.34}$$

The integrals in G^{ii} require special handling. For a two dimensional element for instance they are

$$G^{ii} = \int_{\Gamma_i} u^* \, d\Gamma = \frac{1}{2\pi} \int_{\Gamma_i} \ln\left(\frac{1}{r}\right) d\Gamma \tag{2.35}$$

In order to integrate easily the above expression one can change coordinates to a homogeneous ξ coordinate over the element (figure 2.4) such that,

$$r = \left|\xi \frac{l}{2}\right| \tag{2.36}$$

where l is the element length.

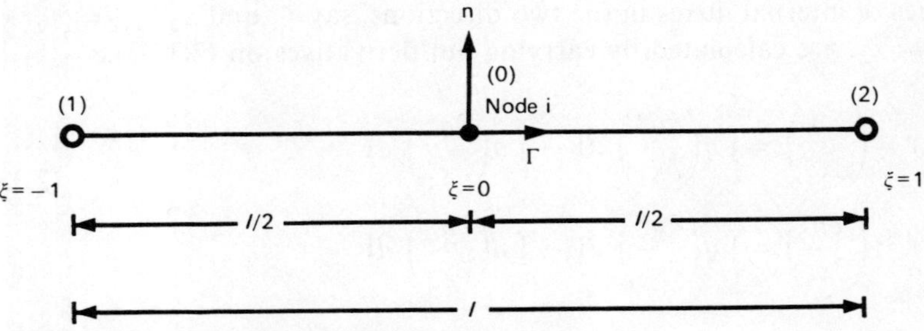

Figure 2.4 Element coordinate system

Hence taking into account symmetry (2.35) can be written

$$G^{ii} = \frac{1}{2\pi} \int_{\text{Point 1}}^{\text{Point 2}} \ln\left(\frac{1}{r}\right) d\Gamma = \frac{1}{\pi} \int_{\text{Node } i}^{\text{Point 2}} \ln\left(\frac{1}{r}\right) dr$$

$$= \frac{1}{\pi}\left(\frac{l}{2}\right) \int_0^1 \ln\left(\frac{1}{\xi l/2}\right) d\xi$$

$$= \frac{1}{\pi}\left(\frac{l}{2}\right)\left[\ln\left(\frac{1}{l/2}\right) + \int_0^1 \ln\left(\frac{1}{\xi}\right) d\xi\right] \quad (2.37)$$

The last integral is equal to 1, i.e.

$$G^{ii} = \frac{1}{\pi}\left(\frac{l}{2}\right)\left[\ln\left(\frac{1}{l/2}\right) + 1\right] \quad (2.38)$$

For more complex cases special weighted formulae are used. The other integrals (i.e. for $i \neq j$) can be calculated using simple Gauss quadrature rules. In the two dimensional codes described in this chapter a 4 points rule has been used (see Appendix A).

2.4 Computer Code for Potential Problems using Constant Elements (POCONBE)

In what follows the above theory will be employed to produce a simple computer code written in FORTRAN for solving Laplace type problems. The code is valid for isotropic materials and uses constant elements. The program can be run in any IBM/PC type XT or AT or compatibles.

Boundary Element codes are substantially different from Finite Element programs. Their internal organization is somewhat simpler as they do not require an assembler. They also produce all the boundary values (u's and q's) and give generally very precise solutions.

Main Program

The macro-flow diagram for the POCONBE boundary element code is shown in figure 2.5. The main program defines the maximum dimensions of the system of equations, which in this case is 100 and allocates the input channel 5 and the output channel 6. It calls the following five routines

INPUTPC – This routine reads the input to the program
GHMATPC – It forms the system matrices **H** and **G** and rearranges them in accordance with the boundary conditions into a matrix **A**. It also creates the right hand side vector **F**.
SLNPD – This is a subroutine for solving systems of equations, with pivoting.
INTERPC – This routine computes the values of potential at internal points.
OUTPTPC: Outputs the results.

The main routine also reads and open files for input and output.

The general integer variables used by the program are defined as follows.

N: Number of elements (equal number of nodes for constant elements).
L: Number of internal points where the function is calculated.

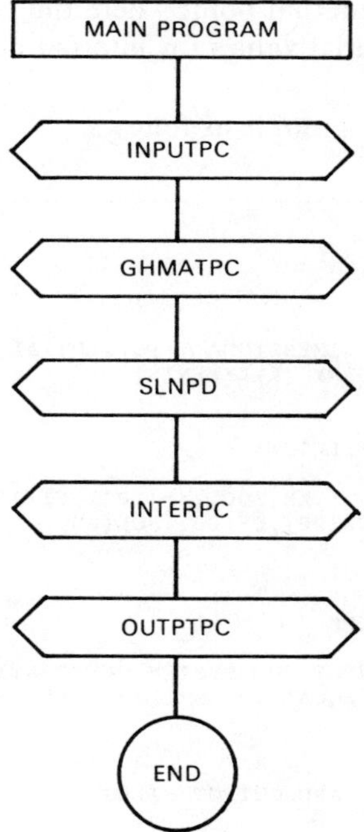

Figure 2.5 Macro flow diagram

KODE: One dimensional array indicating the type of boundary conditions at the nodes. KODE(J) = 0 means that the value of the potential is known at node J and KODE(J) = 1 signifies that the value of q is known at the corresponding boundary node.

The following real arrays are used to store data and results.

X: One dimensional array of x_1 coordinates extreme point of boundary elements.
Y: One dimensional array of x_2 coordinates extreme point of boundary elements.
XM: x_1 coordinates of the nodes. XM(J) contains the x_1 coordinate of node J.
YM: x_2 coordinates of the nodes. YM(J) contains the x_2 coordinates of node J.
G: Matrix defined in equation (2.29). After application of boundary conditions the matrix **A** is stored in the same location.
H: Matrix defined in equation (2.29).
FI: Prescribed value of boundary conditions. FI(J) contains the prescribed value of the condition at node J. If KODE(J) = 0 it means that the potential is prescribed and if KODE(J) = 1 that the q is given for the element associated with the location in those vectors.
DFI: Right hand side vector in equation (2.30). After solution it contains the values of the u's and q's unknowns.
CX: x_1 coordinate for internal point where the value of u is required.
CY: x_2 coordinate for internal point where the value of u is required.
SOL: Vector of the potential values for internal points.

The listing of the MAIN program is as follows:

```
C----------------------------------------------------------------------
      PROGRAM POCONBE
C
C  PROGRAM 1
C
C
C  THIS PROGRAM SOLVES TWO DIMENSIONAL (PO)TENTIAL PROBLEMS
C  USING (CON)STANT (B)OUNDARY (E)LEMENTS
C
C
      CHARACTER*10 FILEIN,FILEOUT
C
      DIMENSION X(101),Y(101),XM(100),YM(100),FI(100),DFI(100)
      DIMENSION KODE(100),CX(20),CY(20),SOL(20)
C
      COMMON/MATG/ G(100,100)
      COMMON/MATH/ H(100,100)
      COMMON       N,L,INP,IPR
C
C     SET MAXIMUN DIMENSION OF THE SYSTEM OF EQUATIONS (NX)
C     (THIS NUMBER MUST BE EQUAL OR SMALLER THAN THE DIMENSION OF XM, ETC...)
C
      NX=100
C
C  ASSIGN NUMBERS FOR INPUT AND OUTPUT FILES
C
      INP=5
      IPR=6
C
```

```
C  READ NAMES AND OPEN  FILES FOR INPUT AND OUTPUT
C
      WRITE(*,' (A) ') ' NAME OF INPUT FILE (MAX. 10 CHART.)'
      READ(*,'  (A) ')FILEIN
      OPEN(INP,FILE=FILEIN,STATUS='OLD')
      WRITE(*,' (A) ') ' NAME OF OUTPUT FILE (MAX. 10 CHART.)'
      READ(*,' (A) ')FILEOUT
      OPEN(IPR,FILE=FILEOUT,STATUS='NEW')
C
C  READ DATA
C
      CALL INPUTPC(CX,CY,X,Y,KODE,FI)
C
C  COMPUTE H AND G MATRICES AND FORM SYSTEM (A X = F)
C
      CALL GHMATPC(X,Y,XM,YM,G,H,FI,DFI,KODE,NX)
C
C  SOLVE SYSTEM OF EQUATIONS
C
      CALL SLNPD(G,DFI,D,N,NX)
C
C  COMPUTE  POTENTIAL VALUES AT INTERNAL POINTS
C
      CALL INTERPC(FI,DFI,KODE,CX,CY,X,Y,SOL)
C
C  PRINT RESULTS AT BOUNDARY NODES AND INTERNAL POINTS
C
      CALL OUTPTPC(XM,YM,FI,DFI,CX,CY,SOL)
C
C   CLOSE INPUT AND OUTPUT FILES
C
      CLOSE (INP)
      CLOSE (IPR)
      STOP
      END
```

Routine INPUTPC

All the input required by the program is read in the program INPUTPC and contained in a file whose name is requested by the main program. The file should contain the following input lines:

(1) *Title Line* One line containing the title of the problem.
(2) *Basic Parameter Line* One line containing the number of boundary elements and the number of internal points where the function is required.
(3) *Extreme Points of Boundary Elements Lines* The coordinates of the extreme of the elements read in counterclockwise direction for the case shown in figure 2.6(a) and in clockwise direction for 2.6(b).
(4) *Boundary Conditions Lines* As many lines as nodes giving the values of KODE and the value of the potential at the node if KODE = 0 or the value of the potential derivative if KODE = 1.
(5) *Internal Points Coordinates Lines* The $x_1 x_2$ coordinates of the internal points are read in free FORMAT in one or more lines.

This subroutine prints the title, the basic parameters the extreme points of the boundary elements and the boundary conditions. The internal point coordinates are printed in the OUTPTPC routine.

Notice that all input is written in free FORMAT.

```
C-----------------------------------------------------------------------
      SUBROUTINE INPUTPC(CX,CY,X,Y,KODE,FI)
C
C  PROGRAM 2
C
      CHARACTER*80 TITLE
      COMMON N,L,INP,IPR
      DIMENSION CX(1),CY(1),X(1),Y(1),KODE(1),FI(1)
C
C  N= NUMBER OF BOUNDARY NODES (=NUMBER OF ELEMENTS)
C  L= NUMBER OF INTERNAL POINTS WHERE THE FUNCTION IS CALCULATED
C
      WRITE(IPR,100)
  100 FORMAT(' ',79('*'))
C
C  READ JOB TITLE
C
      READ(INP,'(A)') TITLE
      WRITE(IPR,'(A)') TITLE
C
C  READ NUMBER OF BOUNDARY ELEMENTS AND INTERNAL POINTS
C
      READ(INP,*)N,L
      WRITE(IPR,300)N,L
  300 FORMAT(//' DATA'//2X,'NUMBER OF BOUNDARY ELEMENTS =',I3/2X,'NUMBER
     1 OF INTERNAL POINTS WHERE THE FUNCTION IS CALCULATED =',I3)
C
C
C  READ COORDINATES OF EXTREME POINTS OF THE BOUNDARY ELEMENTS
C  IN ARRAYS X AND Y
C
      WRITE(IPR,500)
  500 FORMAT(//2X,'COORDINATES OF THE EXTREME POINTS OF THE BOUNDARY ELE
     1MENTS',//1X,'POINT',7X,'X',15X,'Y')
      READ(INP,*) (X(I),Y(I),I=1,N)
      DO 10 I=1,N
   10 WRITE(IPR,700)I,X(I),Y(I)
  700 FORMAT(2X,I3,2(2X,E14.5))
C
C  READ BOUNDARY CONDITIONS IN FI(I) VECTOR, IF KODE(I)=0 THE FI(I)
C  VALUE IS A KNOWN POTENTIAL;IF KODE(I)=1 THE FI(I) VALUE IS A
C  KNOWN POTENTIAL DERIVATIVE (FLUX).
C
      WRITE(IPR,800)
  800 FORMAT(//2X,'BOUNDARY CONDITIONS'//2X,'NODE',6X,'CODE',7X,'PRESCRI
     1BED VALUE')
      DO 20 I=1,N
      READ(INP,*) KODE(I),FI(I)
   20 WRITE(IPR,950)I,KODE(I),FI(I)
  950 FORMAT(2X,I3,9X,I1,8X,E14.5)
C
C  READ COORDINATES OF THE INTERNAL POINTS
C
      IF(L.EQ.0) GO TO 30
      READ(INP,*) (CX(I),CY(I),I=1,L)
   30 RETURN
      END
```

Routine GHMATPC

The routine GHMATPC forms the **G** and **H** matrices of equation (2.29) through its subroutines EXTINPC and LOCINPC. It then rearranges their columns to form **A** matrix and **F** vector of (2.30).

The subroutines EXTINPC and LOCINPC perform the following functions.

EXTINPC: This subroutine computes the **H** and **G** matrix elements by means

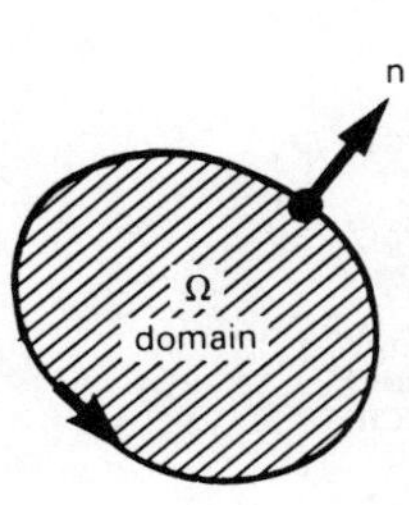

(a) Numbering direction for external surface. CLOSE DOMAIN (anticlockwise)

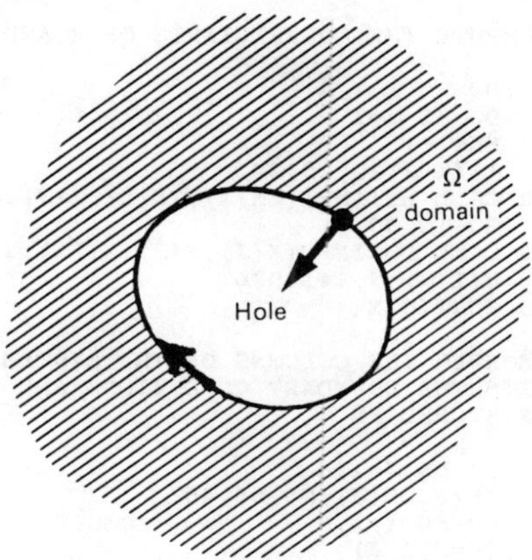

(b) Numbering direction for internal surface – OPEN DOMAIN (clockwise)

Figure 2.6 Numbering directions for external and internal surfaces

of numerical integration along the boundary elements (using 4 points Gauss quadrature).

It calculates all elements except those on the diagonal.

LOCINPC: Only calculates the diagonal elements of **G** matrix, given by equation (2.38).

Notice that as $\hat{H}_{ii} \equiv 0$ the diagonal elements of **H** are simply $\frac{1}{2}$.

It is also important to point out that as the fundamental solution has been taken as $\ln\left(\frac{1}{r}\right)$ – without $\frac{1}{2}\pi$ – all terms in **G** and **H** are effectively multiplied by 2π.

Rearranging the columns of **G** and **H** produces the matrix **A**, which is stored in the original space used by **G**. The columns of this matrix are the columns of **H** or **G** which are multiplied by unknown values of u or q. The right hand side vector **F** is called DFI in the code and is obtained by multiplying the columns of **G** or **H** by the known values (i.e. boundary condition values) of q or u respectively.

```
C-----------------------------------------------------------------------
      SUBROUTINE GHMATPC(X,Y,XM,YM,G,H,FI,DFI,KODE,NX)
C
C  PROGRAM 3
C
C  THIS SUBROUTINE COMPUTES THE G AND H MATRICES
C  AND FORMS THE SYSTEM OF EQUATIONS A X = F
C
      COMMON N,L,INP,IPR
      DIMENSION X(1),Y(1),XM(1),YM(1),FI(1),KODE(1)
      DIMENSION DFI(1),G(NX,NX),H(NX,NX)
C
C  COMPUTE THE NODAL COORDINATES AND STORE IN ARRAYS XM AND YM
C
      X(N+1)=X(1)
      Y(N+1)=Y(1)
      DO 10 I=1,N
      XM(I)=(X(I)+X(I+1))/2
   10 YM(I)=(Y(I)+Y(I+1))/2
C
```

```
C   COMPUTE THE COEFFICIENTS OF G AND H MATRICES
C
        DO 30 I=1,N
        DO 30 J=1,N
        KK=J+1
        IF(I-J)20,25,20
    20  CALL EXTINPC(XM(I),YM(I),X(J),Y(J),X(KK),Y(KK),H(I,J),G(I,J))
        GO TO 30
    25  CALL LOCINPC(X(J),Y(J),X(KK),Y(KK),G(I,J))
        H(I,J)=3.1415926
    30  CONTINUE
C
C   REORDER THE COLUMNS OF THE SYSTEM OF EQUATIONS IN ACCORDANCE
C   WITH THE BOUNDARY CONDITIONS AND FORM SYSTEM MATRIX A WHICH
C   IS STORED IN G
C
        DO 55 J=1,N
        IF(KODE(J))55,55,40
    40  DO 50 I=1,N
        CH=G(I,J)
        G(I,J)=-H(I,J)
        H(I,J)=-CH
    50  CONTINUE
    55  CONTINUE
C
C   FORM THE RIGHT HAND SIDE VECTOR F WHICH IS STORED IN DFI
C
        DO 60 I=1,N
        DFI(I)=0.
        DO 60 J=1,N
        DFI(I)=DFI(I)+H(I,J)*FI(J)
    60  CONTINUE
        RETURN
        END
```

Routine EXTINPC

The values of the off-diagonal coefficients of **H** and **G** are calculated using a 4-point Gauss integration formula (see Appendix A).

The variable *DIST* is the distance from the point under consideration to the line along the boundary element as shown in figure 2.7. If its direction is the same as that of the normal it is defined as positive, otherwise it is negative.

Consider now that instead of the system $x_1 - x_2$ we use an x–y system of coordinates. In this case $X1$, $X2$, $Y1$, $Y2$ are going to be the coordinates of the extreme points of each element considering them in clockwise (open domain) or anticlockwise manner (close domain).

Using numerical integration and changing to a dimensionless system of coordinates the G^{ij} and H^{ij} terms for each element and collocation point can be written as,

$$G^{ij} = \sum_{k=1}^{4} \ln\left(\frac{1}{(RA)_k}\right) w_k \frac{\sqrt{(X1-X2)^2+(Y1-Y2)^2}}{2}$$

$$H^{ij} = \sum_{k=1}^{4} \frac{d}{dn}\left(\ln\frac{1}{(RA)_k}\right) w_k \frac{\sqrt{(X1-X2)^2+(Y1-Y2)^2}}{2}$$

$$= \sum_{k=1}^{4} -\frac{1}{(RA)_k^2}(DIST) w_k \frac{\sqrt{(X1-X2)^2+(Y1-Y2)^2}}{2}$$

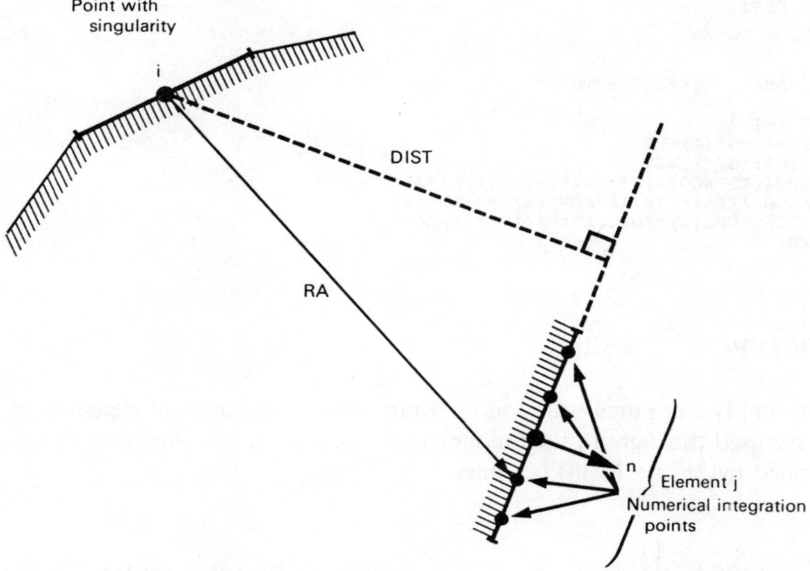

Figure 2.7 Geometrical definitions for numerical integration in routine EXTINPC

w_k are the weighting for each point. Their values and the location of the k points over the element are given in Appendix A.

```
C-----------------------------------------------------------------------
      SUBROUTINE EXTINPC(XP,YP,X1,Y1,X2,Y2,H,G)
C
C  PROGRAM 4
C
C  THIS SUBROUTINE COMPUTES THE OFF DIAGONAL COEFFICIENTS OF
C  H AND G MATRICES USING GAUSS QUADRATURE
C
C  RA=DISTANCE FROM THE COLLOCATION POINT TO THE
C  GAUSS INTEGRATION POINTS ON THE BOUNDARY ELEMENT
C  DIST=DISTANCE FROM THE COLLOCATION POINT TO THE LINE
C  TANGENT TO THE ELEMENT.
C
      DIMENSION XCO(4),YCO(4),GI(4),OME(4)
      DATA GI/0.86113631,-0.86113631,0.33998104,-0.33998104/
      DATA OME/0.34785485,0.34785485,0.65214515,0.65214515/
C
      AX=(X2-X1)/2.
      BX=(X2+X1)/2.
      AY=(Y2-Y1)/2.
      BY=(Y2+Y1)/2.
C
C  COMPUTE THE DISTANCE FROM THE POINT TO THE LINE OF THE ELEMENT
C
      IF(AX)10,20,10
   10 TA=AY/AX
      DIST=ABS((TA*XP-YP+Y1-TA*X1)/SQRT(TA**2+1.))
      GO TO 30
   20 DIST=ABS(XP-X1)
C
C  DETERMINE THE DIRECTION OF THE OUTWARD NORMAL
C
   30 SIG=(X1-XP)*(Y2-YP)-(X2-XP)*(Y1-YP)
      IF(SIG)31,32,32
```

```
   31 DIST=-DIST
   32 G=0.
      H=0.
C
C COMPUTE G AND H COEFFITIENTS
C
      DO 40 I=1,4
      XCO(I)=AX*GI(I)+BX
      YCO(I)=AY*GI(I)+BY
      RA=SQRT((XP-XCO(I))**2+(YP-YCO(I))**2)
      G=G+ALOG(1/RA)*OME(I)*SQRT(AX**2+AY**2)
   40 H=H-(DIST*OME(I)*SQRT(AX**2+AY**2)/RA**2)
      RETURN
      END
```

Routine LOCINPC

This routine simply computes equation (2.38) to obtain the diagonal elements of **G.** As we have used throughout the fundamental solution $\ln(1/r)$, the formula has to be multiplied by 2π, i.e. (2.38) becomes,

$$G^{ii} = l\left\{\ln\frac{1}{l/2} + 1\right\}$$

```
C-----------------------------------------------------------------------
      SUBROUTINE LOCINPC(X1,Y1,X2,Y2,G)
C
C   PROGRAM 5
C
C   THIS SUBROUTINE COMPUTES THE VALUES OF THE DIAGONAL
C   COEFFITIENTS OF THE G MATRIX
C
      AX=(X2-X1)/2.
      AY=(Y2-Y1)/2.
      SR=SQRT(AX**2+AY**2)
      G=2*SR*(1.-ALOG(SR))
      RETURN
      END
```

Routine SLNPD

This is a standard routine given in reference [10] which can solve the system of equations using pivoting if needed. If the matrix **A** has a zero in the diagonal it will interchange rows, deciding that the system matrix is singular only when no row interchange will produce a non-zero diagonal coefficient. If this happens it will give a message indicating a singularity in that row.

After elimination the results are stored in the same right hand side vector DFI.

```
C-----------------------------------------------------------------------
      SUBROUTINE SLNPD(A,B,D,N,NX)
C
C   PROGRAM 6
C
C SOLUTION OF LINEAR SYSTEMS OF EQUATIONS
C BY THE GAUSS ELIMINATION METHOD PROVIDING
C FOR INTERCHANGING ROWS WHEN ENCOUNTERING A
C ZERO DIAGONAL COEFICIENT
C
C A : SYSTEM MATRIX
```

```
C B : ORIGINALLY IT CONTAINS THE INDEPENDENT
C     COEFFICIENTS. AFTER SOLUTION IT CONTAINS
C     THE VALUES OF THE SYSTEM UNKNOWNS.
C
C N : ACTUAL NUMBER OF UNKNOWNS
C NX: ROW AND COLUMN DIMENSION OF A
C
      DIMENSION B(NX),A(NX,NX)
C
      TOL=1.E-6
C
      N1=N-1
      DO 100 K=1,N1
      K1=K+1
      C=A(K,K)
      IF(ABS(C)-TOL)1,1,3
    1 DO 7 J=K1,N
C
C TRY TO INTERCHANGE ROWS TO GET NON ZERO DIAGONAL COEFFICIENT
C
      IF(ABS((A(J,K)))-TOL)7,7,5
    5 DO 6 L=K,N
      C=A(K,L)
      A(K,L)=A(J,L)
    6 A(J,L)=C
      C=B(K)
      B(K)=B(J)
      B(J)=C
      C=A(K,K)
      GO TO 3
    7 CONTINUE
      GO TO 8
C
C DIVIDE ROW BY DIAGONAL COEFFICIENT
C
    3 C=A(K,K)
      DO 4 J=K1,N
    4 A(K,J)=A(K,J)/C
      B(K)=B(K)/C
C
C ELIMINATE UNKNOWN X(K) FROM ROW I
C
      DO 10 I=K1,N
      C=A(I,K)
      DO 9 J=K1,N
    9 A(I,J)=A(I,J)-C*A(K,J)
   10 B(I)=B(I)-C*B(K)
  100 CONTINUE
C
C COMPUTE LAST UNKNOWN
C
      IF(ABS((A(N,N)))-TOL)8,8,101
  101 B(N)=B(N)/A(N,N)
C
C APPLY BACKSUBSTITUTION PROCESS TO COMPUTE REMAINING UNKNOWNS
C
      DO 200 L=1,N1
      K=N-L
      K1=K+1
      DO 200 J=K1,N
  200 B(K)=B(K)-A(K,J)*B(J)
C
C COMPUTE VALUE OF DETERMINANT
C
      D=1.
      DO 250 I=1,N
  250 D=D*A(I,I)
      GO TO 300
    8 WRITE(IPR,2) K
    2 FORMAT(' **** SINGULARITY IN ROW',I5)
      D=0.
  300 RETURN
      END
```

Routine INTERPC

Subroutine INTERPC reorders FI (boundary condition vector) and DFI (unknown vector) in such a way that all the values of the potential are stored in FI and all the values of the derivatives or fluxes in DFI.

This subroutine also computes the potential values for the internal points using formula (2.32). Note that because all the H and G terms appear multiplied by 2π the solution for the internal points is also multiplied by 2π.

```
C-----------------------------------------------------------------------
      SUBROUTINE INTERPC(FI,DFI,KODE,CX,CY,X,Y,SOL)
C
C  PROGRAM 7
C
C  THIS SUBROUTINE COMPUTES THE VALUES OF POTENTIAL AT INTERNAL POINTS
C
      COMMON N,L,INP,IPR
      DIMENSION FI(1),DFI(1),KODE(1),CX(1),CY(1),X(1),Y(1),SOL(1)
C
C  REARRANGE THE FI AND DFI ARRAYS TO STORE ALL THE VALUES OF THE
C  POTENTIAL IN FI AND ALL THE VALUES OF THE DERIVATIVE IN DFI
C
      DO 20 I=1,N
      IF(KODE(I)) 20,20,10
   10 CH=FI(I)
      FI(I)=DFI(I)
      DFI(I)=CH
   20 CONTINUE
C
C  COMPUTE THE VALUES OF POTENTIAL AT INTERNAL POINTS
C
      IF(L.EQ.0) GO TO 50
      DO 40 K=1,L
      SOL(K)=0.
      DO 30 J=1,N
      KK=J+1
      CALL EXTINPC(CX(K),CY(K),X(J),Y(J),X(KK),Y(KK),A,B)
   30 SOL(K)=SOL(K)+DFI(J)*B-FI(J)*A
   40 SOL(K)=SOL(K)/(2*3.1415926)
   50 RETURN
      END
```

Routine OUTPTPC

This routine outputs the results. It first lists the coordinates of the boundary nodes and the corresponding values of potential and its derivatives (or fluxes). It also prints the values of potentials at internal points if any have been requested.

```
C-----------------------------------------------------------------------
      SUBROUTINE OUTPTPC(XM,YM,FI,DFI,CX,CY,SOL)
C
C  PROGRAM 8
C
C  THIS SUBROUTINE PRINTS THE VALUES OF THE POTENTIAL AND ITS NORMAL
C  DERIVATIVE AT BOUNDARY NODES. IT ALSO PRINTS THE VALUES OF THE
C  POTENTIAL AT INTERNAL POINTS
C

      COMMON N,L,INP,IPR
      DIMENSION XM(1),YM(1),FI(1),DFI(1),CX(1),CY(1),SOL(1)
C
      WRITE(IPR,100)
  100 FORMAT(' ',79('*')//1X,'RESULTS'//2X,'BOUNDARY NODES'//8X,'X',15
     1X,'Y',13X,'POTENTIAL',3X,'POTENTIAL DERIVATIVE'/)
```

```
      DO 10 I=1,N
   10 WRITE(IPR,200) XM(I),YM(I),FI(I),DFI(I)
  200 FORMAT(4(2X,E14.5))
C
      IF(L.EQ.0) GO TO 30
      WRITE(IPR,300)
  300 FORMAT(//,2X,'INTERNAL POINTS',//8X,'X',15X,'Y',13X,'POTENTIAL',/
     1)
      DO 20 K=1,L
   20 WRITE(IPR,400)CX(K),CY(K),SOL(K)
  400 FORMAT(3(2X,E14.5))
   30 WRITE(IPR,500)
  500 FORMAT(' ',79('*'))
      RETURN
       END
```

Example 2.1

The following example illustrates how the code can be used to analyse a simple potential problem. Consider the case of a square close domain of the type shown in figure 2.8, where the boundary has been discretized into 12 constant elements with 5 internal points.

The input statements are as follows:

HEAT FLOW EXAMPLE (DATA)

```
 HEAT FLOW EXAMPLE (12 CONSTANT ELEMENTS)
12 5
0. 0. 2. 0. 4. 0. 6. 0. 6. 2. 6. 4.
6. 6. 4. 6. 2. 6. 0. 6. 0. 4. 0. 2.
1 0
1 0
1 0
0 0
0 0
0 0
1 0
1 0
1 0
0 300
0 300
0 300
2. 2. 2. 4. 3. 3. 4. 2. 4. 4.
```

The results are printed out as follows.

HEAT FLOW EXAMPLE (OUTPUT)

```
*******************************************************************************
HEAT FLOW EXAMPLE (12 CONSTANT ELEMENTS)

DATA

 NUMBER OF BOUNDARY ELEMENTS = 12
 NUMBER OF INTERNAL POINTS WHERE THE FUNCTION IS CALCULATED =  5
```

```
COORDINATES OF THE EXTREME POINTS OF THE BOUNDARY ELEMENTS

POINT        X                  Y
   1     0.00000E+00       0.00000E+00
   2     0.20000E+01       0.00000E+00
   3     0.40000E+01       0.00000E+00
   4     0.60000E+01       0.00000E+00
   5     0.60000E+01       0.20000E+01
   6     0.60000E+01       0.40000E+01
   7     0.60000E+01       0.60000E+01
   8     0.40000E+01       0.60000E+01
   9     0.20000E+01       0.60000E+01
  10     0.00000E+00       0.60000E+01
  11     0.00000E+00       0.40000E+01
  12     0.00000E+00       0.20000E+01

BOUNDARY CONDITIONS

NODE      CODE        PRESCRIBED VALUE
  1         1            0.00000E+00
  2         1            0.00000E+00
  3         1            0.00000E+00
  4         0            0.00000E+00
  5         0            0.00000E+00
  6         0            0.00000E+00
  7         1            0.00000E+00
  8         1            0.00000E+00
  9         1            0.00000E+00
 10         0            0.30000E+03
 11         0            0.30000E+03
 12         0            0.30000E+03
*******************************************************************************

RESULTS
 BOUNDARY NODES

       X                 Y                 POTENTIAL     POTENTIAL DERIVATIVE

  0.10000E+01      0.00000E+00      0.25225E+03      0.00000E+00
  0.30000E+01      0.00000E+00      0.15002E+03      0.00000E+00
  0.50000E+01      0.00000E+00      0.47750E+02      0.00000E+00
  0.60000E+01      0.10000E+01      0.00000E+00     -0.52962E+02
  0.60000E+01      0.30000E+01      0.00000E+00     -0.48771E+02
  0.60000E+01      0.50000E+01      0.00000E+00     -0.52962E+02
  0.50000E+01      0.60000E+01      0.47750E+02      0.00000E+00
  0.30000E+01      0.60000E+01      0.15002E+03      0.00000E+00
  0.10000E+01      0.60000E+01      0.25225E+03      0.00000E+00
  0.00000E+00      0.50000E+01      0.30000E+03      0.52969E+02
  0.00000E+00      0.30000E+01      0.30000E+03      0.48737E+02
  0.00000E+00      0.10000E+01      0.30000E+03      0.52969E+02

 INTERNAL POINTS

       X                 Y                 POTENTIAL

  0.20000E+01      0.20000E+01      0.20028E+03
  0.20000E+01      0.40000E+01      0.20028E+03
  0.30000E+01      0.30000E+01      0.15001E+03
  0.40000E+01      0.20000E+01      0.99740E+02
  0.40000E+01      0.40000E+01      0.99740E+02
*******************************************************************************
```

Notice the excellent agreement of the results with the exact solution given in figure 2.8(a)), when the coarseness of the mesh and the simplicity of the model are considered. On the two vertical sides the fluxes are close to -50 and 50 as expected and on the horizontal sides the value of the potential is similar to the analytical solution which varies linearly from 300 on the left hand side to 0 on the right. The accuracy of the internal point results is however even more

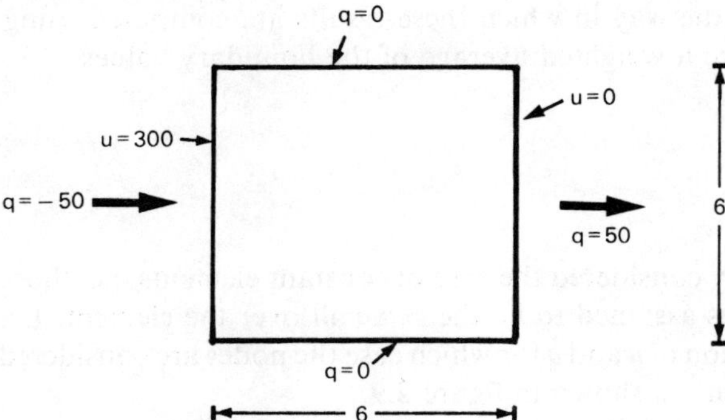

(a) Definition of the Problem

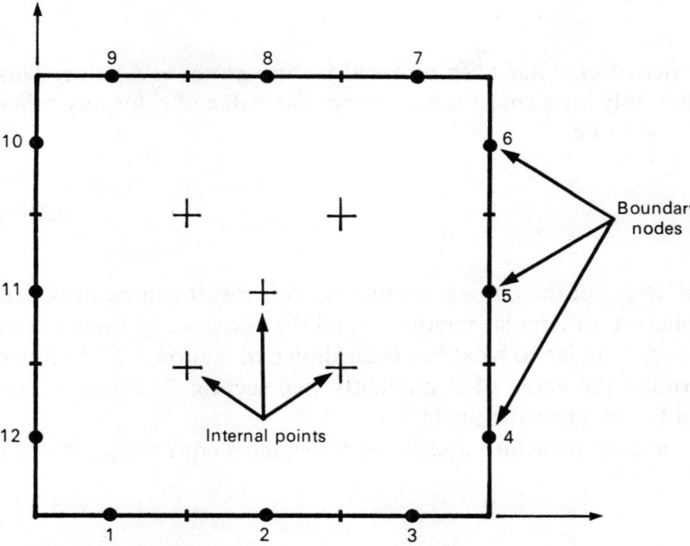

(b) Discretization into elements and internal nodes

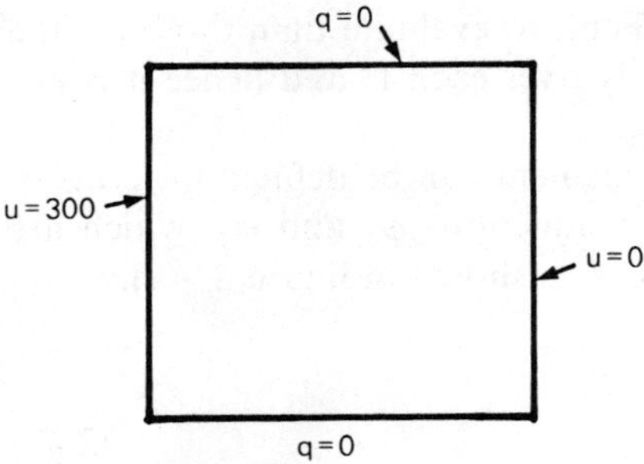

(c) Boundary conditions

Figure 2.8 Simple potential problem

remarkable and this is due to the way in which these results are computed using formula (2.32), i.e. they are like a weighted average of the boundary values.

2.5 Linear Elements

Up to this section we have only considered the case of constant elements, i.e. those with the values of the variables assumed to be the same all over the element. Let us now consider a linear variation of u and q for which case the nodes are considered to be at the ends of the element as shown in figure 2.9.

The governing integral statement can now be written as,

$$c^i u^i + \int_\Gamma u q^* \, d\Gamma = \int_\Gamma u^* q \, d\Gamma \tag{2.39}$$

Notice that the $\frac{1}{2}$ coefficient of u^i has been replaced by an unknown c^i value. This is because $c^i = \frac{1}{2}$ applies only for a smooth boundary. The value of c^i for any other boundary can be proved to be,

$$c^i = \frac{\theta}{2\pi} \tag{2.40}$$

where θ is the internal angle of the corner in radians. This result can be obtained by defining a small spherical or circular region around the corners and then taking the radius of them to zero (similar to what has been shown in section 2.2). Another possibility is to determine the value of c^i implicitly (see section 2.6) and in this case it is not required to calculate the angle.

After discretizing the boundary into a series of N elements equation (2.39) can be written

$$c^i u^i + \sum_{j=1}^{N} \int_{\Gamma_j} u q^* \, d\Gamma = \sum_{j=1}^{N} \int_{\Gamma_j} u^* q \, d\Gamma \tag{2.41}$$

The integrals in this equation are more difficult to evaluate than those for the constant element as the u's and q's vary linearly over each Γ_j and hence it is not possible to take them out of the integrals.

The values of u and q at any point on the element can be defined in terms of their nodal values and two linear interpolation functions ϕ_1 and ϕ_2, which are given in terms of the homogeneous coordinate ξ as shown in figure 2.9(a), i.e.

$$\begin{aligned} u(\xi) &= \phi_1 u^1 + \phi_2 u^2 = [\phi_1 \phi_2] \begin{Bmatrix} u^1 \\ u^2 \end{Bmatrix} \\ q(\xi) &= \phi_1 q^1 + \phi_2 q^2 = [\phi_1 \phi_2] \begin{Bmatrix} q^1 \\ q^2 \end{Bmatrix} \end{aligned} \tag{2.42}$$

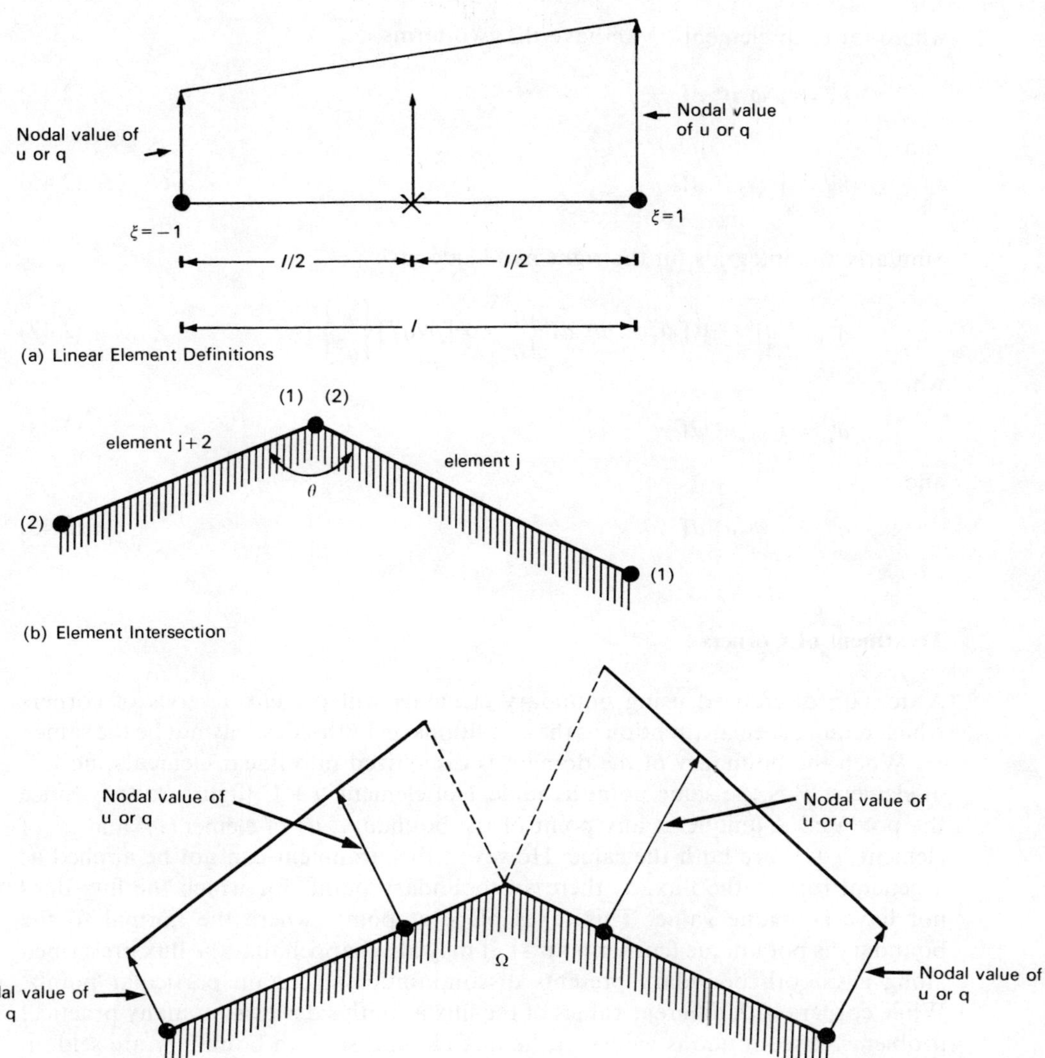

Figure 2.9 Linear element. Basic definitions and corner treatment

ξ is the dimensionless coordinate varying from -1 to $+1$ and the two interpolation functions are

$$\phi_1 = \tfrac{1}{2}(1-\xi); \qquad \phi_2 = \tfrac{1}{2}(1+\xi) \tag{2.43}$$

Let us consider the integrals over an element 'j'. Those on the left hand side can be written as,

$$\int_{\Gamma_j} uq^* \, d\Gamma = \int_{\Gamma_j} [\phi_1 \phi_2] q^* \, d\Gamma \begin{Bmatrix} u^1 \\ u^2 \end{Bmatrix} = [h_1^{ij} h_2^{ij}] \begin{Bmatrix} u^1 \\ u^2 \end{Bmatrix} \tag{2.44}$$

where for each element 'j' we have the two terms,

$$h_1^{ij} = \int_{\Gamma_j} \phi_1 q^* \, d\Gamma \tag{2.45}$$

and

$$h_2^{ij} = \int_{\Gamma_j} \phi_2 q^* \, d\Gamma \tag{2.46}$$

Similarly the integrals on the right hand side give

$$\int_{\Gamma_j} qu^* \, d\Gamma = \int_{\Gamma_j} [\phi_1 \phi_2] u^* \, d\Gamma \begin{Bmatrix} q^1 \\ q^2 \end{Bmatrix} = [g_1^{ij} g_2^{ij}] \begin{Bmatrix} q^1 \\ q^2 \end{Bmatrix} \tag{2.47}$$

where

$$g_1^{ij} = \int_{\Gamma_j} \phi_1 u^* \, d\Gamma \tag{2.48}$$

and

$$g_2^{ij} = \int_{\Gamma_j} \phi_2 u^* \, d\Gamma \tag{2.49}$$

Treatment of Corners

A domain discretized using boundary elements will present a series of corners which require special attention as the conditions on both sides may not be the same.

When the boundary of the domain is discretized into linear elements, node 2 of element 'j' is the same point as node 1 of element '$j+1$' (figure 2.9(b)). Since the potential is unique at any point of the boundary, u^2 of element 'j' and u^1 of element '$j+1$' are both the same. However, this argument can not be applied as a general rule to the flux, as there are boundary points for which the flux does not have a unique value. This takes place at points where the normal to the boundary is not unique (corner points). It may also happen that the flux prescribed along a smooth boundary presents discontinuities at certain particular points. While corners with different values of the flux at both sides exist in many practical problems, discontinuous values of the flux along a smooth boundary are seldom prescribed.

To take into account the possibility that the flux of node 2 of an element may be different from the flux of node 1 of the next element, the fluxes can be arranged in a $2n$ array.

Substituting equations (2.44) and (2.47) for all 'j' elements into (2.41) one obtains the following equation for node 'i'.

$$c^i u^i + [\hat{H}^{i1} \hat{H}^{i2} \ldots \hat{H}^{iN}] \begin{Bmatrix} u^1 \\ u^2 \\ \vdots \\ u^N \end{Bmatrix} = [G^{i1} G^{i2} \ldots G^{i2N}] \begin{Bmatrix} q^1 \\ q^2 \\ \vdots \\ q^{2N} \end{Bmatrix} \tag{2.50}$$

where $\hat{H}^{ij}$ is equal to the h_1^{ij} term of element 'j' plus the h_2^{ij-1} term of element '$j-1$'. Hence formula (2.50) represents the assembled equation for node 'i'. Note the simplicity of this approach. Equation (2.50) can be written as,

$$c^i u^i + \sum_{j=1}^{N} \hat{H}^{ij} u^j = \sum_{j=1}^{2N} G^{ij} q^j \tag{2.51}$$

Similarly, as was previously shown (equation (2.28)), this formula can be written as

$$\sum_{j=1}^{N} H^{ij} u^j = \sum_{j=1}^{2N} G^{ij} q^j \tag{2.52}$$

and the whole set in matrix form becomes

$$\mathbf{HU} = \mathbf{GQ} \tag{2.53}$$

where $\mathbf{G}$ is now an $N \times 2N$ rectangular matrix.

Several situations may occur at a boundary node: First that the boundary be smooth at the node. In such a case both fluxes 'before' and 'after' the node are the same unless they are prescribed as different, but in any case, only one variable will be unknown either the potential or the unique flux. Second, that the node is at a corner point. In this case four different cases are possible depending on the boundary conditions:

(a) Known values: fluxes 'before' and 'after' the corner.
Unknown value: potential
(b) Known values: potential, and flux 'before' the corner.
Unknown value: flux 'after' the corner
(c) Known values: potential, and flux 'after' the corner.
Unknown value: flux 'before' the corner
(d) Known values: potential.
Unknown values: flux 'before' and 'after' the corner.

There is only one unknown per node for the first three cases, and two unknowns for case (d). As long as there is only one unknown per node, system (2.53) can be reordered in such a way that all the unknowns are taken to the left hand side and obtain the usual system of $N \times N$ equations, i.e.

$$\mathbf{AX} = \mathbf{F} \tag{2.54}$$

where $\mathbf{X}$ is the (N) vector of unknowns; $\mathbf{A}$ is the $(N \times N)$ matrix of coefficients which columns are columns of the matrix $\mathbf{H}$, and columns of the matrix $\mathbf{G}$ after a change of sign or sum of two consecutive columns of $\mathbf{G}$ with opposite sign when the unknown is the unique value of the flux at the corresponding node. $\mathbf{F}$ is the known vector computed by the product of the known boundary conditions and the corresponding coefficients of the $\mathbf{G}$ or $\mathbf{H}$ matrices.

When the number of unknowns at a corner node is two (case (d)), one extra equation is needed for the node. The problem can be solved using the idea of 'discontinuous' elements [11]. The nodes here at the end of element 'j' and start of element '$j+1$' are shifted inside the two linear elements which meet at the corner and remain as two distinct nodes instead of joining into one at the corner (see figure 2.9(c)). Thus one equation can be written for each node. The potential and the flux are represented by linear functions along the whole elements in terms of their nodal values but they are in principle always discontinuous at the corner.

2.6 Computer Code for Potential Problems using Linear Elements (POLINBE)

Although this code has many routines which are similar to those developed for the constant element case (POCONBE), there are some parts which require modification.

Main Program

The integer variables have the same meaning as in the constant elements program. The same can be said for the real arrays except for the mid-point coordinates XM and YM that are not needed as now the nodes are at the inter-element junction. Arrays FI and DFI now have a different meaning. The dimension of FI is (N) while the dimension of DFI is ($2N$). Prescribed boundary conditions are read in DFI (two per element). FI is used as the right hand side vector that after solution contains the values of the unknowns. Finally both vectors are reordered to put all the values of the potentials in FI and all the values of the fluxes in DFI as was done for constant elements. Now, however FI contains one potential and DFI two fluxes, per node.

The program allows for the flux 'before' and 'after' any node to be different. When two, equal or different, fluxes are prescribed at a node the potential is computed; if the potential and one flux are prescribed, the other flux is computed; and in the case that only the potential is prescribed, both fluxes are considered to be equal. It should be noticed that in problems with only one uniform region, the case of potential prescribed and two different unknown values of the flux will only take place in a corner where the potential is prescribed along the two elements that join at that corner. This situation is not frequent and since the potential would be known along two different directions emerging from the corner, the potential derivatives along these two directions would be known and consequently the flux along any direction would also be known. Thus, the three variables would be known at that corner and hence any two of them can be prescribed and the third will be computed. Notice that only for the case of a singularity on the corner would one require replacing the corner node by two different nodes inside each of the two adjacent elements.

The listing is as follows:

```
C------------------------------------------------------------------------
C
      PROGRAM POLINBE
C
C  PROGRAM 9
C
C  THIS PROGRAM SOLVES TWO DIMENSIONAL (PO)TENTIAL PROBLEMS
C  USING (LIN)EAR (B)OUNDARY (E)LEMENTS
C
      CHARACTER*10 FILEIN,FILEOUT
      COMMON/MATG/ G(80,160)
      COMMON/MATH/ H(80,80)
      COMMON N,L,INP,IPR
      DIMENSION X(81),Y(81),FI(80),DFI(160)
      DIMENSION KODE(160),CX(20),CY(20),SOL(20)
C
C  SET MAXIMUN DIMENSION OF THE SYSTEM OF EQUATIONS (NX)
C  NX   = MAXIMUN NUMBER OF NODES = MAXIMUN NUMBER OF ELEMENTS
C
      NX=80
      NX1=2*NX
C
C  ASSIGN NUMBERS FOR INPUT AND OUTPUT FILES
C
      INP=5
      IPR=6
C
C  READ NAMES AND OPEN FILES FOR INPUT AND OUTPUT
C
      WRITE(*,' (A) ') ' NAME OF THE INPUT FILE (MAX. 10 CHART.)'
      READ(*,' (A) ') FILEIN
      OPEN(INP,FILE=FILEIN,STATUS='OLD')
      WRITE(*,' (A) ') ' NAME OF THE OUTPUT FILE (MAX.10 CHART.)'
      READ(*,' (A) ') FILEOUT
      OPEN(IPR,FILE=FILEOUT,STATUS='NEW')
C
C  READ DATA
C
      CALL INPUTPL(CX,CY,X,Y,KODE,DFI)
C
C  COMPUTE G AND H MATRICES AND FORM SYSTEM (A X = F)
C
      CALL GHMATPL(X,Y,G,H,FI,DFI,KODE,NX,NX1)
C
C  SOLVE SYSTEM OF EQUATIONS
C
      CALL SLNPD(H,FI,D,N,NX)
C
C  COMPUTE POTENTIAL VALUES AT INTERNAL POINTS
C
      CALL INTERPL(FI,DFI,KODE,CX,CY,X,Y,SOL)
C
C  PRINT RESULTS AT BOUNDARY NODES AND INTERNAL POINTS
C
      CALL OUTPTPL(X,Y,FI,DFI,CX,CY,SOL)
C
      CLOSE (INP)
      CLOSE (IPR)
      STOP
      END
```

Routine INPUTPL

The input subroutine is similar to INPUTPC in POCONBE. Only the boundary conditions are prescribed in a different way. Now, two boundary conditions per element are read in array DFI. Thus, each node 'j' may have a different value of the flux, one as the end node of element '$j-1$' and the other as the start node of element 'j'.

```
C-----------------------------------------------------------------------
      SUBROUTINE INPUTPL(CX,CY,X,Y,KODE,DFI)
C
C  PROGRAM 10
C
C  N= NUMBER OF BOUNDARY ELEMENTS
C  L= NUMBER OF INTERNAL POINTS
C
      CHARACTER*80 TITLE
      COMMON N,L,INP,IPR
      DIMENSION CX(1),CY(1),X(1),Y(1),KODE(1),DFI(1)
      WRITE(IPR,100)
  100 FORMAT(' ',79('*'))
C
C  READ JOB TITLE
C
      READ(INP,'(A)') TITLE
      WRITE(IPR,'(A)') TITLE
C
C  READ NUMBER OF ELEMENTS AND INTERNAL POINTS
C
      READ(INP,*)N,L
      WRITE(IPR,300)N,L
  300 FORMAT(//' DATA'//2X,'NUMBER OF BOUNDARY ELEMENTS =',I3/2X,'NUMBER
     1 OF INTERNAL POINTS WHERE THE FUNCTION IS CALCULATED =',I3)
C
C  READ BOUNDARY NODES COORDINATES IN ARRAYS X AND Y
C
      WRITE(IPR,500)
  500 FORMAT(//2X,'COORDINATES OF THE EXTREME POINTS OF THE BOUNDARY ELE
     1MENTS',//2X,'POINT',10X,'X',18X,'Y')
      READ(INP,*) (X(I),Y(I),I=1,N)
      DO 10 I=1,N
   10 WRITE(IPR,700)I,X(I),Y(I)
  700 FORMAT(3X,I3,2(5X,E14.5))
C
C  READ BOUNDARY CONDITIONS IN DFI(I) VECTOR.IF KODE(I)=0
C  THE DFI(I) VALUE IS A KNOWN POTENTIAL; IF KODE(I)=1 THE
C  DFI(I) VALUE IS A KNOWN POTENTIAL DERIVATIVE (FLUX).
C  TWO BOUNDARY CONDITIONS ARE READ PER ELEMENT.
C  ONE NODE MAY HAVE TWO DIFFERENT VALUES OF THE
C  POTENTIAL DERIVATIVE BUT ONLY ONE VALUE OF THE POTENTIAL
C
      WRITE(IPR,800)
  800 FORMAT(//2X,'BOUNDARY CONDITIONS'//15X,'------FIRST NODE-----',
     19X,'-----SECOND NODE-----'/17X,'PRESCRIBED',20X,'PRESCRIBED'/
     2,1X,'ELEMENT',12X,'VALUE',7X,'CODE',14X,'VALUE',7X,'CODE')
      DO 20 I=1,N
      READ(INP,*) KODE(2*I-1),DFI(2*I-1),KODE(2*I),DFI(2*I)
   20 WRITE(IPR,950)I,DFI(2*I-1),KODE(2*I-1),DFI(2*I),KODE(2*I)
  950 FORMAT(2X,I3,2(10X,E14.7,5X,I1))
C
C  READ COORDINATES OF THE INTERNAL POINTS
C
      IF(L.EQ.0) GO TO 30
      READ(INP,*) (CX(I),CY(I),I=1,L)
   30 RETURN
      END
```

Routine GHMATPL

Notice that this routine is similar to the one described in POCONBE with the main difference that the g^{ij} elements are assembled in a $N \times 2N$ matrix instead of an $N \times N$ one as was previously the case. This is because two possible values of flux are considered at each node, one to the left and the other to the right of it. Then the boundary conditions are applied as described earlier to rearrange the system of equations and prepare it for solving.

The diagonal terms in **H** are computed implicitly. Assuming a constant potential over the whole boundary the flux must be zero and hence

$$\mathbf{HI} = 0 \tag{2.55}$$

where **I** is a vector that for all nodes has a unit potential. Since (2.55) has to be satisfied

$$H^{ii} = -\sum_{j=1}^{N} H^{ij} \qquad (\text{for } j \neq i) \tag{2.56}$$

which gives the diagonal coefficients in terms of the rest of the terms of the **H** matrix.

The above considerations are strictly valid for close domains. When dealing with infinite or semi-infinite regions, equation (2.56) must be modified. If a unit potential is prescribed for a boundless domain the integral

$$\int_{\Gamma_\infty} p^* \, d\Gamma \tag{2.57}$$

over the external boundary Γ_∞ at infinity will now be zero and since p^* is due to a unit source, this integral must be (see equation (2.16) when $r \equiv \varepsilon \to \infty$)

$$\int_{\Gamma_\infty} p^* \, d\Gamma = -1 \tag{2.58}$$

The diagonal terms for this case are,

$$H^{ii} = 1 - \sum_{j=1}^{N} H^{ij} \qquad (\text{for } j \neq i) \tag{2.59}$$

```
C-----------------------------------------------------------------------
      SUBROUTINE GHMATPL(X,Y,G,H,FI,DFI,KODE,NX,NX1)
C
C  PROGRAM 11
C
C  THIS SUBROUTINE COMPUTES THE G AND H MATRICES
C  AND FORMS THE SYSTEM OF EQUATIONS A X = F
C  H IS A SQUARE MATRIX (N,N); G IS RECTANGULAR (N,2*N)
C
      COMMON N,L,INP,IPR
      DIMENSION X(1),Y(1),G(NX,NX1),H(NX,NX),FI(1),KODE(1),DFI(1)
      NN=2*N
      DO 10 I=1,N
      DO 6 J=1,N
    6 H(I,J)=0.
      DO 10 J=1,NN
   10 G(I,J)=0.
C
C  COMPUTE THE COEFFICIENTS OF G AND H MATRICES
C
      X(N+1)=X(1)
      Y(N+1)=Y(1)
      DO 100 I=1,N
      NF=I+1
      NS=I+N-2
      DO 50 JJ=NF,NS
      IF(JJ-N)30,30,20
```

```
   20 J=JJ-N
      GO TO 40
   30 J=JJ
   40 CALL EXTINPL(X(I),Y(I),X(J),Y(J),X(J+1),Y(J+1),A1,A2,B1,B2)
      IF(J-N)42,43,43
   42 H(I,J+1)=H(I,J+1)+A2
      GO TO 44
   43 H(I,1)=H(I,1)+A2
   44 H(I,J)=H(I,J)+A1
      G(I,2*J-1)=B1
      G(I,2*J)=B2
   50 H(I,I)=H(I,I)-A1-A2
      NF=I+N-1
      NS=I+N
      DO 95 JJ=NF,NS
      IF(JJ-N)70,70,60
   60 J=JJ-N
      GO TO 80
   70 J=JJ
   80 CALL LOCINPL(X(J),Y(J),X(J+1),Y(J+1),B1,B2)
      IF(JJ-NF)82,82,83
   82 CH=B1
      B1=B2
      B2=CH
   83 G(I,2*J-1)=B1
   95 G(I,2*J)=B2
C
C  ADD ONE TO THE DIAGONAL COEFFICIENTS
C  FOR EXTERNAL PROBLEMS.
C
      IF(H(I,I)) 98,100,100
   98 H(I,I)=1.+H(I,I)
  100 CONTINUE
C
C  REORDER THE COLUMNS OF THE SYSTEM OF EQUATIONS IN ACCORDANCE
C  WITH THE BOUNDARY CONDITIONS AND FORM THE SYSTEM MATRIX A
C  WHICH IS STORED IN H
C
      DO 155 I=1,N
      DO 150 J=1,2
      IF(KODE(2*I-2+J))110,110,150
  110 IF(I.NE.N .OR. J.NE.2) GO TO 125
      IF(KODE(1)) 115,115,113
  113 DO 114 K=1,N
      CH=H(K,1)
      H(K,1)=-G(K,2*N)
  114 G(K,2*N)=-CH
      GO TO 150
  115 DO 116 K=1,N
      H(K,1)=H(K,1)-G(K,2*N)
  116 G(K,2*N)=0.
      GO TO 150
  125 IF(I.EQ.1 .OR. J.GT.1 .OR. KODE(2*I-2).EQ.1) GO TO 130
      DO 129 K=1,N
      H(K,I)=H(K,I)-G(K,2*I-1)
  129 G(K,2*I-1)=0.
      GO TO 150
  130 DO 132 K=1,N
      CH=H(K,I-1+J)
      H(K,I-1+J)=-G(K,2*I-2+J)
  132 G(K,2*I-2+J)=-CH
  150 CONTINUE
  155 CONTINUE
C
C  FORM THE RIGHT HAND SIDE VECTOR F WHICH IS STORED IN FI
C
      DO 160 I=1,N
      FI(I)=0.
      DO 160 J=1,NN
      FI(I)=FI(I)+G(I,J)*DFI(J)
  160 CONTINUE
      RETURN
      END
```

Routine EXTINPL

This routine is similar to the one in POCONBE but instead of computing only one value per element for *G* and *H* coefficients as in POCONBE, it now computes two values per element, i.e. the part of *G* and *H* coefficients corresponding to the adjacent nodes.

```
C-----------------------------------------------------------------------
      SUBROUTINE EXTINPL(XP,YP,X1,Y1,X2,Y2,A1,A2,B1,B2)
C
C  PROGRAM 12
C
C  THIS SUBROUTINE COMPUTES THE G AND H COEFFITIENTS
C  THAT RELATE A NODE (XP,YP) WITH A BOUNDARY ELEMENT
C  USING GAUSS QUADRATURE
C
C  DIST=DISTANCE FROM THE COLOCATION POINT TO THE
C  LINE TANGENT TO THE ELEMENT.
C  RA=DISTANCE FROM THE COLOCATION POINT TO THE
C  GAUSS INTEGRATION POINTS ON THE BOUNDARY ELEMENTS
C
      DIMENSION XCO(4),YCO(4),GI(4),OME(4)
      DATA GI/0.86113631,-0.86113631,0.33998104,-0.33998104/
      DATA OME/0.34785485,0.34785485,0.65214515,0.65214515/
C
      AX=(X2-X1)/2
      BX=(X2+X1)/2
      AY=(Y2-Y1)/2
      BY=(Y2+Y1)/2
C
C  COMPUTE THE DISTANCE FROM THE POINT TO THE LINE OF THE ELEMENT
C
      IF(AX)10,20,10
   10 TA=AY/AX
      DIST=ABS((TA*XP-YP+Y1-TA*X1)/SQRT(TA**2+1))
      GO TO 30
   20 DIST=ABS(XP-X1)
C
C  DETERMINE THE DIRECTION OF THE OUTWARD NORMAL
C
   30 SIG=(X1-XP)*(Y2-YP)-(X2-XP)*(Y1-YP)
      IF(SIG)31,32,32
   31 DIST=-DIST
   32 A1=0.
      A2=0.
      B1=0.
      B2=0.
C
C  COMPUTE THE TERMS TO BE ADDED TO THE G AND H COEFFITIENTS
C
      DO 40 I=1,4
      XCO(I)=AX*GI(I)+BX
      YCO(I)=AY*GI(I)+BY
      RA=SQRT((XP-XCO(I))**2+(YP-YCO(I))**2)
      H=DIST*OME(I)*SQRT(AX**2+AY**2)/RA**2
      G=ALOG(1/RA)*OME(I)*SQRT(AX**2+AY**2)
      A1=A1+(GI(I)-1)*H/2
      A2=A2-(GI(I)+1)*H/2
      B1=B1-(GI(I)-1)*G/2
   40 B2=B2+(GI(I)+1)*G/2
      RETURN
      END
```

Routine LOCINPL

This routine computes now the part of the elements of the matrix **G** corresponding to the integrals along the elements which include the singularity.

```
C-----------------------------------------------------------------------
      SUBROUTINE LOCINPL(X1,Y1,X2,Y2,B1,B2)
C
C  PROGRAM 13
C
C  THIS SUBROUTINE COMPUTES THE PARTS OF THE G MATRIX
C  COEFFICIENTS CORRESPONDING TO INTEGRALS ALONG AN ELEMENT
C  THAT INCLUDES THE COLLOCATION POINT.
C
      SEP=SQRT((X2-X1)**2+(Y2-Y1)**2)
      B1=SEP*(1.5-ALOG(SEP))/2
      B2=SEP*(0.5-ALOG(SEP))/2
      RETURN
      END
```

Routine INTERPL

This routine replaces the INTERPC program used in POCONBE. It first arranges all potentials in FI and all their derivatives (or fluxes) in DFI and then computes the values of potentials at the internal points if requested.

```
C-----------------------------------------------------------------------
      SUBROUTINE INTERPL(FI,DFI,KODE,CX,CY,X,Y,SOL)
C
C  PROGRAM 14
C
C  THIS SUBROUTINE COMPUTES THE VALUES OF POTENTIAL AT
C  INTERNAL POINTS
C
      COMMON N,L,INP,IPR
      DIMENSION FI(1),DFI(1),KODE(1),CX(1),CY(1),X(1),Y(1),SOL(1)
C
C  REARRANGE THE FI AND DFI ARRAYS TO STORE ALL THE VALUES OF THE
C  POTENTIAL IN FI AND ALL THE VALUES OF THE DERIVATIVE IN DFI
C
      DO 155 I=1,N
      DO 150 J=1,2
      IF(KODE(2*I-2+J))110,110,150
  110 IF(I.NE.N .OR. J.NE.2) GO TO 125
      IF(KODE(1)) 114,114,113
  113 CH=FI(1)
      FI(1)=DFI(2*N)
      DFI(2*N)=CH
      GO TO 150
  114 DFI(2*N)=DFI(1)
      GO TO 150
  125 IF(I.EQ.1 .OR. J.EQ.2 .OR. KODE(2*I-2).EQ.1) GO TO 130
      DFI(2*I-1)=DFI(2*I-2)
      GO TO 150
  130 CH=FI(I-1+J)
      FI(I-1+J)=DFI(2*I-2+J)
      DFI(2*I-2+J)=CH
  150 CONTINUE
  155 CONTINUE
C
C  COMPUTE THE VALUES OF POTENTIAL AT INTERNAL POINTS
C
      IF(L.EQ.0) GO TO 50
      DO 40 K=1,L
      SOL(K)=0.
      DO 30 J=1,N
      CALL EXTINPL(CX(K),CY(K),X(J),Y(J),X(J+1),Y(J+1),A1,A2,B1,B2)
      IF(J-N)32,33,33
   32 SOL(K)=SOL(K)+DFI(2*J-1)*B1+DFI(2*J)*B2-FI(J)*A1-FI(J+1)*A2
      GO TO 30
   33 SOL(K)=SOL(K)+DFI(2*J-1)*B1+DFI(2*J)*B2-FI(J)*A1-FI(1)*A2
   30 CONTINUE
   40 SOL(K)=SOL(K)/(2*3.1415926)
   50 RETURN
      END
```

Routine OUTPTPL

This routine is similar to the one described in POCONBE but instead of printing the mid-point coordinates it now gives directly the values of the coordinates in the *X* and *Y* arrays. It also gives two values for the flux at each boundary node; one 'before' and the other 'after' the node.

```
C-------------------------------------------------------------------------
      SUBROUTINE OUTPTPL(X,Y,FI,DFI,CX,CY,SOL)
C
C  PROGRAM 15
C
C  THIS SUROUTINE PRINTS THE VALUES OF THE POTENTIAL AND ITS
C  NORMAL DERIVATIVE AT BOUNDARY NODES. IT ALSO PRINTS THE
C  VALUES OF THE POTENTIAL AT INTERNAL POINTS.
C
      COMMON N,L,INP,IPR
      DIMENSION X(1),Y(1),FI(1),DFI(1),CX(1),CY(1),SOL(1)
C
      WRITE(IPR,100)
  100 FORMAT(' ',79('*')//2X,'RESULTS'//2X,'BOUNDARY NODES'//
     156X,'POTENTIAL DERIVATIVE'/
     29X,'X',15X,'Y',12X,'POTENTIAL',6X,'BEFORE NODE',6X,'AFTER NODE'/)
      WRITE(IPR,200) X(1),Y(1),FI(1),DFI(2*N),DFI(1)
      DO 10 I=2,N
   10 WRITE(IPR,200) X(I),Y(I),FI(I),DFI(2*I-2),DFI(2*I-1)
  200 FORMAT(5(2X,E14.5))
C
      IF(L.EQ.0) GO TO 30
      WRITE(IPR,300)
  300 FORMAT(//,2X,'INTERNAL POINTS',//9X,'X',15X,'Y',12X,'POTENTIAL',/)
      DO 20 K=1,L
   20 WRITE(IPR,400)CX(K),CY(K),SOL(K)
  400 FORMAT(3(2X,E14.5))
   30 WRITE(IPR,500)
  500 FORMAT(' ',79('*'))
      RETURN
      END
```

Example 2.2

The potential problem solved with POCONBE will now be analysed using linear elements as shown in figure 2.10. The number of elements is still 12 for the linear (figure 2.10(b)) as well as the constant (figure 2.10(a)). Although the constant element solution gives reasonable agreement with the known results, the linear solution is identical to the analytical one within the precision limits of the computer. This result was to be expected since the exact solution varies linearly.

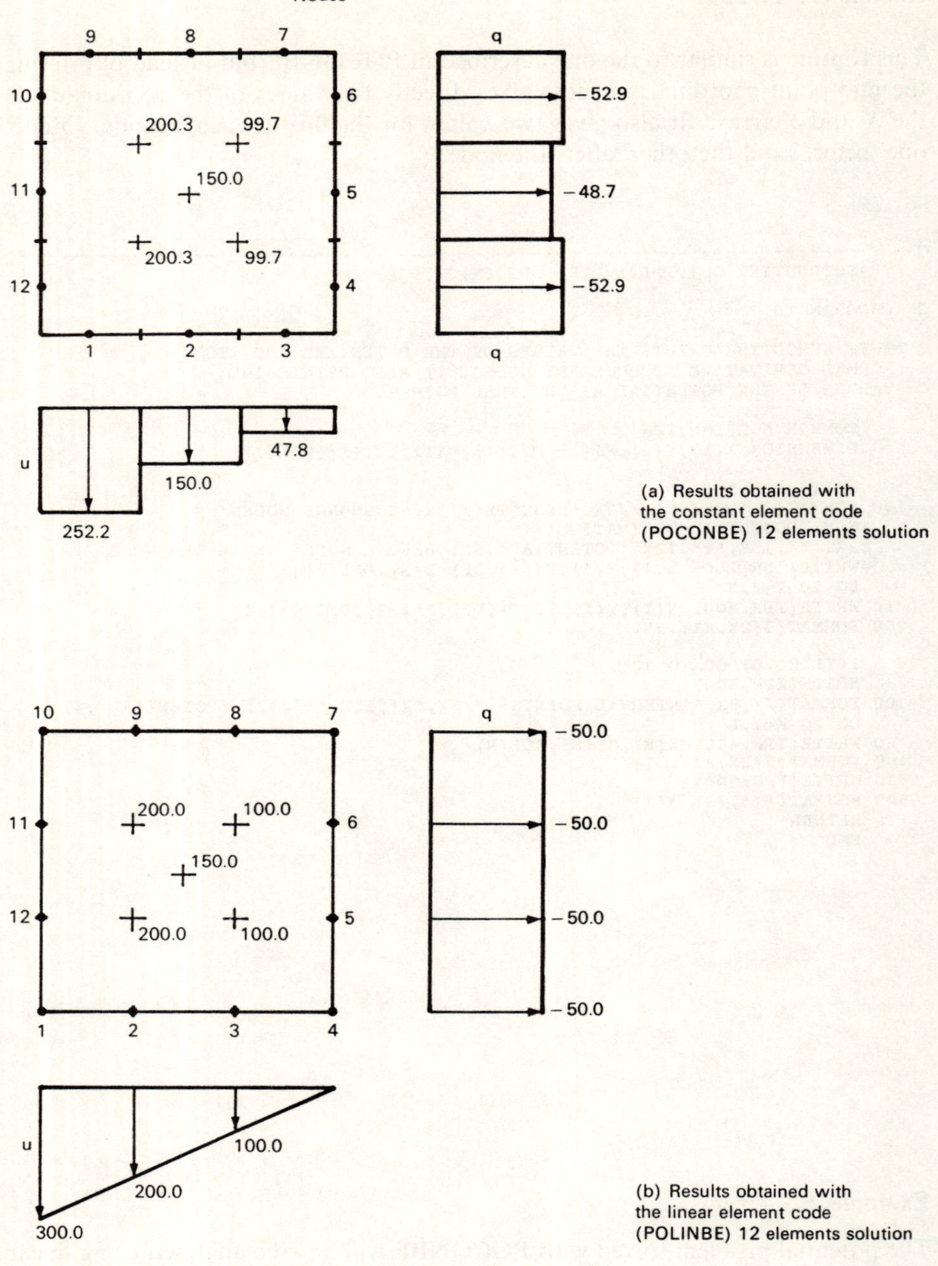

Figure 2.10 Results obtained using constant and linear elements for the heat flow example

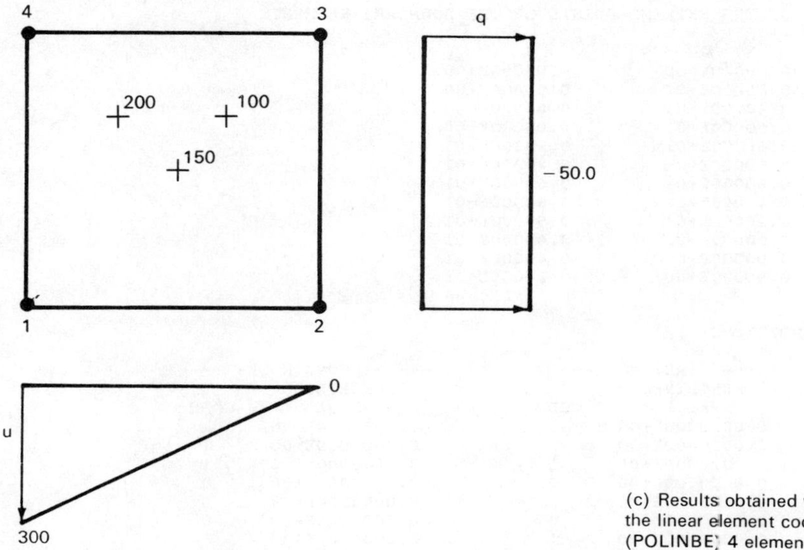

(c) Results obtained with the linear element code (POLINBE) 4 elements solution

Figure 2.10 *continued*

The data corresponding to the 12 element problem is as follows.

HEAT FLOW EXAMPLE (DATA)

```
  HEAT FLOW EXAMPLE (12 LINEAR ELEMENTS)
12 5
0. 0. 2. 0. 4. 0. 6. 0. 6. 2. 6. 4. 6. 6.
4. 6. 2. 6. 0. 6. 0. 4. 0. 2.
1 0. 1 0.
1 0. 1 0.
1 0. 1 0.
0 0. 0 0.
0 0. 0 0.
0 0. 0 0.
1 0. 1 0.
1 0. 1 0.
1 0. 1 0.
0 300. 0 300.
0 300. 0 300.
0 300. 0 300.
2. 2. 2. 4. 3. 3. 4. 2. 4. 4.
```

and the output is given by

HEAT FLOW EXAMPLE (OUTPUT)

```
*******************************************************************************
 HEAT FLOW EXAMPLE (12 LINEAR ELEMENTS)


DATA

 NUMBER OF BOUNDARY ELEMENTS = 12
 NUMBER OF INTERNAL POINTS WHERE THE FUNCTION IS CALCULATED =  5
```

```
COORDINATES OF THE EXTREME POINTS OF THE BOUNDARY ELEMENTS

POINT          X                 Y
   1      0.00000E+00      0.00000E+00
   2      0.20000E+01      0.00000E+00
   3      0.40000E+01      0.00000E+00
   4      0.60000E+01      0.00000E+00
   5      0.60000E+01      0.20000E+01
   6      0.60000E+01      0.40000E+01
   7      0.60000E+01      0.60000E+01
   8      0.40000E+01      0.60000E+01
   9      0.20000E+01      0.60000E+01
  10      0.00000E+00      0.60000E+01
  11      0.00000E+00      0.40000E+01
  12      0.00000E+00      0.20000E+01

BOUNDARY CONDITIONS

             ------FIRST NODE-----            -----SECOND NODE-----
               PRESCRIBED                       PRESCRIBED
ELEMENT          VALUE          CODE              VALUE          CODE
   1        0.0000000E+00        1           0.0000000E+00        1
   2        0.0000000E+00        1           0.0000000E+00        1
   3        0.0000000E+00        1           0.0000000E+00        1
   4        0.0000000E+00        0           0.0000000E+00        0
   5        0.0000000E+00        0           0.0000000E+00        0
   6        0.0000000E+00        0           0.0000000E+00        0
   7        0.0000000E+00        1           0.0000000E+00        1
   8        0.0000000E+00        1           0.0000000E+00        1
   9        0.0000000E+00        1           0.0000000E+00        1
  10        0.3000000E+03        0           0.3000000E+03        0
  11        0.3000000E+03        0           0.3000000E+03        0
  12        0.3000000E+03        0           0.3000000E+03        0
******************************************************************************
RESULTS

BOUNDARY NODES

                                                     POTENTIAL DERIVATIVE
      X               Y            POTENTIAL       BEFORE NODE      AFTER NODE

 0.00000E+00     0.00000E+00     0.30000E+03     0.50000E+02     0.00000E+00
 0.20000E+01     0.00000E+00     0.20000E+03     0.00000E+00     0.00000E+00
 0.40000E+01     0.00000E+00     0.10000E+03     0.00000E+00     0.00000E+00
 0.60000E+01     0.00000E+00     0.00000E+00     0.00000E+00    -0.50000E+02
 0.60000E+01     0.20000E+01     0.00000E+00    -0.50000E+02    -0.50000E+02
 0.60000E+01     0.40000E+01     0.00000E+00    -0.50000E+02    -0.50000E+02
 0.60000E+01     0.60000E+01     0.00000E+00    -0.50000E+02     0.00000E+00
 0.40000E+01     0.60000E+01     0.10000E+03     0.00000E+00     0.00000E+00
 0.20000E+01     0.60000E+01     0.20000E+03     0.00000E+00     0.00000E+00
 0.00000E+00     0.60000E+01     0.30000E+03     0.00000E+00     0.50000E+02
 0.00000E+00     0.40000E+01     0.30000E+03     0.50000E+02     0.50000E+02
 0.00000E+00     0.20000E+01     0.30000E+03     0.50000E+02     0.50000E+02

INTERNAL POINTS

      X               Y            POTENTIAL

 0.20000E+01     0.20000E+01     0.20000E+03
 0.20000E+01     0.40000E+01     0.20000E+03
 0.30000E+01     0.30000E+01     0.15000E+03
 0.40000E+01     0.20000E+01     0.10000E+03
 0.40000E+01     0.40000E+01     0.10000E+03
******************************************************************************
```

Example 2.3

It is interesting to note that in this case even a simple four elements representation can give exact results using double value of the flux at the corners (figure 2.10(c)).

The input in this case is

HEAT FLOW EXAMPLE (DATA)

```
 HEAT FLOW EXAMPLE (4 LINEAR ELEMENTS)
4 3
0. 0. 6. 0. 6. 6. 0. 6.
1 0. 1 0.
0 0. 0 0.
1 0. 1 0.
0 300. 0 300.
2. 4. 3. 3. 4. 4.
```

The corresponding output is very accurate taking into consideration the simplicity of the mesh.

HEAT FLOW EXAMPLE (OUTPUT)

```
*****************************************************************************
HEAT FLOW EXAMPLE (4 LINEAR ELEMENTS)


DATA

 NUMBER OF BOUNDARY ELEMENTS =  4
 NUMBER OF INTERNAL POINTS WHERE THE FUNCTION IS CALCULATED =   3

 POINT          X                  Y
    1       0.00000E+00        0.00000E+00
    2       0.60000E+01        0.00000E+00
    3       0.60000E+01        0.60000E+01
    4       0.00000E+00        0.60000E+01

 BOUNDARY CONDITIONS

             ------FIRST NODE-----              -----SECOND NODE-----
               PRESCRIBED                          PRESCRIBED
ELEMENT           VALUE           CODE               VALUE          CODE
   1          0.0000000E+00        1            0.0000000E+00        1
   2          0.0000000E+00        0            0.0000000E+00        0
   3          0.0000000E+00        1            0.0000000E+00        1
   4          0.3000000E+03        0            0.3000000E+03        0
*****************************************************************************


 RESULTS

 BOUNDARY NODES

                                                         POTENTIAL DERIVATIVE
        X                Y               POTENTIAL      BEFORE NODE       AFTER NO

    0.00000E+00     0.00000E+00     0.30000E+03     0.50000E+02      0.00000E+
    0.60000E+01     0.00000E+00     0.00000E+00     0.00000E+00     -0.50000E+
    0.60000E+01     0.60000E+01     0.00000E+00    -0.50000E+02      0.00000E+
    0.00000E+00     0.60000E+01     0.30000E+03     0.00000E+00      0.50000E+


 INTERNAL POINTS

        X                Y               POTENTIAL

    0.20000E+01     0.40000E+01     0.20138E+03
    0.30000E+01     0.30000E+01     0.14979E+03
    0.40000E+01     0.40000E+01     0.10001E+03
*****************************************************************************
```

2.7 Quadratic and Higher Order Elements

It is usually more convenient for arbitrary geometries to implement some type of curvilinear elements. The simplest of these are the three noded quadratic elements which require working with transformations. Consider the curved boundary shown in figure 2.11 where Γ is defined along the boundary and the $\vec{r}$ position vector is a function of the cartesian system, x_1, x_2. The variables u or q can be written in terms of interpolation functions which are functions of the homogeneous coordinate ξ, i.e.

$$u(\xi) = \phi_1 u^1 + \phi_2 u^2 + \phi_3 u^3 = [\phi_1 \phi_2 \phi_3] \begin{Bmatrix} u^1 \\ u^2 \\ u^3 \end{Bmatrix}$$

$$q(\xi) = \phi_1 q^1 + \phi_2 q^2 + \phi_3 q^3 = [\phi_1 \phi_2 \phi_3] \begin{Bmatrix} q^1 \\ q^2 \\ q^3 \end{Bmatrix} \tag{2.60}$$

where the interpolation functions are

$$\phi_1 = \tfrac{1}{2}\xi(\xi - 1); \qquad \phi_2 = (1 - \xi)(1 + \xi); \qquad \phi_3 = \tfrac{1}{2}\xi(1 + \xi) \tag{2.61}$$

These functions are quadratic in ξ and give the nodal values of the variable u or q when specialized for the nodes; i.e. with reference to figure 2.12

Node	ξ	ϕ_1	ϕ_2	ϕ_3
1	-1	1	0	0
2	0	0	1	0
3	$+1$	0	0	1

(2.62)

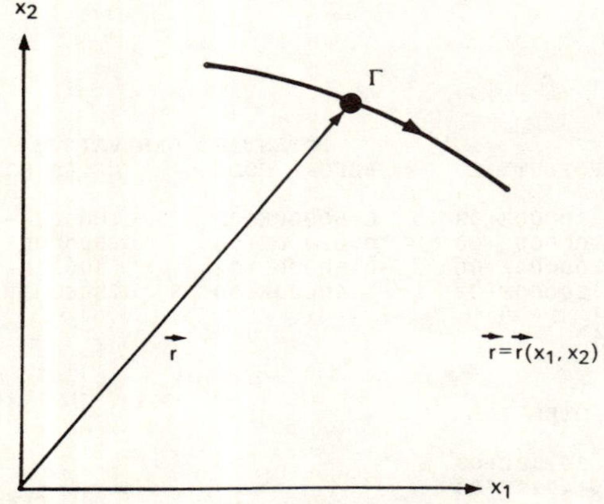

Figure 2.11 Curved boundary

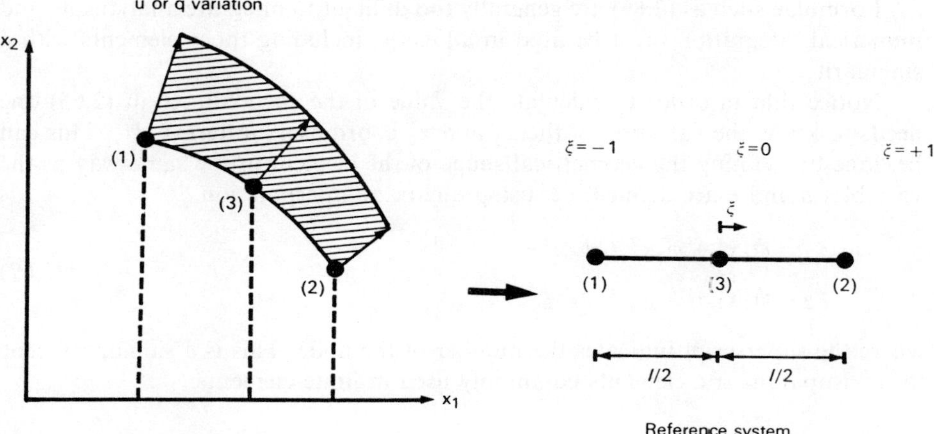

Figure 2.12 Quadratic element

The integrals along any 'j' element are similar to those computed for the linear element, but there are now three nodal unknowns and the integration requires the use of a Jacobian. Consider for instance an integral for the H type terms, i.e.

$$\int_{\Gamma_j} u(\xi) q^* \, d\Gamma = \int_{\Gamma_j} [\phi_1 \phi_2 \phi_3] q^* \, d\Gamma \begin{Bmatrix} u^1 \\ u^2 \\ u^3 \end{Bmatrix} = [h_1^{ij} h_2^{ij} h_3^{ij}] \begin{Bmatrix} u^1 \\ u^2 \\ u^3 \end{Bmatrix} \tag{2.63}$$

where

$$h_1^{ij} = \int_{\Gamma_j} \phi_1 q^* \, d\Gamma; \qquad h_2^{ij} = \int_{\Gamma_j} \phi_2 q^* \, d\Gamma; \qquad h_3^{ij} = \int_{\Gamma_j} \phi_3 q^* \, d\Gamma \tag{2.64}$$

The evaluation of these terms requires the use of a Jacobian as the ϕ_i functions are expressed in terms of ξ, but the integrals are functions of Γ. For a curve such as given in figure 2.11, the transformation is simple,

$$d\Gamma = \left\{ \sqrt{\left(\frac{dx_1}{d\xi}\right)^2 + \left(\frac{dx_2}{d\xi}\right)^2} \right\} d\xi = |G| \, d\xi \tag{2.65}$$

where $|G|$ is the Jacobian. Hence one can write,

$$h_k^{ij} = \int_{\Gamma_j} \phi_k(\xi) q^* \, d\Gamma = \int_{\text{Node 1}}^{\text{Node 2}} \phi_k(\xi) q^* |G| \, d\xi \tag{2.66}$$

Formulae such as (2.66) are generally too difficult to integrate analytically and numerical integration must be used in all cases, including those elements with a singularity.

Notice that in order to calculate the value of the Jacobian $|G|$ in (2.65) one needs to know the variation of the x_1 and x_2 coordinates in terms of ξ. This can be done by defining the geometrical shape of the element in the same way as the variables u and q are defined, i.e. using quadratic interpolation,

$$\begin{aligned} x_1 &= \phi_1 x_1^1 + \phi_2 x_1^2 + \phi_3 x_1^3 \\ x_2 &= \phi_2 x_2^1 + \phi_2 x_2^2 + \phi_2 x_2^3 \end{aligned} \tag{2.67}$$

where the superscript indicates the number of the node. This is a similar concept to the isoparametric elements commonly used in finite elements.

Cubic Elements

Elements of order higher than quadratic are seldom used in practice, but they may be interesting in some particular applications. Because of this we will briefly describe the case of elements with cubic variation of geometry, and u or q variables. In this case the functions are described by taking four nodes over each element (figure 2.13).

$$\begin{aligned} u &= \phi_1 u^1 + \phi_2 u^2 + \phi_3 u^3 + \phi_4 u^4 \\ q &= \phi_1 q^1 + \phi_2 q^2 + \phi_3 q^3 + \phi_4 q^4 \end{aligned} \tag{2.68}$$

and similarly

$$\begin{aligned} x_1 &= \phi_1 x_1^1 + \phi_2 x_1^2 + \phi_3 x_1^3 + \phi_4 x_1^4 \\ x_2 &= \phi_1 x_2^1 + \phi_2 x_2^2 + \phi_3 x_2^3 + \phi_4 x_2^4 \end{aligned} \tag{2.69}$$

where the interpolation functions are,

$$\begin{aligned} \phi_1 &= \tfrac{1}{16}(1-\xi)[-10+9(\xi^2+1)] \qquad & \phi_2 &= \tfrac{1}{16}(1+\xi)[-10+9(\xi^2+1)] \\ \phi_3 &= \tfrac{9}{16}(1-\xi^2)(1-3\xi) & \phi_4 &= \tfrac{9}{16}(1-\xi^2)(1+3\xi) \end{aligned} \tag{2.70}$$

which can be specialized at the nodes as follows,

Node	ξ	ϕ_1	ϕ_2	ϕ_3	ϕ_4
1	-1	1	0	0	0
2	$-1/3$	0	1	0	0
3	$1/3$	0	0	1	0
4	1	0	0	0	1

(2.71)

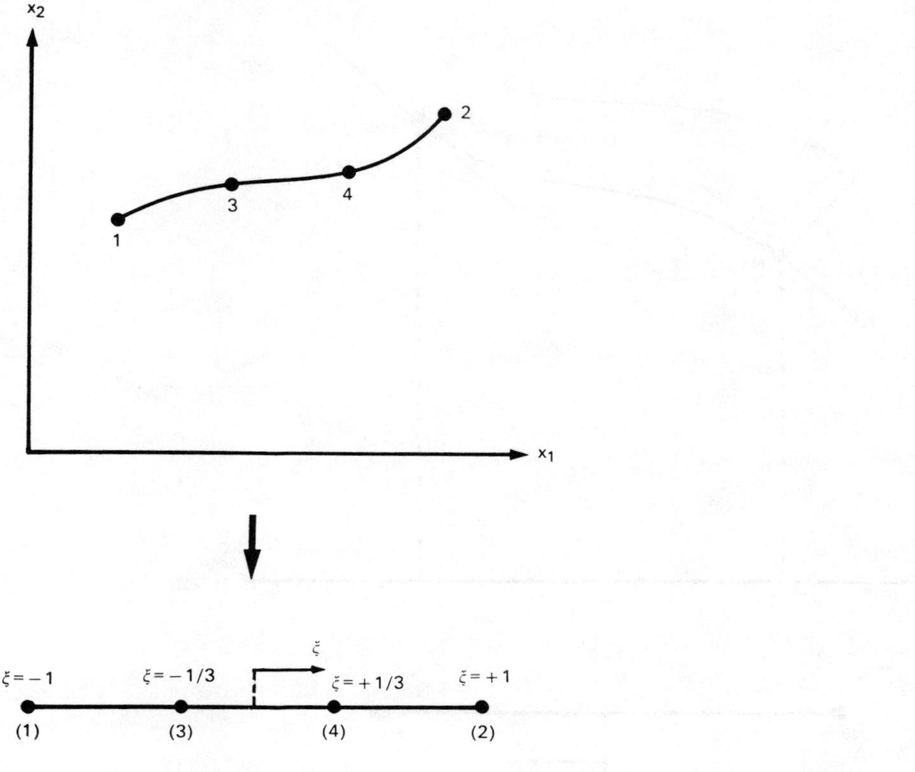

Figure 2.13 Cubic elements with four nodes

Another possibility with cubic elements is to define the variation of u or q in terms of the function and its derivative along the element, at the two end points – i.e nodes 1 and 2 – as shown in figure 2.14. The corresponding function for u (same applies for q and $x_1 x_2$) is then given by

$$u = \phi_1 u^1 + \phi_2 \left(\frac{\partial u}{\partial \Gamma}\right)^1 + \phi_3 u^2 + \phi_4 \left(\frac{\partial u}{\partial \Gamma}\right)^2 \tag{2.72}$$

with

$$\begin{aligned} \phi_1 &= \tfrac{1}{4}(\xi - 1)^2(\xi + 2) \qquad & \phi_2 &= -\tfrac{1}{4}l(\xi - 1)^2(\xi + 1) \\ \phi_3 &= \tfrac{1}{4}(\xi + 1)^2(\xi - 2) \qquad & \phi_4 &= -\tfrac{1}{4}l(\xi + 1)^2(\xi - 1) \end{aligned} \tag{2.73}$$

where l is the element length.

This last type of cubic element could be used in cases where we wish to have a correct definition of the derivative along Γ, for instance to calculate fluxes in that direction, or if we prefer to reduce the number of nodes along the element. In some cases it may still be better to continue defining the geometry with four nodes as it is generally more difficult to have accurate results for the slopes.

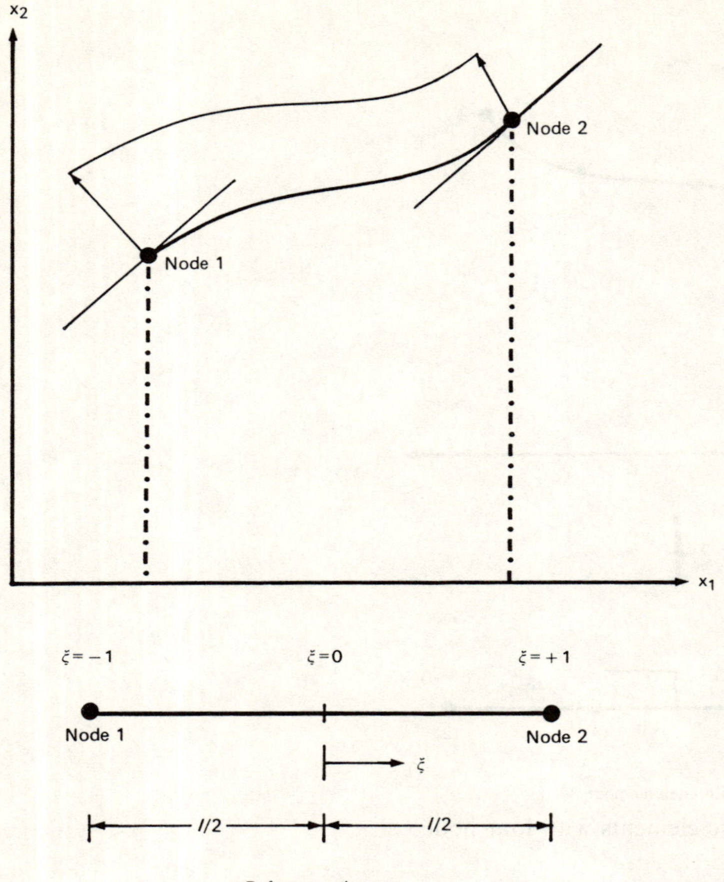

Figure 2.14 Cubic elements with only two nodes

2.8 Computer Code for Potential Problems using Quadratic Elements (POQUABE)

In what follows a FORTRAN code for potential problems using quadratic elements is described. The program has the same organization as the two previously studied.

All variables in the code have the same meaning as for the linear element program (POLINBE). FI and DFI have a slightly different form. The dimension of FI is (N), N being the number of nodes and that of DFI is $(3NE)$, where NE is the number of elements. The prescribed boundary conditions are read in DFI (three per element).

The program allows for the values of flux at both sides of the nodes connecting two elements to be different. Then, (i) when both fluxes are prescribed as different at both sides of the node, the potential is the only unknown; (ii) when the potential and one flux are prescribed, the other flux is the unknown and (iii) if only the potential is prescribed, one value of the flux is the unknown and will be the same on both sides of the node. Thus, the situation at corner nodes of quadratic elements is the same as for linear elements.

Main Program

The program follows the same structure of the constant and linear potential codes.

The listing is as follows.

```
C-------------------------------------------------------------------------
      PROGRAM POQUABE
C
C  PROGRAM 16
C
C
C  THIS PROGRAM SOLVES TWO DIMENSIONAL (PO)TENTIAL PROBLEMS
C  USING (QUA)DRATIC (B)OUNDARY (E)LEMENTS.
C
      CHARACTER*10 FILEIN,FILEOUT
      COMMON/MATG/G(100,150)
      COMMON/MATH/H(100,100)
      COMMON N,L,INP,IPR
      DIMENSION X(101),Y(101)
      DIMENSION DFI(150),FI(100),KODE(150)
      DIMENSION CX(20),CY(20),SOL(20)
C
C  SET MAXIMUN DIMENSION OF THE SYSTEM OF EQUATIONS (NX)
C  NX= MAXIMUN NUMBER OF NODES= 2*MAXIMUN NUMBER OF ELEMENTS
C  NX1= 3*MAXIMUN NUMBER ELEMENTS
C
      NX=100
      NX1=150
C
C  ASSIGN NUMBERS FOR INPUT AND OUTPUT FILES
C
      INP=5
      IPR=6
C
C  READ NAMES AND OPEN  FILES FOR INPUT AND OUTPUT
C
      WRITE(*,' (A) ') ' NAME OF INPUT FILE (MAX. 10 CHART.)'
      READ(*,'  (A)  ')FILEIN
      OPEN(INP,FILE=FILEIN,STATUS='OLD')
      WRITE(*,' (A)  ') ' NAME OF OUTPUT FILE (MAX. 10 CHART.)'
      READ(*,' (A)  ')FILEOUT
      OPEN(IPR,FILE=FILEOUT,STATUS='NEW')
C
C  READ DATA
C
      CALL INPUTPQ(CX,CY,X,Y,KODE,DFI)
C
C  COMPUTE H AND G MATRICES AND FORM SYSTEM (A X = F)
C
      CALL GHMATPQ(X,Y,G,H,FI,DFI,KODE,NX,NX1)
C
C  SOLVE SYSTEM OF EQUATIONS
C
      CALL SLNPD(H,FI,D,N,NX)
C
C  COMPUTE POTENTIAL VALUES AT INTERNAL POINTS
C
      CALL INTERPQ(FI,DFI,KODE,CX,CY,X,Y,SOL)
C
C  PRINT RESULTS AT BOUNDARY NODES AND INTERNAL POINTS
C
```

```
      CALL OUTPTPQ(X,Y,FI,DFI,CX,CY,SOL)
C
C   CLOSE INPUT AND OUTPUT FILES
C
      CLOSE (INP)
      CLOSE (IPR)
      STOP
      END
```

Routine INPUTPQ

This subroutine reads all the input required by the program and requests a file from the user containing the following lines:

(i) *Title Line* Contains the title of the problem
(ii) *Basic Parameter Line* Contains the number of elements and the number of internal points.
(iii) *Boundary Nodes Coordinates Lines* Contains $x_1 x_2$ coordinates read counter-clockwise for external boundaries and clockwise for internal ones. The lines are organized in free format.
(iv) *Boundary Conditions Lines* As many lines as boundary elements. Three values of KODE and the known variables are read for each element, corresponding to the three nodes. In this way a value of the flux may be prescribed for an extreme node as part of one element and a different value as part of the other element. The potential however must be unique for any node and the flux must also be unique for the mid-node of any element. The known variables are the potential if KODE = 0 and the flux if KODE = 1. The order of reading is first KODE(I) and then value of the variable (I) for $I = 1, 2, 3$.
(v) *Internal Points Coordinates Lines* Contain $x_1 x_2$ coordinates of the internal points organized in free format. There will be one or more lines if necessary.

This subroutine first prints the name of the run and the basic parameters. Then the coordinates of the nodes and the boundary conditions for each element, with codes and prescribed values are printed. The internal points coordinates are only printed in the subroutine OUTPTPQ.

The FORTRAN listing of INPUTPQ is as follows:

```
C------------------------------------------------------------------------
      SUBROUTINE INPUTPQ(CX,CY,X,Y,KODE,DFI)
C
C  PROGRAM 17
C
C  NE= NUMBER OF BOUNDARY ELEMENTS
C  N = NUMBER OF BOUNDARY NODES = 2 * NE
C  L = NUMBER OF INTERNAL POINTS
C
      CHARACTER*80 TITLE
      COMMON N,L,INP,IPR
      DIMENSION KODE(1),X(1),Y(1),CX(1),CY(1),DFI(1)
      WRITE(IPR,100)
  100 FORMAT(' ',79('*'))
C
C  READ JOB TITLE
```

```
C
      READ(INP,'(A)') TITLE
      WRITE(IPR,'(A)') TITLE
C
C  READ NUMBER OF BOUNDARY ELEMENTS AND INTERNAL POINTS
C
      READ(INP,*)NE,L
      WRITE(IPR,210)NE,L
  210 FORMAT(//2X,'DATA'/2X,'NUMBER OF BOUNDARY ELEMENTS=',
     1I3/2X,'NUMBER OF INTERNAL POINTS=',I3)
      N=2*NE
C
C  READ BOUNDARY NODES COORDINATES IN ARRAYS X AND Y
C
      WRITE(IPR,500)
      READ(INP,*) (X(I),Y(I),I=1,N)
      DO 10 I=1,N
   10 WRITE(IPR,240) I,X(I),Y(I)
  500 FORMAT(//2X,'BOUNDARY NODES COORDINATES'///4X,
     1'NODE',10X,'X',18X,'Y'/)
  240 FORMAT(5X,I3,2(5X,E14.7))
C
C  READ BOUNDARY CONDITIONS IN DFI(I) VECTOR,IF KODE(I)=0
C  THE DFI(I) VALUE IS A KNOWN POTENTIAL; IF KODE(I)=1 THE
C  DFI(I) VALUE IS A KNOWN POTENTIAL DERIVATIVE (FLUX).
C  THREE BOUNDARY CONDITIONS ARE READ PER ELEMENT.
C  NODES BETWEEN TWO ELEMENTS MAY HAVE TWO DIFFERENT VALUES
C  OF THE POTENTIAL DERIVATIVE BUT ONLY ONE VALUE OF THE
C  POTENTIAL.
C
      WRITE(IPR,800)
  800 FORMAT(//2X,'BOUNDARY CONDITIONS'//11X,
     1'-----FIRST NODE------',3X,'-----SECOND NODE-----',3X,
     2'-----THIRD NODE------'/13X,'PRESCRIBED',14X,
     3'PRESCRIBED',14X,'PRESCRIBED'/1X,'ELEMENT',8X,'VALUE',
     47X,'CODE',8X,'VALUE',7X,'CODE',8X,'VALUE',7X,'CODE')
      DO 20 I=1,NE
      READ(INP,*) (KODE(3*I-3+J),DFI(3*I-3+J),J=1,3)
   20 WRITE(IPR,950)I,(DFI(3*I-3+J),KODE(3*I-3+J),J=1,3)
  950 FORMAT(3X,I3,2X,3(4X,E14.7,5X,I1))
C
C  READ COORDINATES OF THE INTERNAL POINTS
C
      IF(L.EQ.0) GO TO 30
      READ(INP,*) (CX(I),CY(I),I=1,L)
   30 RETURN
      END
```

Routine GHMATPQ

This subroutine computes the **G** and **H** system matrices by calling routines EXTINPQ and LOCINPQ.

EXTINPQ: Computes the **GW** and **HW** (3) submatrices which relate a collocation point with an element as defined by its three nodes. The collocation point is not any one of the element nodes.

LOCINPQ: Computes the **GW** (3) submatrix for the case when the collocation point is one of the nodes within the element under consideration (i.e. the singularity is in the same element). Notice that the corresponding **HW** (3) is computed using EXTINPQ because the singularity will occur only on the diagonal term and this is computed later on by adding the off-diagonal terms of the row.

The resulting **GW** and **HW** submatrices are assembled in the **G** and **H** system matrices. Matrix **G** is now rectangular since each extreme node of an element may have different fluxes, i.e. one 'before' and another 'after' the node. The diagonal terms in **H** are computed using constant potential considerations, which results in adding row coefficients together.

Once the matrices **H** and **G** are assembled, the system of equations needs to be reordered in accordance with the boundary conditions to form

$$\mathbf{AX} = \mathbf{F}$$

where **X** is a (N) vector of unknowns, N being the number of nodes; **A** is a $(N \times N)$ matrix whose columns are a combination of columns of **H** or **G** depending on the boundary conditions or of two consecutive columns of **G** when the unknown is the *unique* value of the tractions at both sides of the extreme node of an element; **F** is a known vector computed by multiplying the prescribed boundary conditions by the corresponding row terms of **G** or **H**.

At the end of the subroutine GHMATPQ and after rearranging **H** contains the matrix **A**, and **FI** the **F** vector.

The FORTRAN listing of GHMATPQ is as follows.

```
C-----------------------------------------------------------------------
      SUBROUTINE GHMATPQ(X,Y,G,H,FI,DFI,KODE,NX,NX1)
C
C     PROGRAM 18
C
C     THIS SUBROUTINE COMPUTES THE G AND H MATRICES AND FORMS
C     THE SYSTEM OF EQUATIONS A X = F
C     H IS A SQUARE MATRIX (2*NE,2*NE); G IS RECTANGULAR (2*NE,3*NE)
C
      DIMENSION X(1),Y(1),G(NX,NX1),H(NX,NX),HW(3),GW(3)
      DIMENSION FI(1),DFI(1),KODE(1)
      COMMON N,L,INP,IPR
      NE=N/2
      DO 20 I=1,N
      DO 11 J=1,N
   11 H(I,J)=0.
      DO 12 J=1,3*NE
   12 G(I,J)=0.
   20 CONTINUE
      X(N+1)=X(1)
      Y(N+1)=Y(1)
C
C    COMPUTE THE GW AND HW MATRICES FOR EACH COLLOCATION
C    POINT AND EACH BOUNDARY ELEMENT
C
      DO 40 LL=1,N
      DO 40 I=1,N-1,2
      IF((LL-I)*(LL-I-1)*(LL-I-2)*(LL-I+N-2)) 22,21,22
   21 NODO=LL-I+1
      IF((LL.EQ.1).AND.(I.EQ.N-1)) NODO=NODO+N
      CALL EXTINPQ(X(LL),Y(LL),X(I),Y(I),X(I+1),Y(I+1),X(I+2),Y(I+2)
     *,HW,GW)
      CALL LOCINPQ(X(I),Y(I),X(I+1),Y(I+1),X(I+2),Y(I+2),GW,NODO)
      GO TO 34
   22 CALL EXTINPQ(X(LL),Y(LL),X(I),Y(I),X(I+1),Y(I+1),X(I+2),Y(I+2)
     *,HW,GW)
C
C    PLUG THE GW AND HW MATRICES INTO THE GENERAL G AND H MATRICES.
C
   34 DO 38 J=1,3
      K=3*(I-1)/2
      G(LL,K+J)=G(LL,K+J)+GW(J)
      IF(I-N+1) 37,35,37
```

```
   35 IF(J-3) 37,36,36
   36 H(LL,1)=H(LL,1)+HW(J)
      GO TO 38
   37 H(LL,I-1+J)=H(LL,I-1+J)+HW(J)
   38 CONTINUE
   40 CONTINUE
C
C  COMPUTE THE DIAGONAL COEFFICIENTS OF THE H MATRIX
C
      DO 70 I=1,N
      H(I,I)=0.
      DO 60 J=1,N
      IF(I.EQ.J) GO TO 60
      H(I,I)=H(I,I)-H(I,J)
   60 CONTINUE
C
C  ADD ONE TO THE DIAGONAL COEFFICIENTS FOR
C  EXTERNAL PROBLEMS.
C
      IF(H(I,I)) 65,70,70
   65 H(I,I)=1.+H(I,I)
   70 CONTINUE
C
C  REORDER THE COLUMNS OF THE SYSTEM OF EQUATIONS IN ACCORDANCE
C  WITH THE BOUNDARY CONDITIONS AND FORM SYSTEM MATRIX A WHICH
C  IS STORED IN H
C
      DO 180 I=1,NE
      DO 170 J=1,3
      IF(KODE(3*I-3+J)) 110,110,170
  110 IF((I-NE).NE.0 .OR. J.NE.3) GO TO 125
      IF(KODE(1)) 115,115,113
  113 DO 114 K=1,N
      CH=H(K,1)
      H(K,1)=-G(K,3*I)
  114 G(K,3*I)=-CH
      GO TO 170
  115 DO 116 K=1,N
      H(K,1)=H(K,1)-G(K,3*I)
  116 G(K,3*I)=0.
      GO TO 170
  125 IF(I.EQ.1 .OR. J.GT.1 .OR. KODE(3*I-3).EQ.1) GO TO 130
      DO 129 K=1,N
      H(K,2*I-1)=H(K,2*I-1)-G(K,3*I-2)
  129 G(K,3*I-2)=0.
      GO TO 170
  130 DO 132 K=1,N
      CH=H(K,2*I-2+J)
      H(K,2*I-2+J)=-G(K,3*I-3+J)
  132 G(K,3*I-3+J)=-CH
  170 CONTINUE
  180 CONTINUE
C
C  FORM THE RIGHT HAND SIDE VECTOR F WHICH IS STORED IN FI
C
      DO 190 I=1,N
      FI(I)= 0.
      DO 185 J=1,3*NE
  185 FI(I)=FI(I)+G(I,J)*DFI(J)
  190 CONTINUE
      RETURN
      END
```

Routine EXTINPQ

This subroutine computes using numerical integration, the (3) submatrices **GW** and **HW** that correspond to an element when the collocation point is at a node other than any of those 3 in the element. The correlation of the collocation points are XP and YP. The integrals are of the type (2.63), i.e.

$$\mathbf{HW} = \int_{\Gamma_j} \phi \mathbf{q}^* \, d\Gamma = \int_{-1}^{+1} \phi \mathbf{q}^* |G| \, d\xi \tag{2.74}$$

$$\mathbf{GW} = \int_{\Gamma_j} \phi \mathbf{u}^* \, d\Gamma = \int_{-1}^{+1} \phi \mathbf{u}^* |G| \, d\xi \tag{2.75}$$

The Jacobians are calculated by taking derivatives of the expressions for the x_1 and x_2 coordinates, which are defined as follows (equation (2.67)),

$$\begin{aligned} x_1 &= \phi_1 x_1^1 + \phi_2 x_1^2 + \phi_3 x_1^3 \\ x_2 &= \phi_1 x_2^1 + \phi_2 x_2^2 + \phi_3 x_2^3 \end{aligned} \tag{2.76}$$

After substituting the ϕ_i expression (equation (2.61)) the above relationships can be written as,

$$\begin{aligned} x_1 &= \tfrac{1}{2}\xi^2(x_1^1 - 2x_1^2 + x_1^3) + \tfrac{1}{2}\xi(x_1^3 - x_1^1) + x_1^2 \\ x_2 &= \tfrac{1}{2}\xi^2(x_2^1 - 2x_2^2 + x_2^3) + \tfrac{1}{2}\xi(x_2^3 - x_2^1) + x_2^2 \end{aligned} \tag{2.77}$$

The Jacobian is obtained by substituting (2.77) into (2.65) which gives

$$\begin{aligned} |G| = [\{&(x_1^3 - 2x_1^2 + x_1^1)\xi + \tfrac{1}{2}(x_1^3 - x_1^1)\}^2 \\ &+ \{(x_2^3 - 2x_2^2 + x_2^1)\xi + \tfrac{1}{2}(x_2^3 - x_2^1)^2\}^2] \end{aligned} \tag{2.78}$$

A Gauss quadrature formula with ten points has been taken instead of the four points formula applied in the constant and linear element cases. The reason is twofold: (i) The variation of potential and flux is quadratic and hence more points should be taken and (ii) the element geometry is also quadratic.

```
C-----------------------------------------------------------------------
      SUBROUTINE EXTINPQ(XP,YP,X1,Y1,X2,Y2,X3,Y3,HW,GW)
C
C   PROGRAM 19
C
C   THIS SUBRUOTINE COMPUTES THE  HW AND GW MATRICES
C   THAT RELATE A NODE (XP,YP) WITH A BOUNDARY
C   ELEMENT USING GAUSS QUADRATURE
C
C   RA             = RADIUS
C   RD1,RD2,RDN    = RADIUS DERIVATIVES
C   ETA1,ETA2      = COMPONENTS OF THE UNIT NORMAL TO THE ELEMENT
C   XCO,YCO        = INTEGRATION POINT ALONG THE ELEMENT
C   XJA            = JACOBIAN
C
      COMMON N,L,INP,IPR
      DIMENSION GW(3),HW(3)
      DIMENSION GI(10),OME(10)
      DATA GI/0.9739065285,-0.9739065285,0.8650633666,-0.8650633666
     @,0.6794095683,-0.6794095682,0.4333953941,-0.4333953941,
     @0.1488743389,-0.1488743389/
      DATA OME/0.0666713443,0.0666713443,0.1494513491,0.1494513491
     @,0.2190863625,0.2190863625,0.2692667193,0.2692667193,
     @0.2955242247,0.2955242247/
```

```
      DO 20 J=1,3
      HW(J)=0.
   20 GW(J)=0.
      A=X3-2*X2+X1
      B=(X3-X1)/2
      C=Y3-2*Y2+Y1
      D=(Y3-Y1)/2
      DO 40 I=1,10
C
C   COMPUTE THE VALUES OF THE SHAPE FUNCTIONS AT THE
C   INTEGRATION POINTS
C
      F1=GI(I)*(GI(I)-1)*0.5
      F2=1.-GI(I)**2
      F3=GI(I)*(GI(I)+1)*0.5
C
C   COMPUTE GEOMETRICAL PROPERTIES AT THE INTEGRATION POINTS
C
      XCO=X1*F1+X2*F2+X3*F3
      YCO=Y1*F1+Y2*F2+Y3*F3
      XJA=SQRT((GI(I)*A+B)**2+(GI(I)*C+D)**2)
      ETA1=(GI(I)*C+D)/XJA
      ETA2=-(GI(I)*A+B)/XJA
      RA=SQRT((XP-XCO)**2+(YP-YCO)**2)
      RD1=(XCO-XP)/RA
      RD2=(YCO-YP)/RA
      RDN=RD1*ETA1+RD2*ETA2
C
C   COMPUTE GW AND HW MATRICES
C
      GW(1)=GW(1)+ALOG(1./RA)*OME(I)*XJA*F1
      GW(2)=GW(2)+ALOG(1./RA)*OME(I)*XJA*F2
      GW(3)=GW(3)+ALOG(1./RA)*OME(I)*XJA*F3
      HW(1)=HW(1)-RDN/RA*OME(I)*XJA*F1
      HW(2)=HW(2)-RDN/RA*OME(I)*XJA*F2
40    HW(3)=HW(3)-RDN/RA*OME(I)*XJA*F3
      RETURN
      END
```

Routine LOCINPQ

This subroutine computes using numerical integration, the submatrix **GW** that corresponds to an element when the collocation point is one of those three in the element. The integrals are

$$\mathbf{GW} = \int_{\Gamma_j} \phi \mathbf{u}^* \, d\Gamma \tag{2.79}$$

Three cases are considered depending on the position of the collocation point, i.e. NODE = 1, 2 or 3 (see figure 2.15).

(i) *Collocation point at Node* (*1*)

First a change of coordinates from $x_1 x_2$ to ξ is defined in the same way that was done in subroutine EXTINPQ. Then, in order to integrate the singularity a new change of variables is carried out, i.e.

$$\eta = \frac{\xi + 1}{2} \tag{2.80}$$

The integral then gives two parts, one with a singular term ln $1/\eta$ and the other with no singularity. The first part is integrated by means of a special integration formula of the type (see Appendix A)

$$I = \int_0^1 \ln\left(\frac{1}{\eta}\right) f(\eta)\, d\eta \cong \sum_{i=1}^{n} w_i f(\eta_i) \tag{2.81}$$

(in the program $\eta_i \equiv$ GIL(I), $w_i \equiv$ OMEL(I)). The second part is integrated by the standard Gauss quadrature formula in terms of the variable ξ (in the program $\xi_i =$ GI(I), $w_i =$ OME(I)). The shape functions ϕ_1, ϕ_2, ϕ_3 are given by F1, F2, F3 in terms of ξ and by FL1, FL2, FL3 in terms of η.

XJA1 is the Jacobian for the special integration and XJA2 is the Jacobian for the standard Gauss quadrature.

(ii) *Collocation point at Node* (2)

In order to integrate the two singularities that appear at both sides of the nodes, the integral is divided into two parts,

$$\mathbf{GW} = \int_{(1)}^{(3)} \boldsymbol{\phi}\mathbf{u}^*|G|\, d\xi = \int_{(1)}^{(2)} \boldsymbol{\phi}\mathbf{u}^*|G|\, d\xi + \int_{(2)}^{(3)} \boldsymbol{\phi}\mathbf{u}^*|G|\, d\xi \tag{2.82}$$

Then, the first part is changed to the variable (see figure 2.15) $\eta' = -\xi$ and the second part ot the variable $\eta = \xi$. Now each singular part of these two integrals is computed using the special integration formula, and the two non-singular parts together are integrated using standard 10 points Gauss quadrature.

In the program XJA1 and XJA11 are the Jacobians for the two special logarithmic integrations and XJA2 is the Jacobian for the standard Gauss quadrature. The functions ϕ_1, ϕ_2 and ϕ_3 in terms of the new variables η are FLN1, FLN2 and FLN3.

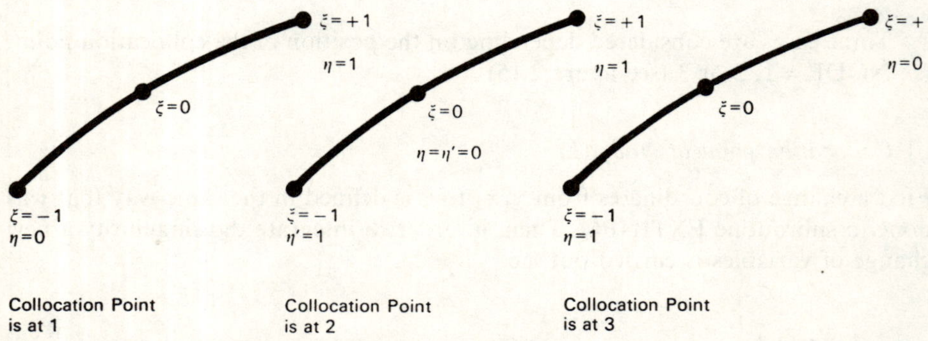

Figure 2.15 Geometrical coordinates systems for numerical integration

(iii) *Collocation point at Node (3)*

This case is similar to the first one with the logarithmic integration variable being now

$$\eta = \frac{1-\xi}{2} \tag{2.83}$$

```
C-----------------------------------------------------------------------
      SUBROUTINE LOCINPQ(XG1,YG1,XG2,YG2,XG3,YG3,GW,NODO)
C
C   PROGRAM 20
C
C   THIS SUBROUTINE COMPUTES THE GW MATRIX WHEN THE COLLOCATION
C   POINT IS ONE OF THE NODES OF THE INTEGRATION ELEMENT.
C   THE COEFFICIENTS ARE COMPUTED BY NUMERICAL INTEGRATION:
C   THE NON SINGULAR PART IS COMPUTED USING STANDARD GAUSS QUADRATURE,
C   THE LOGARITHMIC PART IS COMPUTED USING A SPECIAL QUADRATURE FORMULA.
C
      COMMON N,L,INP,IPR
      DIMENSION GI(10),OME(10),GIL(10),OMEL(10),GW(3)
C
C   DATA FOR THE GAUSS QUADRATURE
C
      DATA GI/0.9739065285,-0.9739065285,0.8650633666,-0.8650633666
     @,0.6794095682,-0.6794095682,0.4333953941,-0.4333953941,
     @0.1488743389,-0.1488743389/
      DATA OME/0.0666713443,0.0666713443,0.1494513491,0.1494513491
     @,0.2190863625,0.2190863625,0.2692667193,0.2692667193,
     @0.2955242247,0.2955242247/
C
C   DATA FOR THE SPECIAL QUADRATURE
C
      DATA GIL/0.0090426309,0.0539712662,0.1353118246,0.2470524162
     @,0.3802125396,0.5237923179,0.6657752055,0.7941904160,
     @0.8981610912,0.9688479887/
      DATA OMEL/0.1209551319,0.1863635425,0.1956608732,0.1735771421
     @,0.1356956729,0.0936467585,0.0557877273,0.0271598109,
     @0.0095151826,0.0016381576/
C
C   SET A LOCAL COORDINATES SYSTEM
C
      GO TO(1,2,3),NODO
    1 X3=XG3-XG1
      Y3=YG3-YG1
      X2=XG2-XG1
      Y2=YG2-YG1
      A1=(X3-2*X2)*0.5
      B1=X2
      A2=(Y3-2*Y2)*0.5
      B2=Y2
      GO TO 4
    2 X3=XG3-XG2
      Y3=YG3-YG2
      X1=XG1-XG2
      Y1=YG1-YG2
      A1=X1+X3
      B1=X3-X1
      A2=Y1+Y3
      B2=Y3-Y1
      GO TO 4
    3 X2=XG2-XG3
      Y2=YG2-YG3
      X1=XG1-XG3
      Y1=YG1-YG3
      A1=(X1-2*X2)*0.5
      B1=-X2
      A2=(Y1-2*Y2)*0.5
      B2=-Y2
```

```
    4 CONTINUE
C
      DO 10 J=1,3
   10 GW(J)=0.
C
      DO 250 I=1,10
C
C   COMPUTE SHAPE FUNCTIONS FOR NUMERICAL INTEGRATIONS
C
      F3=0.5*GI(I)*(GI(I)+1.)
      F2=1.-GI(I)**2
      F1=0.5*GI(I)*(GI(I)-1.)
      FL3=GIL(I)*(2.*GIL(I)-1.)
      FL2=4.*GIL(I)*(1.-GIL(I))
      FL1=(GIL(I)-1.)*(2.*GIL(I)-1.)
      FLN3=0.5*GIL(I)*(GIL(I)+1.)
      FLN2=1.-GIL(I)**2
      FLN1=0.5*GIL(I)*(GIL(I)-1.)
C
C   COMPUTE GEOMETRICAL PROPERTIES
C
      GO TO(50,60,70) NODO
C
   50 XJA1=SQRT((4*A1*GIL(I)-2*A1+0.5*X3)**2+(4*A2*GIL(I)-2*A2+0.5*Y3)**
     @  2)*2
      XJA2=SQRT((A1*GI(I)*2+0.5*X3)**2+(A2*GI(I)*2+0.5*Y3)**2)
      XLO=-ALOG(2*SQRT((GI(I)*A1+B1)**2+(GI(I)*A2+B2)**2))
      S3=FL3*XJA1*OMEL(I)+F3*XJA2*XLO*OME(I)
      S2=FL2*XJA1*OMEL(I)+F2*XJA2*XLO*OME(I)
      S1=FL1*XJA1*OMEL(I)+F1*XJA2*XLO*OME(I)
      GO TO 200
   60 XJA1=SQRT((0.5*B1-A1*GIL(I))**2+(0.5*B2-A2*GIL(I))**2)
      XJA11=SQRT((0.5*B1+A1*GIL(I))**2+(0.5*B2+A2*GIL(I))**2)
      XJA2=SQRT((0.5*B1+A1*GI(I))**2+(0.5*B2+A2*GI(I))**2)
      XLO=-0.5*ALOG((GI(I)*A1*0.5+B1*0.5)**2+(GI(I)*A2*0.5+B2*0.5)**2)
      S3=(FLN1*XJA1+FLN3*XJA11)*OMEL(I)+F3*XJA2*XLO*OME(I)
      S2=FLN2*(XJA1+XJA11)*OMEL(I)+F2*XJA2*XLO*OME(I)
      S1=(FLN3*XJA1+FLN1*XJA11)*OMEL(I)+F1*XJA2*XLO*OME(I)
      GO TO 200
   70 XJA1=SQRT((2*A1-4*A1*GIL(I)-0.5*X1)**2+(2*A2-4*A2*GIL(I)-0.5*Y1)**
     @2)*2
      XJA2=SQRT((2*A1*GI(I)-0.5*X1)**2+(2*A2*GI(I)-0.5*Y1)**2)
      XLO=-ALOG(2*SQRT((A1*GI(I)+B1)**2+(A2*GI(I)+B2)**2))
      S3=FL1*XJA1*OMEL(I)+F3*XJA2*XLO*OME(I)
      S2=FL2*XJA1*OMEL(I)+F2*XJA2*XLO*OME(I)
      S1=FL3*XJA1*OMEL(I)+F1*XJA2*XLO*OME(I)
C
C   COMPUTE GW MATRIX
C
  200 GW(3)= GW(3)+S3
      GW(2)= GW(2)+S2
      GW(1)= GW(1)+S1
C
  250 CONTINUE
C
      RETURN
      END
```

Routine INTERPQ

This subroutine first reorders the vectors DFI and FI in such a way that all the boundary fluxes are stored in DFI and all the potentials in FI. It then computes the potentials at internal points.

The potential at any interior point is given by

$$u^i = \sum_{j=0}^{NE} \left\{ \int_{\Gamma_j} u^* \boldsymbol{\phi} \, \mathbf{d}\Gamma \right\} \mathbf{q}^j - \sum_{j=1}^{NE} \left\{ \int_{\Gamma_j} q^* \boldsymbol{\phi} \, d\Gamma \right\} \mathbf{u}^j \tag{2.84}$$

where the integrals along the boundary elements are computed numerically by calling again the subroutine EXTINPQ.

The listing of INTERPQ is as follows:

```
C-----------------------------------------------------------------------
      SUBROUTINE INTERPQ(FI,DFI,KODE,CX,CY,X,Y,SOL)
C
C  PROGRAM 21
C
C  THIS SUBROUTINE COMPUTES THE VALUES OF POTENTIAL AT
C  INTERNAL POINTS.
C
      COMMON N,L,INP,IPR
      DIMENSION FI(1),DFI(1),KODE(1),CX(1),CY(1)
      DIMENSION X(1),Y(1),SOL(1)
      DIMENSION HW(3),GW(3)
C
C  REARRANGE THE FI AND DFI ARRAYS TO STORE ALL THE VALUES OF THE
C  POTENTIAL IN FI AND ALL THE VALUES OF THE DERIVATIVE IN DFI
C
      NE=N/2
      DO 180 I=1,NE
      DO 170 J=1,3
      IF(KODE(3*I-3+J)) 110,110,170
  110 IF((I-NE).NE.0 .OR. J.NE.3) GO TO 125
      IF(KODE(1)) 114,114,113
  113 CH=FI(1)
      FI(1)=DFI(3*I)
      DFI(3*I)=CH
      GO TO 170
  114 DFI(3*I)=DFI(1)
      GO TO 170
  125 IF(I.EQ.1 .OR. J.GT.1 .OR. KODE(3*I-3).EQ.1) GO TO 130
      DFI(3*I-2)=DFI(3*I-3)
      GO TO 170
  130 CH=FI(2*I-2+J)
      FI(2*I-2+J)=DFI(3*I-3+J)
      DFI(3*I-3+J)=CH
  170 CONTINUE
  180 CONTINUE
C
C  COMPUTE THE VALUES OF POTENTIAL AT INTERNAL POINTS
C
      IF(L.EQ.0) GO TO 50
      DO 240 K=1,L
      SOL(K)=0.
      DO 230 I=1,NE
      CALL EXTINPQ(CX(K),CY(K),X(2*I-1),Y(2*I-1),X(2*I),Y(2*I),X(2*I+1)
     1,Y(2*I+1),HW,GW)
      DO 220 J=1,3
      IJ2=2*I-2+J
      IF(IJ2.GT.(2*NE)) IJ2=J-2
  220 SOL(K)=SOL(K)+GW(J)*DFI(3*I-3+J)-HW(J)*FI(IJ2)
  230 CONTINUE
      SOL(K)=SOL(K)/(2.*3.1415926536)
  240 CONTINUE
   50 RETURN
      END
```

Routine OUTPTPQ

This subroutine prints the results in th following order.

(i) Potentials and fluxes at boundary nodes (fluxes 'before' and 'after' each node are printed. Mid-nodes always have the same flux at both sides).

(ii) Internal points potentials.

The listing is as follows:

```
C-----------------------------------------------------------------------
      SUBROUTINE OUTPTPQ(X,Y,FI,DFI,CX,CY,SOL)
C
C  PROGRAM 22
C
C  THIS SUBROUTINE PRINTS THE VALUES OF THE POTENTIAL AND ITS NORMAL
C  DERIVATIVE AT BOUNDARY NODES. IT ALSO PRINTS THE VALUES OF THE
C  POTENTIAL AT INTERNAL POINTS
C
      COMMON N,L,INP,IPR
      DIMENSION X(1),Y(1),FI(1),DFI(1),CX(1),CY(1),SOL(1)
C
      NE=N/2
      WRITE(IPR,100)
  100 FORMAT(' ',79('*')//2X,'RESULTS'//2X,'BOUNDARY NODES'//
     156X,'POTENTIAL DERIVATIVE'/
     29X,'X',15X,'Y',12X,'POTENTIAL',6X,'BEFORE NODE',6X,'AFTER NODE'/)
      WRITE(IPR,200) X(1),Y(1),FI(1),DFI(3*NE),DFI(1)
      WRITE(IPR,200) X(2),Y(2),FI(2),DFI(2),DFI(2)
      DO 10 I=2,NE
      WRITE(IPR,200) X(2*I-1),Y(2*I-1),FI(2*I-1),DFI(3*I-3),DFI(3*I-2)
   10 WRITE(IPR,200) X(2*I),Y(2*I),FI(2*I),DFI(3*I-1),DFI(3*I-1)
  200 FORMAT(5(2X,E14.5))
C
      IF(L.EQ.0) GO TO 30
      WRITE(IPR,300)
  300 FORMAT(//,2X,'INTERNAL POINTS',//9X,'X',15X,'Y',12X,'POTENTIAL',/)
      DO 20 K=1,L
   20 WRITE(IPR,400)CX(K),CY(K),SOL(K)
  400 FORMAT(3(2X,E14.5))
   30 WRITE(IPR,500)
  500 FORMAT(' ',79('*'))
      RETURN
      END
C-----------------------------------------------------------------------
```

Example 2.4

The problem of an elliptical bar under torsion (figure 2.16(a)) is analysed using program POQUABE. Under Saint-Venant type torsion the displacements are given by

$$\begin{aligned} u_1 &= -\theta x_3 x_2 \\ u_2 &= \theta x_3 x_1 \\ u_3 &= \theta\phi \end{aligned} \tag{a}$$

where θ is the torsion angle per unit length and $\phi(x, y)$ is the warping function given by

$$\nabla^2\phi = 0 \tag{b}$$

The boundary conditions are as follows.

Tractions normal to the boundary are identically zero, hence

$$\frac{\partial\phi}{\partial n} = |r| \cos(\vec{r}, \vec{t}) \tag{c}$$

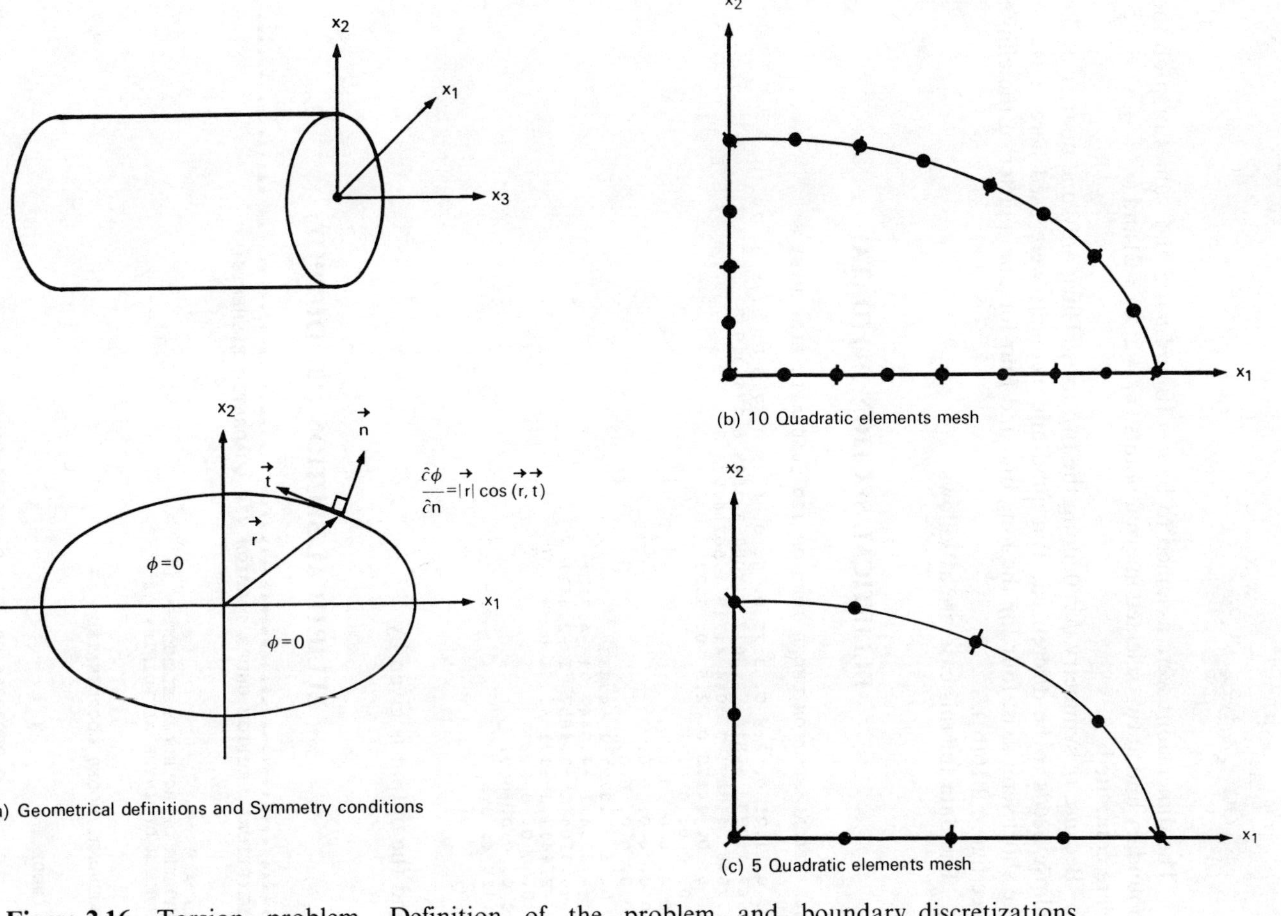

Figure 2.16 Torsion problem. Definition of the problem and boundary discretizations

For the case of an ellipse this becomes,

$$\frac{\partial \phi}{\partial n} = \frac{a^2 - b^2}{\sqrt{a^4 x_2^2 + b^4 x_1^2}} x_1 x_2 \tag{d}$$

The dimensions were assumed to be $a = 10$ and $b = 5$ and values of ϕ on the boundary and at two selected internal points ($x_1 = 2, x_2 = 2$) and ($x_1 = 4, x_2 = 3.5$) were computed.

Because of symmetry $\phi \equiv 0$ along the two axes. Thus, only one quarter of the ellipse needs to be discretized. Ten quadratic elements were used here. Two for the short semi-axis, four for the long one and four for one quarter of the ellipse (see figure 2.16(b)).

The data for this case are as follows:

ELLIPTICAL SECTION (10) (DATA)

```
ELLIPTICAL SECTION UNDER TORSION (10 QUADRATIC ELEMENTS)
10  2
0. 0. 1.25 0. 2.5 0. 3.75 0. 5. 0. 6.25 0. 7.5 0. 8.75 0. 10. 0.
9.67 1.273  8.814  2.3617 7.7008 3.1898 6.174 3.933
4.7898  4.3891 3.3044 4.719 1.557 4.939
0. 5. 0. 3.375 0. 2.5  0. 1.25
0 0. 0 0. 0 0.
 0 0. 0 0. 0 0.
 0 0. 0 0. 0 0.
 0 0. 0 0. 0 0.
1 0. 1 -3.379   1 -4.8334
1 -4.8334 1 -4.9447  1 -4.3104
1 -4.3104 1 -3.4657  1 -2.4411
1 -2.4411 1 -1.1643  1 0.
0 0. 0 0. 0 0.
 0 0. 0 0. 0 0.
2. 2. 4. 3.5
```

and the output is given by

ELLIPTICAL SECTION (10) (OUTPUT)

```
********************************************************************************
ELLIPTICAL SECTION UNDER TORSION (10 QUADRATIC ELEMENTS)

 DATA
 NUMBER OF BOUNDARY ELEMENTS= 10
 NUMBER OF INTERNAL POINTS=  2


 BOUNDARY NODES COORDINATES


   NODE           X                    Y

     1      0.0000000E+00       0.0000000E+00
     2      0.1250000E+01       0.0000000E+00
     3      0.2500000E+01       0.0000000E+00
     4      0.3750000E+01       0.0000000E+00
     5      0.5000000E+01       0.0000000E+00
```

6	0.6250000E+01	0.0000000E+00
7	0.7500000E+01	0.0000000E+00
8	0.8750000E+01	0.0000000E+00
9	0.1000000E+02	0.0000000E+00
10	0.9670000E+01	0.1273000E+01
11	0.8814000E+01	0.2361700E+01
12	0.7700800E+01	0.3189800E+01
13	0.6174000E+01	0.3933000E+01
14	0.4789800E+01	0.4389100E+01
15	0.3304400E+01	0.4719000E+01
16	0.1557000E+01	0.4939000E+01
17	0.0000000E+00	0.5000000E+01
18	0.0000000E+00	0.3375000E+01
19	0.0000000E+00	0.2500000E+01
20	0.0000000E+00	0.1250000E+01

BOUNDARY CONDITIONS

	-----FIRST NODE------		-----SECOND NODE-----		-----THIRD NODE------	
ELEMENT	PRESCRIBED VALUE	CODE	PRESCRIBED VALUE	CODE	PRESCRIBED VALUE	CODE
1	0.0000000E+00	0	0.0000000E+00	0	0.0000000E+00	0
2	0.0000000E+00	0	0.0000000E+00	0	0.0000000E+00	0
3	0.0000000E+00	0	0.0000000E+00	0	0.0000000E+00	0
4	0.0000000E+00	0	0.0000000E+00	0	0.0000000E+00	0
5	0.0000000E+00	1	-0.3379000E+01	1	-0.4833400E+01	1
6	-0.4833400E+01	1	-0.4944700E+01	1	-0.4310400E+01	1
7	-0.4310400E+01	1	-0.3465700E+01	1	-0.2441100E+01	1
8	-0.2441100E+01	1	-0.1164300E+01	1	0.0000000E+00	1
9	0.0000000E+00	0	0.0000000E+00	0	0.0000000E+00	0
10	0.0000000E+00	0	0.0000000E+00	0	0.0000000E+00	0

****

RESULTS

BOUNDARY NODES

X	Y	POTENTIAL	POTENTIAL DERIVATIVE BEFORE NODE	POTENTIAL DERIVATIVE AFTER NODE
0.00000E+00	0.00000E+00	0.00000E+00	-0.25810E-03	-0.25810E-03
0.12500E+01	0.00000E+00	0.00000E+00	0.74965E+00	0.74965E+00
0.25000E+01	0.00000E+00	0.00000E+00	0.14996E+01	0.14996E+01
0.37500E+01	0.00000E+00	0.00000E+00	0.22502E+01	0.22502E+01
0.50000E+01	0.00000E+00	0.00000E+00	0.29999E+01	0.29999E+01
0.62500E+01	0.00000E+00	0.00000E+00	0.37536E+01	0.37536E+01
0.75000E+01	0.00000E+00	0.00000E+00	0.44915E+01	0.44915E+01
0.87500E+01	0.00000E+00	0.00000E+00	0.53029E+01	0.53029E+01
0.10000E+02	0.00000E+00	0.00000E+00	0.58640E+01	0.00000E+00
0.96700E+01	0.12730E+01	-0.74619E+01	-0.33790E+01	-0.33790E+01
0.88140E+01	0.23617E+01	-0.12506E+02	-0.48334E+01	-0.48334E+01
0.77008E+01	0.31898E+01	-0.14746E+02	-0.49447E+01	-0.49447E+01
0.61740E+01	0.39330E+01	-0.14576E+02	-0.43104E+01	-0.43104E+01
0.47898E+01	0.43891E+01	-0.12616E+02	-0.34657E+01	-0.34657E+01
0.33044E+01	0.47190E+01	-0.93634E+01	-0.24411E+01	-0.24411E+01
0.15570E+01	0.49390E+01	-0.46000E+01	-0.11643E+01	-0.11643E+01
0.00000E+00	0.50000E+01	0.00000E+00	0.00000E+00	0.30154E+01
0.00000E+00	0.33750E+01	0.00000E+00	0.20220E+01	0.20220E+01
0.00000E+00	0.25000E+01	0.00000E+00	0.14999E+01	0.14999E+01
0.00000E+00	0.12500E+01	0.00000E+00	0.74935E+00	0.74935E+00

INTERNAL POINTS

X	Y	POTENTIAL
0.20000E+01	0.20000E+01	-0.23990E+01
0.40000E+01	0.35000E+01	-0.84020E+01

****

$

Results for some representative boundary nodes and the two internal points are compared with the known exact solution in the following table.

Boundary node	Potential using 10 quadratic elements	Exact potential solution
$x_1 = 8.814$, $x_2 = 2.361$	-12.506	-12.489
$x_1 = 6.174$, $x_2 = 3.933$	-14.576	-14.570
$x_1 = 3.304$, $x_2 = 4.719$	-9.363	-9.356
Internal points		
$x_1 = 2.$, $x_2 = 2.$	-2.399	-2.400
$x_1 = 4$, $x_2 = 3.5$	-8.402	-8.400

Example 2.5

This example is the solution of the same elliptical sections as described in figure 2.16(a)) and studied in Example 2.4 but the number of quadratic elements has been reduced to 5. It is interesting to see how the simple model gives results which are in good agreement with the theory.

The input for the five element model (figure 2.16(c)) is as follows:

ELLIPTICAL SECTION (5) (DATA)

```
ELLIPTICAL SECTION UNDER TORSION (5 QUADRATIC ELEMENTS)

5  2
0. 0.   2.5 0. 5. 0. 7.5 0. 10. 0.
  8.814   2.3617   6.174 3.933
 3.3044 4.719
0. 5.   0. 2.5
 0 0. 0 0. 0 0.
 0 0. 0 0. 0 0.
1 0.   1 -4.8334   1 -4.3104
1 -4.3104    1 -2.4411   1 0.
 0 0. 0 0. 0 0.
2. 2. 4. 3.5
```

The corresponding output is given below.

ELLIPTICAL SECTION (5) (OUTPUT)

```
******************************************************************************
ELLIPTICAL SECTION UNDER TORSION (5 QUADRATIC ELEMENTS)


 DATA
 NUMBER OF BOUNDARY ELEMENTS=  5
 NUMBER OF INTERNAL POINTS=  2

BOUNDARY NODES COORDINATES

   NODE          X                  Y

     1       0.0000000E+00      0.0000000E+00
     2       0.2500000E+01      0.0000000E+00
     3       0.5000000E+01      0.0000000E+00
     4       0.7500000E+01      0.0000000E+00
     5       0.1000000E+02      0.0000000E+00
     6       0.8814000E+01      0.2361700E+01
     7       0.6174000E+01      0.3933000E+01
     8       0.3304400E+01      0.4719000E+01
     9       0.0000000E+00      0.5000000E+01
    10       0.0000000E+00      0.2500000E+01


 BOUNDARY CONDITIONS

          -----FIRST NODE------      -----SECOND NODE-----      -----THIRD NODE----
           PRESCRIBED                 PRESCRIBED                 PRESCRIBED
ELEMENT      VALUE        CODE          VALUE       CODE          VALUE        CO
    1    0.0000000E+00     0       0.0000000E+00     0       0.0000000E+00
    2    0.0000000E+00     0       0.0000000E+00     0       0.0000000E+00
    3    0.0000000E+00     1      -0.4833400E+01     1      -0.4310400E+01
    4   -0.4310400E+01     1      -0.2441100E+01     1       0.0000000E+00
    5    0.0000000E+00     0       0.0000000E+00     0       0.0000000E+00
******************************************************************************

 RESULTS

 BOUNDARY NODES

                                                        POTENTIAL DERIVATIVE
       X                 Y               POTENTIAL      BEFORE NODE       AFTER NO

   0.00000E+00      0.00000E+00      0.00000E+00     -0.10336E-01      -0.10336E-
   0.25000E+01      0.00000E+00      0.00000E+00      0.15216E+01       0.15216E+
   0.50000E+01      0.00000E+00      0.00000E+00      0.29757E+01       0.29757E+
   0.75000E+01      0.00000E+00      0.00000E+00      0.46613E+01       0.46613E+
   0.10000E+02      0.00000E+00      0.00000E+00      0.51624E+01       0.00000E+
   0.88140E+01      0.23617E+01     -0.12779E+02     -0.48334E+01      -0.48334E+
   0.61740E+01      0.39330E+01     -0.14839E+02     -0.43104E+01      -0.43104E+
   0.33044E+01      0.47190E+01     -0.94349E+01     -0.24411E+01      -0.24411E+
   0.00000E+00      0.50000E+01      0.00000E+00      0.00000E+00       0.30003E+
   0.00000E+00      0.25000E+01      0.00000E+00      0.15278E+01       0.15278E+


 INTERNAL POINTS

       X                 Y               POTENTIAL

   0.20000E+01      0.20000E+01     -0.24306E+01
   0.40000E+01      0.35000E+01     -0.84719E+01
******************************************************************************
```

The following table compares the 5 element solution with the exact solution.

Boundary node	Potential using 5 quadratic elements	Exact potential solution
$x_1 = 8.814$, $x_2 = 2.361$	−12.779	−12.489
$x_1 = 6.174$, $x_2 = 3.933$	−14.839	−14.570
$x_1 = 3.304$, $x_2 = 4.719$	− 9.435	− 9.356
Internal points		
$x_1 = 2.$, $x_2 = 2.$	− 2.431	− 2.400
$x_1 = 4$, $x_2 = 3.5$	− 8.472	− 8.400

As an exercise, the reader can run the same problem using programs POCONBE and POLINBE for constant and linear elements and compare results with those shown in the above table and in Example 2.4.

2.9 Computer Code for Multiboundary Problems (POMCOBE)

In engineering practice many problems have more than one surface as shown in figure 2.17, with internal and external boundaries. Both types of boundary can be differentiated by identifying the direction of the normals. This can easily be done in two dimensional problems by adopting the rule that the numbering on the external surface is done counterclockwise and the one on the internal surface is carried out in the clockwise direction. From these rules the normal will be well defined in the computer code.

The following computer code is based on the constant element code (POCONBE) of section 4, but while the previous code was applicable to problems with only one surface the following listing applies for multisurface cases.

Main Program

Similar to program 1 already described in POCONBE, but now the Common statement is replaced by

```
COMMON N,L,NC(5),M,LEC,IMP
```

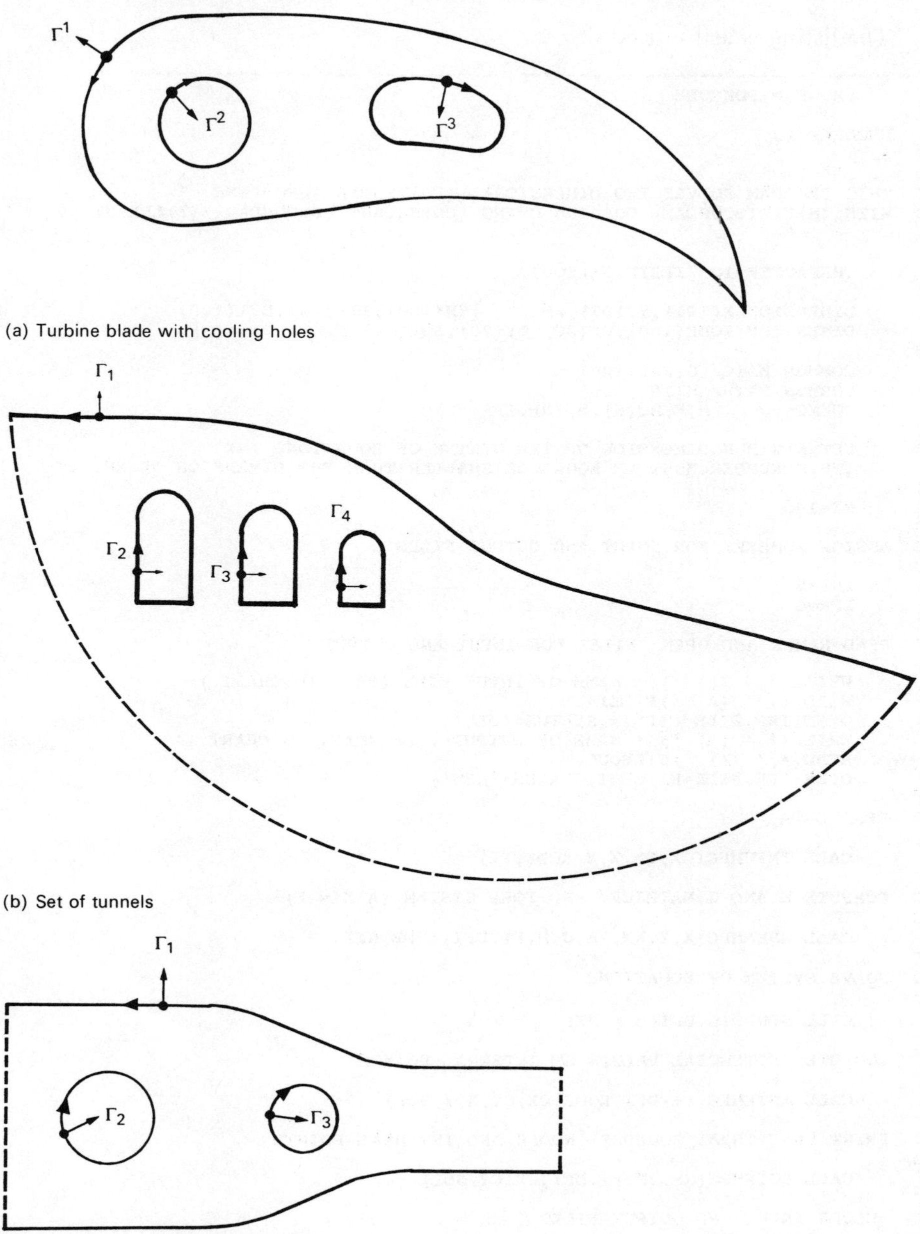

(a) Turbine blade with cooling holes

(b) Set of tunnels

(c) Potential flow around several obstacles

Figure 2.17 Problems with more than one surface

where M defines the number of different surfaces and NC stores the last node of each different surface. The dimensions of NC allow for 5 different surfaces in the present listing but this can easily be extended by the user if required.

The listing is as follows:

```
C-----------------------------------------------------------------------
      PROGRAM POMCOBE
C
C  PROGRAM 23
C
C
C  THIS PROGRAM SOLVES TWO DIMENSIONAL (PO)TENTIAL PROBLEMS
C  WITH (M)ULTIBOUNDARY DOMAINS USING (CO)NSTANT (B)OUNDARY (E)LEMENTS
C
C
      CHARACTER*10 FILEIN,FILEOUT
C
      DIMENSION X(101),Y(101),XM(100),YM(100),FI(100),DFI(100)
      DIMENSION KODE(100),CX(20),CY(20),SOL(20)
C
      COMMON/MATG/ G(100,100)
      COMMON/MATH/ H(100,100)
      COMMON      N,L,NC(5),M,INP,IPR
C
C     SET MAXIMUN DIMENSION OF THE SYSTEM OF EQUATIONS (NX)
C     (THIS NUMBER MUST BE EQUAL OR SMALLER THAN THE DIMENSION OF XM, ETC...)
C
      NX=100
C
C  ASSIGN NUMBERS FOR INPUT AND OUTPUT FILES
C
      INP=5
      IPR=6
C
C  READ NAMES AND OPEN  FILES FOR INPUT AND OUTPUT
C
      WRITE(*,' (A) ') ' NAME OF INPUT FILE (MAX. 10 CHART.)'
      READ(*,'  (A) ')FILEIN
      OPEN(INP,FILE=FILEIN,STATUS='OLD')
      WRITE(*,' (A) ') ' NAME OF OUTPUT FILE (MAX. 10 CHART.)'
      READ(*,' (A) ')FILEOUT
      OPEN(IPR,FILE=FILEOUT,STATUS='NEW')
C
C  READ DATA
C
      CALL INPUMPC(CX,CY,X,Y,KODE,FI)
C
C  COMPUTE H AND G MATRICES AND FORM SYSTEM (A X = F)
C
      CALL GHMAMPC(X,Y,XM,YM,G,H,FI,DFI,KODE,NX)
C
C  SOLVE SYSTEM OF EQUATIONS
C
      CALL SLNPD(G,DFI,D,N,NX)
C
C  COMPUTE  POTENTIAL VALUES AT INTERNAL POINTS
C
      CALL INTEMPC(FI,DFI,KODE,CX,CY,X,Y,SOL)
C
C  PRINT RESULTS AT BOUNDARY NODES AND INTERNAL POINTS
C
      CALL OUTPMPC(XM,YM,FI,DFI,CX,CY,SOL)
C
C   CLOSE INPUT AND OUTPUT FILES
C
      CLOSE (INP)
      CLOSE (IPR)
      STOP
      END
```

Routine INPUMPC

The input required is the same as in Program 2 of POCONBE with the exception of M and NC which are read in the same line as N and L (i.e. number of elements and number of internal points where the function u is required).

The listing is as follows:

```
C-----------------------------------------------------------------------
      SUBROUTINE INPUMPC(CX,CY,X,Y,KODE,FI)
C
C  PROGRAM 24
C
      CHARACTER*80 TITLE
      COMMON N,L,NC(5),M,INP,IPR
      DIMENSION CX(1),CY(1),X(1),Y(1),KODE(1),FI(1)
C
C  N= NUMBER OF BOUNDARY NODES (=NUMBER OF ELEMENTS)
C  L= NUMBER OF INTERNAL POINTS WHERE THE FUNCTION IS CALCULATED
C
      WRITE(IPR,100)
  100 FORMAT(' ',79('*'))
C
C  READ JOB TITLE
C
      READ(INP,'(A)') TITLE
      WRITE(IPR,'(A)') TITLE
C
C  READ NUMBER OF BOUNDARY ELEMENTS,NUMBER OF INTERNAL POINTS,
C  NUMBER OF DIFFERENT BOUNDARIES AND LAST NODE OF EACH BCUNDARY
C
      READ(INP,*)N,L,M,(NC(K),K=1,M)
      WRITE(IPR,300)N,L
  300 FORMAT(//' DATA'//2X,'NUMBER OF BOUNDARY ELEMENTS =',I3/2X,'NUMBER
     1 OF INTERNAL POINTS WHERE THE FUNCTION IS CALCULATEC =',I3)
C
      IF(M)40,40,30
   30 WRITE(IPR,999)M,(NC(K),K=1,M)
  999 FORMAT(/2X,'NUMBER OF DIFFERENT BOUNDARIES =',I3/2X,
     1'LAST NODE OF EACH BOUNDARY =',5(I3,','))
C
C
C  READ COORDINATES OF EXTREME POINTS OF THE BOUNDARY ELEMENTS
C  IN ARRAYS X AND Y
C
   40 WRITE(IPR,500)
  500 FORMAT(//2X,'COORDINATES OF THE EXTREME POINTS OF THE BOUNDARY ELE
     1MENTS',//1X,'POINT',7X,'X',15X,'Y')
      READ(INP,*) (X(I),Y(I),I=1,N)
      DO 10 I=1,N
   10 WRITE(IPR,700)I,X(I),Y(I)
  700 FORMAT(2X,I3,2(2X,E14.5))
C
C  READ BOUNDARY CONDITIONS IN FI(I) VECTOR, IF KODE(I)=0 THE FI(I)
C  VALUE IS A KNOWN POTENTIAL;IF KODE(I)=1 THE FI(I) VALUE IS A
C  KNOWN POTENTIAL DERIVATIVE (FLUX).
C
      WRITE(IPR,800)
  800 FORMAT(//2X,'BOUNDARY CONDITIONS'//2X,'NODE',6X,'CODE',7X,'PRESCRI
     1BED VALUE')
      DO 20 I=1,N
      READ(INP,*) KODE(I),FI(I)
   20 WRITE(IPR,950)I,KODE(I),FI(I)
  950 FORMAT(2X,I3,9X,I1,8X,E14.5)
C
C  READ COORDINATES OF THE INTERNAL POINTS
C
      IF(L.EQ.0) GO TO 50
      READ(INP,*) (CX(I),CY(I),I=1,L)
   50 RETURN
      END
```

Routine GHMAMPC

In this routine the COMMON needs to be changed as in the MAIN program. In addition some extra commands have been included to differentiate the points on each of the surfaces. These are required in order to compute the mid-point coordinates XM and YM. Each surface has to close and the last node of each surface is in the mid-point between the last extreme point to the first point on that surface.

The listing of GHMAMPC is now as follows.

```
C-----------------------------------------------------------------------
      SUBROUTINE GHMAMPC(X,Y,XM,YM,G,H,FI,DFI,KODE,NX)
C
C  PROGRAM 25
C
C  THIS SUBROUTINE COMPUTES THE G AND H MATRICES
C  AND FORMS THE SYSTEM OF EQUATIONS A X = F
C
      COMMON N,L,NC(5),M,INP,IPR
      DIMENSION X(1),Y(1),XM(1),YM(1),FI(1),KODE(1)
      DIMENSION DFI(1),G(NX,NX),H(NX,NX)
C
C  COMPUTE THE NODAL COORDINATES AND STORE IN ARRAYS XM AND YM
C
      X(N+1)=X(1)
      Y(N+1)=Y(1)
      DO 10 I=1,N
      XM(I)=(X(I)+X(I+1))/2
   10 YM(I)=(Y(I)+Y(I+1))/2
      IF(M-1)15,15,12
   12 XM(NC(1))=(X(NC(1))+X(1))/2
      YM(NC(1))=(Y(NC(1))+Y(1))/2
      DO 13 K=2,M
      XM(NC(K))=(X(NC(K))+X(NC(K-1)+1))/2
   13 YM(NC(K))=(Y(NC(K))+Y(NC(K-1)+1))/2
C
C  COMPUTE THE COEFFICIENTS OF G AND H MATRICES
C
   15 DO 30 I=1,N
      DO 30 J=1,N
      IF(M-1)16,16,17
   17 IF(J-NC(1))19,18,19
   18 KK=1
      GO TO 23
   19 DO 22 K=2,M
      IF(J-NC(K))22,21,22
   21 KK=NC(K-1)+1
      GO TO 23
   22 CONTINUE
   16 KK=J+1
   23 IF(I-J)20,25,20
   20 CALL EXTINPC(XM(I),YM(I),X(J),Y(J),X(KK),Y(KK),H(I,J),G(I,J))
      GO TO 30
   25 CALL LOCINPC(X(J),Y(J),X(KK),Y(KK),G(I,J))
      H(I,J)=3.1415926
   30 CONTINUE
C
C  REORDER THE COLUMNS OF THE SYSTEM OF EQUATIONS IN ACCORDANCE
C  WITH THE BOUNDARY CONDITIONS AND FORM SYSTEM MATRIX A WHICH
C  IS STORED IN G
C
      DO 55 J=1,N
      IF(KODE(J))55,55,40
   40 DO 50 I=1,N
      CH=G(I,J)
      G(I,J)=-H(I,J)
      H(I,J)=-CH
   50 CONTINUE
```

```
   55 CONTINUE
C
C  FORM THE RIGHT HAND SIDE VECTOR F WHICH IS STORED IN DFI
C
      DO 60 I=1,N
      DFI(I)=0.
      DO 60 J=1,N
      DFI(I)=DFI(I)+H(I,J)*FI(J)
   60 CONTINUE
      RETURN
      END
```

Routine EXTINPC

As in POCONBE (Program 4).

Routine LOCINPC

As in POCONBE (Program 5).

Routine SLNPD

As in POCONBE (Program 6).

Routine INTEMPC

This routine varies from Program 7 by a few statements to take into account the different surfaces.

```
C-----------------------------------------------------------------------
      SUBROUTINE INTEMPC(FI,DFI,KODE,CX,CY,X,Y,SOL)
C
C  PROGRAM 26
C
C  THIS SUBROUTINE COMPUTES THE VALUES OF POTENTIAL AT INTERNAL POINTS
C
      COMMON N,L,NC(5),M,INP,IPR
      DIMENSION FI(1),DFI(1),KODE(1),CX(1),CY(1),X(1),Y(1),SOL(1)
C
C  REARRANGE THE FI AND DFI ARRAYS TO STORE ALL THE VALUES OF THE
C  POTENTIAL IN FI AND ALL THE VALUES OF THE DERIVATIVE IN DFI
C
      DO 20 I=1,N
      IF(KODE(I)) 20,20,10
   10 CH=FI(I)
      FI(I)=DFI(I)
      DFI(I)=CH
   20 CONTINUE
C
C  COMPUTE THE VALUES OF POTENTIAL AT INTERNAL POINTS
C
      IF(L.EQ.0) GO TO 50
      DO 40 K=1,L
      SOL(K)=0.
      DO 30 J=1,N
      IF(M-1)28,28,22
   22 IF(J-NC(1))24,23,24
```

```
   23 KK=1
      GO TO 29
   24 DO 26 LK=2,M
      IF(J-NC(LK))26,25,26
   25 KK=NC(LK-1)+1
      GO TO 29
   26 CONTINUE
   28 KK=J+1
   29 CALL EXTINPC(CX(K),CY(K),X(J),Y(J),X(KK),Y(KK),A,B)
   30 SOL(K)=SOL(K)+DFI(J)*B-FI(J)*A
   40 SOL(K)=SOL(K)/(2*3.1415926)
   50 RETURN
      END
```

Routine OUTPMPC

Same as Program 8 but with the new COMMON.

```
C-----------------------------------------------------------------------
      SUBROUTINE OUTPMPC(XM,YM,FI,DFI,CX,CY,SOL)
C
C  PROGRAM 27
C
C  THIS SUBROUTINE PRINTS THE VALUES OF THE POTENTIAL AND ITS NORMAL
C  DERIVATIVE AT BOUNDARY NODES. IT ALSO PRINTS THE VALUES OF THE
C  POTENTIAL AT INTERNAL POINTS
C
      COMMON N,L,NC(5),M,INP,IPR
      DIMENSION XM(1),YM(1),FI(1),DFI(1),CX(1),CY(1),SOL(1)
C
      WRITE(IPR,100)
  100 FORMAT(' ',79('*')//1X,'RESULTS'//2X,'BOUNDARY NODES'//8X,'X',15
     1X,'Y',13X,'POTENTIAL',3X,'POTENTIAL DERIVATIVE'/)
      DO 10 I=1,N
   10 WRITE(IPR,200) XM(I),YM(I),FI(I),DFI(I)
  200 FORMAT(4(2X,E14.5))
C
      IF(L.EQ.0) GO TO 30
      WRITE(IPR,300)
  300 FORMAT(//,2X,'INTERNAL POINTS',//8X,'X',15X,'Y',13X,'POTENTIAL',/
     1)
      DO 20 K=1,L
   20 WRITE(IPR,400)CX(K),CY(K),SOL(K)
  400 FORMAT(3(2X,E14.5))
   30 WRITE(IPR,500)
  500 FORMAT(' ',79('*'))
      RETURN
      END
```

2.10 Boundary Elements for Three Dimensional Problems

The elements used in three dimensional problems are surface elements which cover the boundary of the body (figure 2.18). They are usually of two types; triangular or quadrilateral and both can be flat or curved. The functions u and q and those used to describe the geometry can be constant over the element, vary linearly, being second order functions and others which produce a curved element. While the development of constant or flat elements is comparatively simple, curved elements are more important in three dimensions as they can follow better the geometry of actual engineering components and hence they will be described in detail here.

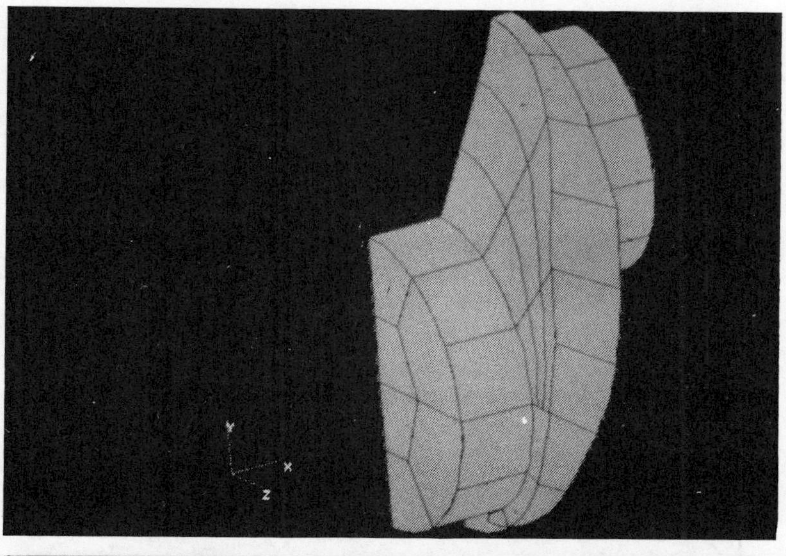

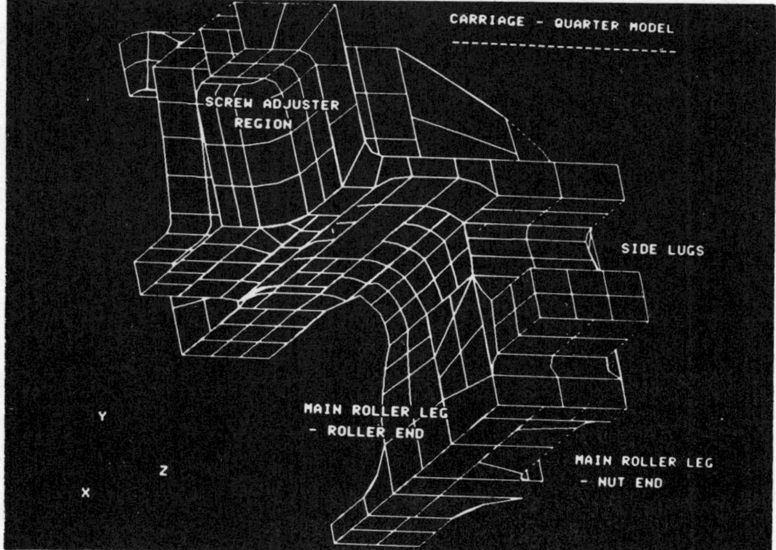

Figure 2.18 Some three dimensional applications (lower figure by courtesy of British Aerospace plc)

To study curved elements first we need to define the way in which we can pass from the $x_1x_2x_3$ global cartesian system to the $\xi_1\xi_2$, η system defined over the element, where $\xi_1\xi_2$ are oblique coordinates and η is in the direction of the normal (see figure 2.19).

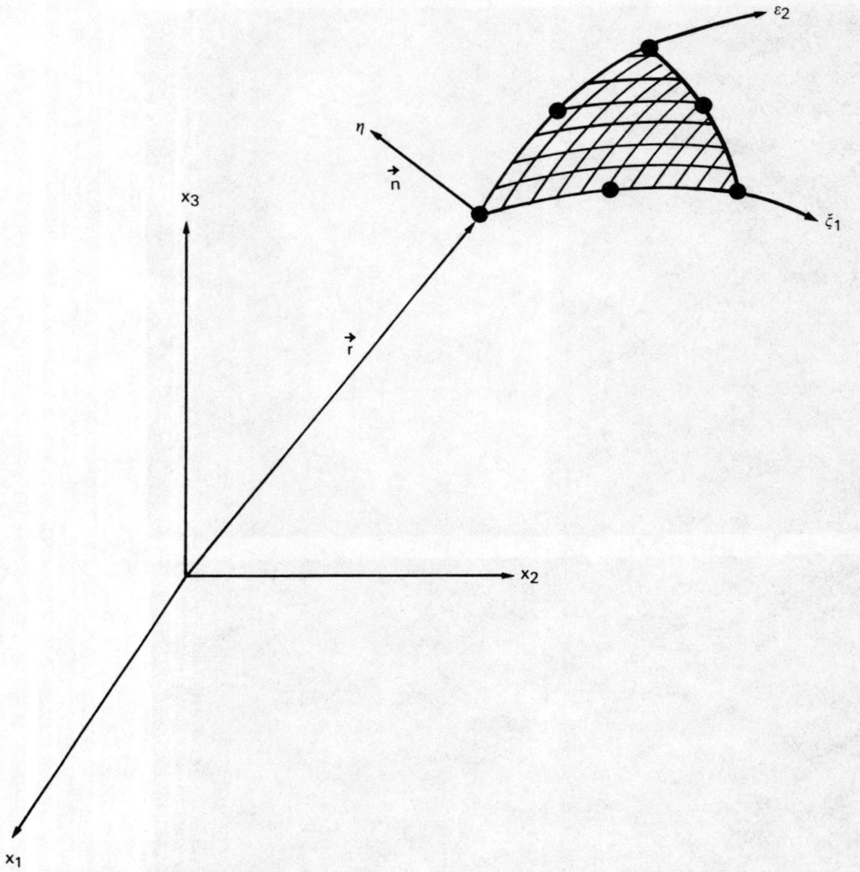

Figure 2.19 a) Triangular curved elements for three dimensional problems

The transformation for a given function – say u – is related through the following,

$$\begin{Bmatrix} \dfrac{\partial u}{\partial \xi_1} \\ \dfrac{\partial u}{\partial \xi_2} \\ \dfrac{\partial u}{\partial \eta} \end{Bmatrix} = \begin{bmatrix} \dfrac{\partial x_1}{\partial \xi_1} & \dfrac{\partial x_2}{\partial \xi_1} & \dfrac{\partial x_3}{\partial \xi_1} \\ \dfrac{\partial x_1}{\partial \xi_2} & \dfrac{\partial x_2}{\partial \xi_2} & \dfrac{\partial x_3}{\partial \xi_2} \\ \dfrac{\partial x_1}{\partial \eta} & \dfrac{\partial x_2}{\partial \eta} & \dfrac{\partial x_3}{\partial \eta} \end{bmatrix} \begin{Bmatrix} \dfrac{\partial u}{\partial x_1} \\ \dfrac{\partial u}{\partial x_2} \\ \dfrac{\partial u}{\partial x_3} \end{Bmatrix} \tag{2.85}$$

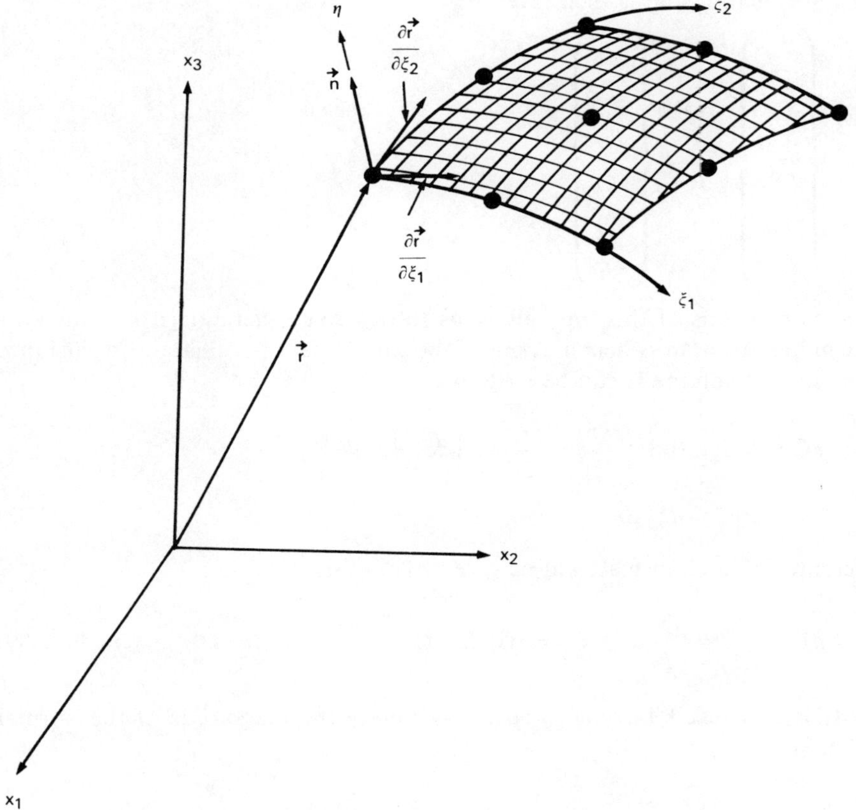

Figure 2.19 b) Quadrilateral curved elements for three dimensional problems

where the square matrix is the Jacobian or **J**. Hence

$$\begin{Bmatrix} \dfrac{\partial u}{\partial \xi_1} \\ \dfrac{\partial u}{\partial \xi_2} \\ \dfrac{\partial u}{\partial \eta} \end{Bmatrix} = \mathbf{J} \begin{Bmatrix} \dfrac{\partial x_1}{\partial x_1} \\ \dfrac{\partial u}{\partial x_2} \\ \dfrac{\partial u}{\partial x_3} \end{Bmatrix} \tag{2.86}$$

The inverse relationship is then given by

$$\begin{Bmatrix} \dfrac{\partial u}{\partial x_1} \\ \dfrac{\partial u}{\partial x_2} \\ \dfrac{\partial u}{\partial x_3} \end{Bmatrix} = \mathbf{J}^{-1} \begin{Bmatrix} \dfrac{\partial u}{\partial \xi_1} \\ \dfrac{\partial u}{\partial \xi_2} \\ \dfrac{\partial u}{\partial \eta} \end{Bmatrix} \tag{2.87}$$

Transformations of this type allow us to describe differentials of volume or surface in the cartesian system in terms of the curvilinear coordinates. For instance a differential of volume Ω can be written as,

$$\begin{aligned} d\Omega &= \text{Magnitude}\left(\frac{\partial \vec{r}}{\partial \xi_1} \times \frac{\partial \vec{r}}{\partial \xi_2} \cdot \frac{\partial \vec{r}}{\partial \eta}\right) d\xi_1\, d\xi_2\, d\eta \\ &= |\mathbf{J}|\, d\xi_1\, d\xi_2\, d\eta \end{aligned} \tag{2.88}$$

A differential of area instead will be given by

$$d\Gamma = \left|\frac{\partial \vec{r}}{\partial \xi_1} \times \frac{\partial \vec{r}}{\partial \xi_2}\right| d\xi_1\, d\xi_2 = |\mathbf{G}|\, d\xi_1\, d\xi_2 \tag{2.89}$$

where $\mathbf{G}$ is a reduced Jacobian and $|\mathbf{G}|$ is simply the magnitude of the normal vector $\vec{\eta}$; i.e.

$$\vec{\eta} = \frac{\partial \vec{r}}{\partial \xi_1} \times \frac{\partial \vec{r}}{\partial \xi_2} = (g_1, g_2, g_3) = \left(\frac{\partial x_1}{\partial \eta}, \frac{\partial x_2}{\partial \eta}, \frac{\partial x_3}{\partial \eta}\right) \tag{2.90}$$

where

$$\frac{\partial \vec{r}}{\partial \xi_1} = \left(\frac{\partial x_1}{\partial \xi_1}, \frac{\partial x_2}{\partial \xi_1}, \frac{\partial x_3}{\partial \xi_1}\right); \qquad \frac{\partial \vec{r}}{\partial \xi_2} = \left(\frac{\partial x_1}{\partial \xi_2}, \frac{\partial x_2}{\partial \xi_2}, \frac{\partial x_3}{\partial \xi_2}\right)$$

Notice that the values of g_1, g_2 and g_3 are given by

$$\begin{aligned} g_1 &= \left(\frac{\partial x_2}{\partial \xi_1}\frac{\partial x_3}{\partial \xi_2} - \frac{\partial x_2}{\partial \xi_2}\frac{\partial x_3}{\partial \xi_1}\right) \\ g_2 &= \left(\frac{\partial x_3}{\partial \xi_1}\frac{\partial x_1}{\partial \xi_2} - \frac{\partial x_1}{\partial \xi_1}\frac{\partial x_3}{\partial \xi_2}\right) \\ g_3 &= \left(\frac{\partial x_1}{\partial \xi_1}\frac{\partial x_2}{\partial \xi_2} - \frac{\partial x_2}{\partial \xi_1}\frac{\partial x_1}{\partial \xi_2}\right) \end{aligned} \tag{2.91}$$

Hence the magnitude of $|\mathbf{G}|$ is given by

$$|\mathbf{G}| = \sqrt{(g_1^2 + g_2^2 + g_3^2)} \tag{2.92}$$

These relationships can be used to integrate any of the boundary integrals terms such as,

$$\int_{\Gamma} u^* q \, d\Gamma \qquad \text{or} \qquad \int_{\Gamma} u q^* \, d\Gamma \tag{2.93}$$

which now become

$$\int_{\Gamma_\xi} u^* q |\mathbf{G}| \, d\xi_1 \, d\xi_2; \qquad \int_{\Gamma_\xi} u q^* |\mathbf{G}| \, d\xi_1 \, d\xi_2 \tag{2.94}$$

A whole range of quadrilateral and triangular elements can be defined. For quadrilateral cases one prefers to use Lagrangian type of elements (i.e. the ones which have in some cases nodes inside) as such elements give better numerical results when used with point collocation. This is simply because the collocation points are better distributed in these cases. Table 2.1 shows some of the elements which can be used.

2.11 Poisson's Equation

Many practical applications are governed by the Poisson's rather than Laplace equations. In these cases sources are distributed in the Ω domain in accordance with a $b(\mathbf{x})$ function. This produces the following governing equation

$$\nabla^2 u = b \qquad \text{in } \Omega \tag{2.95}$$

where b is a known function of position.

Sometimes the above equation can be reduced to Laplace's simply by substituting a particular solution or a change of variables. Care should be taken in these cases also to transform the boundary conditions accordingly.

When the function $b(\mathbf{x})$ has a more complex formulation it may be difficult to find any suitable transformation and one needs to start with (2.95) plus the appropriate boundary conditions in order to deduce the basic integral equation, i.e.

$$\int_{\Omega} (\nabla^2 u - b) u^* \, d\Omega = \int_{\Gamma_2} (q - \bar{q}) u^* \, d\Gamma - \int_{\Gamma_1} (u - \bar{u}) q^* \, d\Gamma \tag{2.96}$$

which integrated by parts twice produces,

$$\begin{aligned} \int_{\Omega} (\nabla^2 u^*) u \, d\Omega - \int_{\Omega} b u^* \, d\Omega = & -\int_{\Gamma_2} \bar{q} u^* \, d\Gamma - \int_{\Gamma_1} q u^* \, d\Gamma \\ & + \int_{\Gamma_2} u q^* \, d\Gamma + \int_{\Gamma_1} \bar{u} q^* \, d\Gamma \end{aligned} \tag{2.97}$$

Table 2.1 Table of Triangular and Quadrilateral Surface Elements for 3D Problems

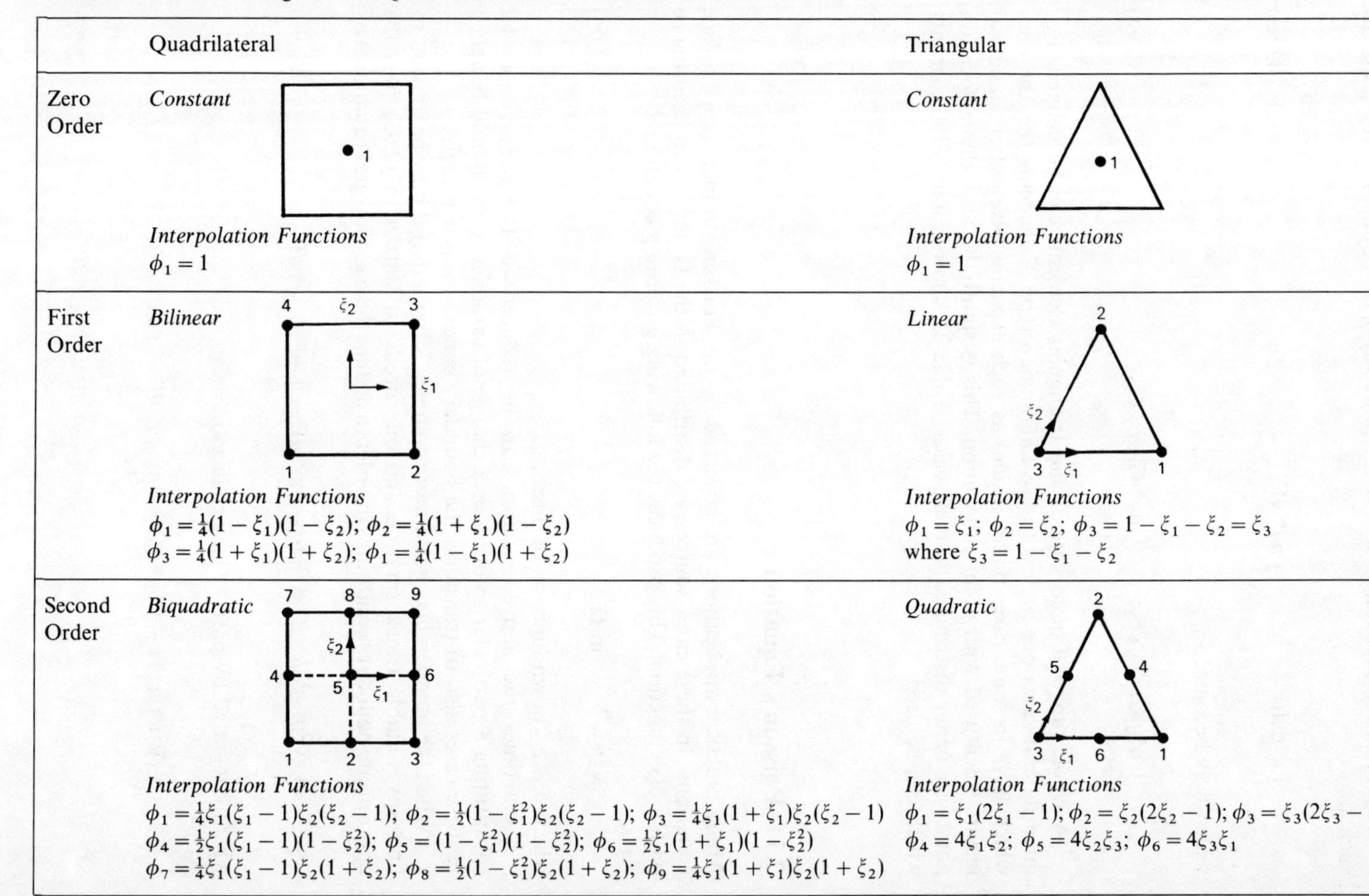

	Quadrilateral	Triangular
Zero Order	*Constant* *Interpolation Functions* $\phi_1 = 1$	*Constant* *Interpolation Functions* $\phi_1 = 1$
First Order	*Bilinear* *Interpolation Functions* $\phi_1 = \frac{1}{4}(1-\xi_1)(1-\xi_2)$; $\phi_2 = \frac{1}{4}(1+\xi_1)(1-\xi_2)$ $\phi_3 = \frac{1}{4}(1+\xi_1)(1+\xi_2)$; $\phi_1 = \frac{1}{4}(1-\xi_1)(1+\xi_2)$	*Linear* *Interpolation Functions* $\phi_1 = \xi_1$; $\phi_2 = \xi_2$; $\phi_3 = 1-\xi_1-\xi_2 = \xi_3$ where $\xi_3 = 1-\xi_1-\xi_2$
Second Order	*Biquadratic* *Interpolation Functions* $\phi_1 = \frac{1}{4}\xi_1(\xi_1-1)\xi_2(\xi_2-1)$; $\phi_2 = \frac{1}{2}(1-\xi_1^2)\xi_2(\xi_2-1)$; $\phi_3 = \frac{1}{4}\xi_1(1+\xi_1)\xi_2(\xi_2-1)$ $\phi_4 = \frac{1}{2}\xi_1(\xi_1-1)(1-\xi_2^2)$; $\phi_5 = (1-\xi_1^2)(1-\xi_2^2)$; $\phi_6 = \frac{1}{2}\xi_1(1+\xi_1)(1-\xi_2^2)$ $\phi_7 = \frac{1}{4}\xi_1(\xi_1-1)\xi_2(1+\xi_2)$; $\phi_8 = \frac{1}{2}(1-\xi_1^2)\xi_2(1+\xi_2)$; $\phi_9 = \frac{1}{4}\xi_1(1+\xi_1)\xi_2(1+\xi_2)$	*Quadratic* *Interpolation Functions* $\phi_1 = \xi_1(2\xi_1-1)$; $\phi_2 = \xi_2(2\xi_2-1)$; $\phi_3 = \xi_3(2\xi_3-1)$ $\phi_4 = 4\xi_1\xi_2$; $\phi_5 = 4\xi_2\xi_3$; $\phi_6 = 4\xi_3\xi_1$

After substituting the Laplace fundamental solution u^* and grouping all boundary terms together (i.e. in $\Gamma = \Gamma_1 + \Gamma_2$), one obtains

$$c^i u^i + \int_\Gamma u q^* \, d\Gamma + \int_\Omega b u^* \, d\Omega = \int_\Gamma q u^* \, d\Gamma \tag{2.98}$$

Notice that although the b functions are known and consequently the integrals in Ω do not introduce any new unknowns, the problem has changed in character as we need now to carry out integrals in the domain as well as on the boundary.

Regions of integration called cells can now be employed to compute the domain integral in (2.98). One usually applies a numerical integration scheme such as Gauss. In this case for each position of the singularity at a boundary point i, the integral in (2.98) can be written as,

$$B^i = \int_\Omega b u^* \, d\Omega = \sum_{e=1}^{M} \left(\sum_{k=1}^{r} w_k (b u^*)_k \right) \Omega_e \tag{2.99}$$

where e are the different cells (e varies from 1 to M, where M is the total number of cells describing Ω domain), w_k are the integration weights, the function (bu^*) needs to be evaluated at r integration points on each cell, Ω_e is the area of the cell 'e'. B^i is the result, different for each 'i' position of the fundamental solution, where i is one of the boundary nodes.

Hence equation (2.98) now becomes

$$c^i u^i + \sum_{j=1}^{N} \hat{H}^{ij} u^j + B^i = \sum_{j=1}^{N} G^{ij} q^j \tag{2.100}$$

or in matrix form

$$\mathbf{HU} + \mathbf{B} = \mathbf{GQ} \tag{2.101}$$

Notice that the domain integrals need to be computed as well when calculating any values of potentials or fluxes at internal points. Hence,

$$u^i = \sum_{j=1}^{N} G^{ij} q^j - \sum_{j=1}^{N} \hat{H}^{ij} u^j - B^i \tag{2.102}$$

where i is now an internal point, at which the singularity is applied.

Concentrated sources are very simple to handle in boundary elements. They are a special case for which the function b at the internal point 'l' becomes,

$$b = Q^l \Delta^l \tag{2.103}$$

where Q^l is the magnitude of the source and Δ^l is a Dirac delta function whose integral is equal to 1 at the point l and zero elsewhere. Assuming that a number of these functions exist one can write,

$$c^i u^i + \int_\Gamma uq^* \, d\Gamma + \int_\Omega bu^* \, d\Omega + \sum_{l=1}^{P} (Q^l u^{*l}) = \int_\Gamma qu^* \, d\Gamma \qquad (2.104)$$

u^{*l} is the value of the fundamental solution at the point l. P are the number of concentrated forces within the domain.

Another way of dealing with volume sources is to transform the domain integral into equivalent boundary integrals. This is possible when the function b is harmonic (although the approach can in principle be extended to harmonic, or bi-harmonic, etc.). Harmonic means that b obeys the following equation,

$$\nabla^2 b = 0 \qquad (2.105)$$

To effect the transformation – which is basically an integration by parts – we need to express the fundamental solution u^* in function of some derivatives. This is done by proposing a function v^* such that,

$$u^* = \nabla^2 v^* \qquad (2.106)$$

and then writing Green's identity, in the form

$$\int_\Omega (b\nabla^2 v^* - v^* \nabla^2 b) \, d\Omega = \int_\Gamma \left(b \frac{\partial v^*}{\partial n} - v^* \frac{\partial b}{\partial n} \right) d\Gamma \qquad (2.107)$$

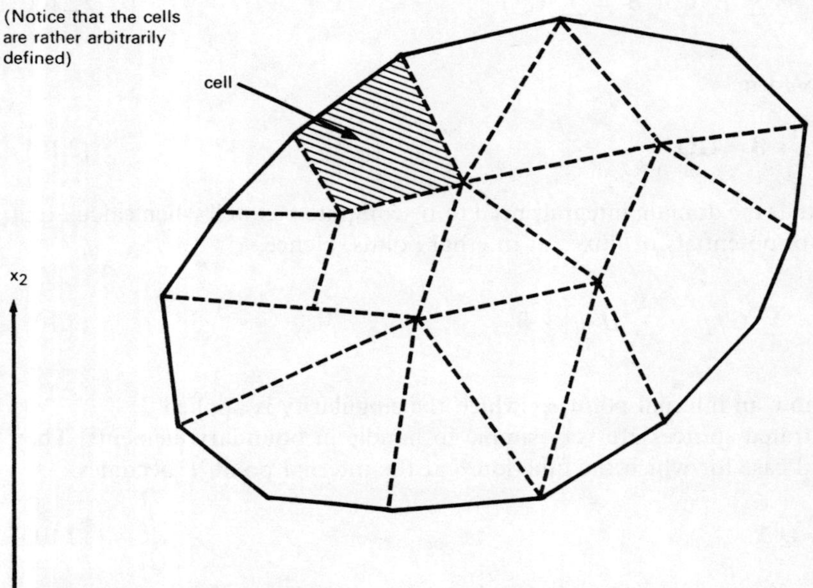

Figure 2.20 Boundary elements and internal cells

which if $\nabla^2 b \equiv 0$ reduces to,

$$\int_\Omega bu^* \, d\Omega = \int_\Gamma \left(b \frac{\partial v^*}{\partial n} - v^* \frac{\partial b}{\partial n} \right) d\Gamma \tag{2.108}$$

Hence we have effectively reduced the domain integrals to two different boundary integrals.

The function v^* required in this case is simply the fundamental solution of the biharmonic equation as,

$$\nabla^2 u^* = \nabla^2(\nabla^2 v^*) = \nabla^4 v^* = -\Delta^i \tag{2.109}$$

which for two dimensions is a well known fundamental solution used in plate bending, i.e.

$$v^* = \frac{r^2}{8\pi} \left[\ln\left(\frac{1}{r}\right) + 1 \right] \tag{2.110}$$

and for three dimensions

$$v^* = \frac{1}{8\pi} r \tag{2.111}$$

Notice that for two dimensions v^* satisfies Laplace's equation as follows,

$$\nabla^2 v^* = \frac{1}{r} \frac{\partial}{\partial r} \left(r \frac{\partial v^*}{\partial r} \right) = \frac{1}{2\pi} \ln \frac{1}{r} = u^* \tag{2.112}$$

and for three dimensions one finds that,

$$\nabla^2 v^* = \frac{1}{r^2} \frac{\partial}{\partial r} \left(r^2 \frac{\partial v^*}{\partial r} \right) = \frac{1}{4\pi r} = u^* \tag{2.113}$$

2.12 Orthotropy and Anisotropy

Up to now we have considered problems with isotropic materials, i.e. those for which the properties are the same in all directions. We will study those materials for which the properties vary in different directions. If one finds the directions of orthotropy (figure 2.21) the governing equation for a two dimensional problem can be written,

$$k_1 \frac{\partial^2 u}{\partial y_1^2} + k_2 \frac{\partial^2 u}{\partial y_2^2} = 0 \tag{2.114}$$

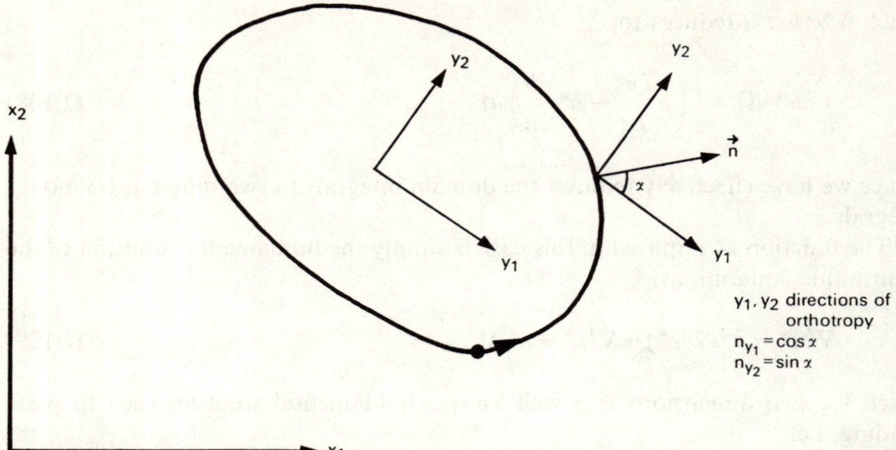

Figure 2.21 Orthotropic medium

k_i is the medium property coefficient in the direction of orthotropy i. For three dimensional cases the equation is

$$k_1 \frac{\partial^2 u}{\partial y_1^2} + k_2 \frac{\partial^2 u}{\partial y_2^2} + k_3 \frac{\partial^2 u}{\partial y_3^2} = 0 \tag{2.115}$$

The fundamental solution corresponding to the above equations can be found by transforming the y_j system of coordinates into a z_j system in which the governing equations (2.114) or (2.115) become a Laplace equation without the k_j coefficients. This is achieved by using the following transformation,

$$z_j = \frac{y_j}{\sqrt{k_j}} \tag{2.116}$$

The fundamental solution of the Laplace equation in the z_j system is known and can then be backtransformed to the x_j coordinates. This gives for the two dimensional case the following solution:

$$u^* = \frac{1}{\sqrt{k_1 k_2}} \ln \frac{1}{r} \tag{2.117}$$

where r is now

$$r = \left\{ \frac{1}{k_1} [y_1^i - y_1]^2 + \frac{1}{k_2} [y_2^i - y_2]^2 \right\}^{1/2} \tag{2.118}$$

where y_1, y_2 are the coordinates of the point under consideration and $y_1^i y_2^i$ are those of the node on which the fundamental solution is applied.

The corresponding normal flux is

$$q^* = k_1 \frac{\partial u^*}{\partial y_1} n_{y1} + k_2 \frac{\partial u^*}{\partial y_2} n_{y2} \tag{2.119}$$

where n_{y1} and n_{y2} are the direction cosines of the normal n to the surface with respect to y_1 and y_2 respectively. Similarly for the actual flux we have

$$q = k_1 \frac{\partial u}{\partial y_1} n_{y1} + k_2 \frac{\partial u}{\partial y_2} n_{y2} \tag{2.120}$$

For a fully anisotropic media, governing equation (2.114) becomes,

$$k_{11} \frac{\partial^2 u}{\partial x_1^2} + 2k_{12} \frac{\partial^2 u}{\partial x_1 \partial x_2} + k_{22} \frac{\partial^2 u}{\partial x_2^2} = 0 \tag{2.121}$$

where the k_{ij} coefficients represent the terms of the properties tensor. The fundamental solution can now be expressed

$$u^* = \frac{1}{\sqrt{|k_{ij}|}} \ln \frac{1}{r} \tag{2.122}$$

where $|k_{ij}|$ is the determinant of the medium properties coefficients matrix, i.e.

$$|k_{ij}| = k_{11}k_{22} - k_{12}^2 \tag{2.123}$$

The r distance is now

$$r = \left\{ \frac{1}{k_{11}} [x_1^i - x_i]^2 + \frac{2}{k_{12}} [x_1^i - x_1][x_2^i - x_2] + \frac{1}{k_{22}} [x_2^i - x_2]^2 \right\}^{1/2} \tag{2.124}$$

The normal flux is

$$q^* = \left(k_{11} \frac{\partial u^*}{\partial x_1} + k_{12} \frac{\partial u^*}{\partial x_2} \right) n_{x1} + \left(k_{12} \frac{\partial u^*}{\partial x_1} + k_{22} \frac{\partial u^*}{\partial x_2} \right) n_{x2} \tag{2.125}$$

and similarly for q.

The fundamental solution for the three dimensional orthotropic case is

$$u^* = \frac{1}{\sqrt{k_1 k_2 k_3}} \frac{1}{4\pi} \frac{1}{r} \tag{2.126}$$

where r is now

$$r = \left\{ \frac{1}{k_1} [y_1^i - y_1]^2 + \frac{1}{k_2} [y_2^i - y_2]^2 + \frac{1}{k_3} [y_3^3 - y_3]^2 \right\} \tag{2.127}$$

Similar considerations as for two dimensional cases apply for the fluxes q^* and q.

2.13 Subregions

In certain cases the region under study may be piecewise homogeneous and then the boundary element procedure can be applied to each subregion in turn as if they were independent of each other. The final set of equations for the whole region can then be obtained by assembling the set of equations for each subregion using compatibility of potentials and fluxes between the common interfaces.

Consider for instance the two subregions shown in figure 2.22 one called Ω^1 and the other Ω^2. Over subregion Ω^1 we define,

$\mathbf{U}^1, \mathbf{Q}^1$; Nodal potential and fluxes at the external boundary Γ^1

$\mathbf{U}_I^1, \mathbf{Q}_I^1$; Nodal potential and fluxes at the interface Γ_I considering it belongs to Ω_1

Over the other subregion, Ω_2, one can define,

$\mathbf{U}^2, \mathbf{Q}^2$; Nodal potential and fluxes at the external boundary Γ^2

$\mathbf{U}_I^2, \mathbf{Q}_I^2$; Nodal potential and fluxes at the interface Γ_I considering it belongs to Ω_2.

The system of equations corresponding to Ω_1 can be written as

$$[\mathbf{H}^1\mathbf{H}_I^1]\begin{Bmatrix}\mathbf{U}^1\\ \mathbf{U}_I^1\end{Bmatrix} = [\mathbf{G}^1\mathbf{G}_I^1]\begin{Bmatrix}\mathbf{Q}^1\\ \mathbf{Q}_I^1\end{Bmatrix} \tag{2.128}$$

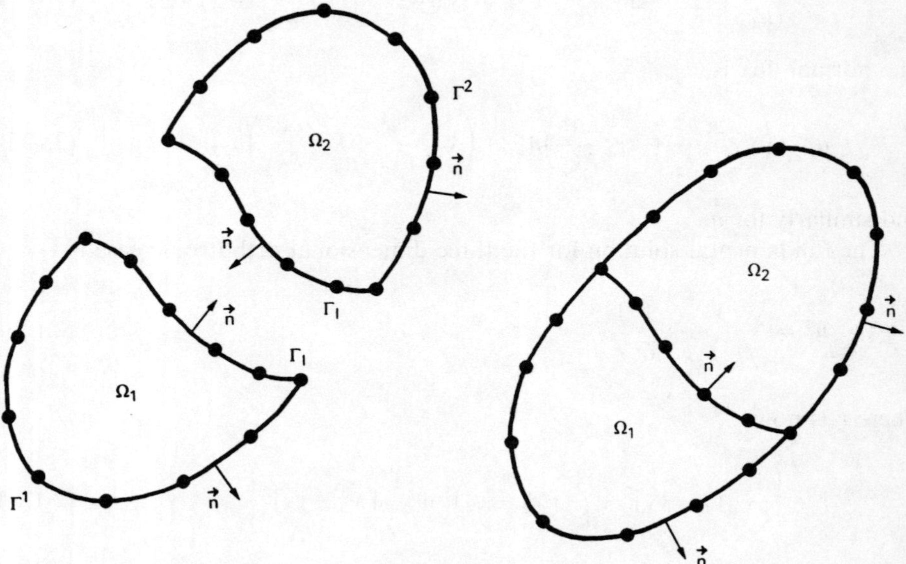

Figure 2.22 Piecewise homogeneous body consisting of two subregions

and the one for subregion Ω_2 gives

$$[\mathbf{H}^2\mathbf{H}_\mathrm{I}^2]\begin{Bmatrix}\mathbf{U}^2\\\mathbf{U}_\mathrm{I}^2\end{Bmatrix}=[\mathbf{G}^2\mathbf{G}_\mathrm{I}^2]\begin{Bmatrix}\mathbf{Q}^2\\\mathbf{Q}_\mathrm{I}^2\end{Bmatrix}\tag{2.129}$$

The compatibility and equilibrium conditions on the interface Γ_I can be expressed as

$$\mathbf{U}_\mathrm{I}^1=\mathbf{U}_\mathrm{I}^2\tag{2.130}$$

and

$$\mathbf{Q}_\mathrm{I}^1+\mathbf{Q}_\mathrm{I}^2=\mathbf{0}\tag{2.131}$$

If one calls the potentials at the interface $\mathbf{U}_\mathrm{I}$ and adapts on the same interface the fluxes of Ω_1 as reference values (which is equivalent to say that the normal on the interface is the normal to Ω_1) one has,

$$\mathbf{U}_\mathrm{I}=\mathbf{U}_\mathrm{I}^1=\mathbf{U}_\mathrm{I}^2\tag{2.132}$$

and

$$\mathbf{Q}_\mathrm{I}=\mathbf{Q}_\mathrm{I}^1=-\mathbf{Q}_\mathrm{I}^2\tag{2.133}$$

These conditions can be introduced in equations (2.128) and (2.129) which can now be written together as follows,

$$\begin{bmatrix}\mathbf{H}^1 & \mathbf{H}_\mathrm{I}^1 & \mathbf{0}\\ \mathbf{0} & \mathbf{H}_\mathrm{I}^2 & \mathbf{H}^2\end{bmatrix}\begin{Bmatrix}\mathbf{U}^1\\\mathbf{U}_\mathrm{I}\\\mathbf{U}^2\end{Bmatrix}=\begin{bmatrix}\mathbf{G}^1 & \mathbf{G}_\mathrm{I}^1 & \mathbf{0}\\ \mathbf{0} & -\mathbf{G}_\mathrm{I}^2 & \mathbf{G}^2\end{bmatrix}\begin{Bmatrix}\mathbf{Q}^1\\\mathbf{Q}_\mathrm{I}\\\mathbf{Q}^2\end{Bmatrix}\tag{2.134}$$

As $\mathbf{U}_\mathrm{I}$ and $\mathbf{Q}_\mathrm{I}$ are unknown at the interface the above system is frequently written as

$$\begin{bmatrix}\mathbf{H}^1 & \mathbf{H}_\mathrm{I}^1 & -\mathbf{G}_\mathrm{I}^1 & \mathbf{0}\\ \mathbf{0} & \mathbf{H}_\mathrm{I}^2 & \mathbf{G}_\mathrm{I}^2 & \mathbf{H}^2\end{bmatrix}\begin{Bmatrix}\mathbf{U}^1\\\mathbf{U}_\mathrm{I}\\\mathbf{Q}_\mathrm{I}\\\mathbf{U}^2\end{Bmatrix}=\begin{bmatrix}\mathbf{G}^1 & \mathbf{0}\\ \mathbf{0} & \mathbf{G}^2\end{bmatrix}\begin{Bmatrix}\mathbf{Q}^1\\\mathbf{Q}^2\end{Bmatrix}\tag{2.135}$$

Notice that we still have to apply the relevant boundary conditions on the external surface of the region, i.e. Γ^1 and Γ^2. It should also be noticed that the matrix on the left hand side of equation (2.135) is square and that the one on the right hand side is not.

It is interesting to point out that when more interfaces are included the system of equations may tend to be banded or have a large number of zero submatrices, which can result in an *A* matrix which is computationally more efficient. For this reason most boundary element codes offer nowadays subregions, particularly for analysing three dimensional problems.

2.14 Helmholtz Equation

Another useful equation in potential problems is the so called Helmholtz or wave equation. Its time dependent version is

$$\nabla^2 u + \lambda^2 u = 0 \qquad \text{in } \Omega \tag{2.136}$$

where ∇^2 is the two or three dimensional Laplacian and λ^2 is a positive and known parameter. Let us consider that the boundary conditions are the normal

$$\begin{aligned} &\text{(i) Essential conditions} \qquad u = \bar{u} \quad \text{on } \Gamma_1 \\ &\text{(ii) Natural conditions} \qquad q = \bar{q} \quad \text{on } \Gamma_2 \end{aligned} \tag{2.137}$$

Mixed boundary conditions can easly be incorporated in the formulation and are important in many practical problems. We will however restrict the discussion now to the above two conditions for simplicity.

The corresponding weighted residual statement for (2.136) and (2.137) is

$$\int_\Omega (\nabla^2 u + \lambda^2 u) u^* \, d\Omega = \int_{\Gamma_2} (q - \bar{q}) u^* \, d\Gamma - \int_{\Gamma_1} (u - \bar{u}) q^* \, d\Gamma \tag{2.138}$$

Integrating twice by parts one obtains.

$$\begin{aligned} \int_\Omega (\nabla^2 u^* + \lambda^2 u^*) u \, d\Omega = &- \int_{\Gamma_2} \bar{q} u^* \, d\Gamma - \int_{\Gamma_1} q u^* \, d\Gamma \\ &+ \int_{\Gamma_1} \bar{u} q^* \, d\Gamma + \int_{\Gamma_2} u q^* \, d\Gamma \end{aligned} \tag{2.139}$$

or in more compact form

$$\int_\Omega (\nabla^2 u^* + \lambda^2 u^*) u \, d\Omega = -\int_\Gamma q u^* \, d\Gamma + \int_\Gamma u q^* \, d\Gamma \tag{2.140}$$

where $\Gamma = \Gamma_1 + \Gamma_2$.

The fundamental solution u^* for the Helmholtz equation now needs to satisfy

$$\nabla^2 u^* + \lambda^2 u^* + \Delta_i = 0 \tag{2.141}$$

For two dimensions this results in

$$u^* = \frac{i}{4} H_0^{(1)}(\lambda r) \tag{2.142}$$

and

$$q^* = \frac{\partial u^*}{\partial r} = -\frac{\lambda i}{4} H_1^{(1)}(\lambda r) \tag{2.143}$$

where r is as usual the distance from the source point to the point under consideration, $i=\sqrt{-1}$, $H_0^{(1)}$ is the Hankel function of the first kind and zero order and $H_1^{(1)}$ a Hankel function of the first kind and first order, i.e.

$$H_0^{(1)}(\lambda r)=J_0(\lambda r)+iY_0(\lambda r) \tag{2.144}$$

$$H_1^{(1)}(\lambda r)=J_1(\lambda r)+iY_1(\lambda r) \tag{2.145}$$

where J and Y are Bessel functions of first and second kind, with the subscripts indicating their order.

The three dimensional fundamental solution is

$$u^*=\frac{1}{4\pi r}e^{i\lambda r} \tag{2.146}$$

and

$$q^*=\frac{1}{4\pi r}\left(\frac{-1}{r}+i\lambda\right)e^{i\lambda r} \tag{2.147}$$

2.15 Axisymmetric Formulation

A way of dealing with the formulation for axisymmetric problems is by integrating the three dimensional solution.

Assume first that all boundary values have axial symmetry and consequently all domain values are axisymmetric. We can write the governing integral equation (2.39) in terms of the cylindrical polar coordinates (ρ,θ,Z) given in figure 2.23 as,

$$c^iu^i+\int_{\hat{\Gamma}}u\int_0^{2\pi}q^*\,d\theta\,\rho\,d\hat{\Gamma}=\int_{\hat{\Gamma}}q\int_0^{2\pi}u^*\,d\theta\,\rho\,d\hat{\Gamma} \tag{2.148}$$

where $\hat{\Gamma}$ is the generating body created by the intersection of the three dimensional Γ surface with the part of the plane defined by ρ and Z positive.

The three dimensional fundamental solution can be written in cylindrical coordinates

$$u^*=\frac{1}{4\pi r}=\frac{1}{4\pi r(\rho,\theta,Z)} \tag{2.149}$$

and then integrated with respect to θ, which gives the type of ring source solution required by axisymmetric problems, i.e.

$$\hat{u}^*=\int_0^{2\pi}u^*\,d\theta=\frac{4K(m)}{(a+b)^{1/2}} \tag{2.150}$$

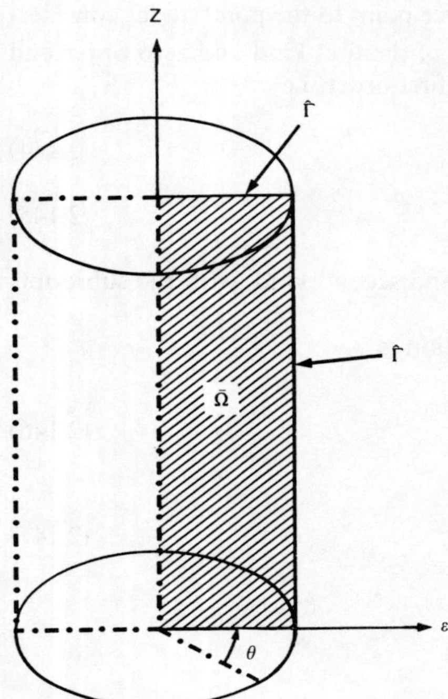

Figure 2.23 Geometrical definition, generating area and boundary contour of solid of revolution

$$\text{where: } m = \frac{2b}{a+b} \qquad (0 \leqslant m \leqslant 1)$$

$$a = \rho_i^2 + \rho^2 + (Z_i - Z)^2; \qquad b = 2\rho_i\rho \tag{2.151}$$

$K(m) =$ is the complete elliptical integral of the first kind.

The subscript 'i' refers to the position of the fundamental solution. Notice that unlike the 2 or 3 dimensional cases, the axisymmetric fundamental solution can not be written simply as a function of the distance between two points but also depends on the distance from the source 'i', the point under consideration to the axis of revolution.

A different formulation of the axisymmetric problem will be presented in section 3.9 where the elliptical integral is avoided by direct numerical integration of the three-dimensional fundamental solution.

2.16 Indirect Formulation

Up to now we have been using the so-called direct formulation which is nowadays the one commonly associated with boundary elements. In the past it was frequent to associate boundary integral solutions with what is now called the 'indirect' approach. In this formulation the boundary solutions are obtained by using sources or sometimes dipoles and it is interesting to point out that this can be interpreted as a particular case of the more general direct approach.

Considering for instance the Poisson's equation, i.e.

$$\nabla^2 u + b = 0 \qquad \text{in } \Omega \tag{2.152}$$

The weighted residual formulation applied on (2.152) produces the direct method integral statement, i.e. for any point in the body.

$$u^i + \int_\Gamma uq^* \, d\Gamma = \int_\Gamma u^* q \, d\Gamma - \int_\Omega bu^* \, d\Omega \tag{2.153}$$

If we now call the region external to Ω by Ω' and assume that u' is the solution to the Laplace equation, $\nabla^2 u' = 0$ within Ω', one can write the following statement,

$$\int_\Gamma q^* u' \, d\Gamma - \int_\Gamma u^* q' \, d\Gamma = 0 \tag{2.154}$$

where the point 'i' has been assumed as external to Ω' (internal to Ω) and the tractions q and q^* have been referred to the normal n to the internal domain so that the two values of q^* in (2.153) and (2.154) become identical. One can now specify u' as the solution in Ω' which generates potentials around Γ identical to those of our initial problem, in Ω, i.e.

$$u \equiv u' \qquad \text{on } \Gamma \tag{2.155}$$

We can now add equation (2.153) and (2.154) to give

$$u^i - \int_\Gamma u^*(q - q') \, d\Gamma + \int_\Omega bu^* \, d\Omega = 0 \tag{2.156}$$

or,

$$u^i = \int_\Gamma u^* \sigma \, d\Gamma - \int_\Omega bu^* \, d\Omega \tag{2.157}$$

$\sigma = q - q'$ represents the initially unknown density distribution of u^* over Γ necessary to generate u_i through equation (2.157). The physical interpretation of σ is the difference between the fluxes generated by the two solutions (internal and external).

We can also deduce another 'indirect' formulation from the same equations (2.153) and (2.154). This is the indirect approach in terms of dipoles. Instead of continuity of potentials between the two fields one can assume continuity of fluxes, i.e.

$$q - q' \equiv 0 \tag{2.158}$$

Then one can subtract the two equations to obtain

$$u^i + \int_\Gamma (u - u')q^* \, d\Gamma = -\int_\Omega bu^* \, d\Omega \tag{2.159}$$

or

$$u^i = \int_\Gamma \mu q^* \, d\Gamma - \int_\Omega bu^* \, d\Omega \tag{2.160}$$

$\mu = u' - u$ are called dipoles.

Notice that in order to apply material conditions on Γ, in both the source or dipole formulation one needs still to compute the derivatives of (2.157) or (2.160). This is rather cumbersome in the dipole formulation as it involves derivatives of the potential.

References

[1] Brebbia, C. A. *The Boundary Element Method for Engineers*. Pentech Press, London, 1978.

[2] Brebbia, C. A. and Dominguez, J. Boundary Element Methods versus Finite Elements. *Proc. Int. Conference on Applied Numerical Modelling, Southampton University, 1977*. Ed. C. A. Brebbia, Pentech Press, London, 1978.

[3] Brebbia, C. A. and Dominguez, J. Boundary Element Methods for Potential Problems. *Applied Mathematical Modelling*, **1**, 7, December 1977.

[4] Hess, J. L. and Smith, A. M. O. Calculation of Potential Flow about Arbitrary Bodies. *Progress in Aeronautical Sciences Vol. 8*. Ed. D. Kuchemann, Pergamon Press, London, 1967.

[5] Harrington, R. F., Pontoppidan, K., Abramhamsen, P. and Albertsen, N. C. Computation of Laplacian Potentials by an Equivalent-source Method. *Proc. IEE*, **116**, 1715–1720 1969.

[6] Mautz, J. R. and Harrington, R. F. Computation of Rotationally Symmetric Laplacian Potentials. *Proc. IEE*, **117**, 850–852, 1970.

[7] Jaswon, M. Integral Equation Methods in Potential Theory, I. *Proc. Roy. Soc. Ser. A.*, **275**, 23–32, 1963.

[8] Symm, G. T. Integral Equation Methods in Potential Theory, II. *Proc. Roy. Soc. Ser. A.*, **275**, 33–46, 1963.

[9] Jaswon, M. and Ponter, A. R. An Integral Equation Solution of the Torsion Problem. *Proc. Roy. Soc. Ser. A.*, **273**, 237–246, 1963.

[10] Brebbia, C. A. and Ferrante, A. *Computational Methods for the Solution of Engineering Problems*. Pentech Press, London, Wiley, USA, 1986.

[11] Danson, D. J., Brebbia, C. A. and Adey, R. A. BEASY. A Boundary Element Analysis System. *Finite Element Systems Handbook*. Springer-Verlag and Computational Mechanics Publications, Southampton, 1982.

Exercises

2.1. Verify that the two-dimensional fundamental solution satisfies Laplace's equation for any point where $r \neq 0$.

2.2. The two-dimensional fundamental solution is $u^* = (1/2\pi)\ln(1/r)$. Since it includes a logarithm of a non-dimensionless quantity its value depends on the scale. Will this fact change the solution of a problem depending on its units?

2.3. Using program POLINBE, solve the example of the figure with 12 linear elements and prescribing values of the potential on the whole boundary: linear variation of u between nodes 1 and 4, $u=0$ between nodes 4 and 5 linear variation of u between nodes 5 and 6, $u=100$ between nodes 6 and 8, linear variation of u between nodes 8 and 10 and $u=300$ between nodes 10 and 1. Check that the results for the fluxes at the corners are very poor.

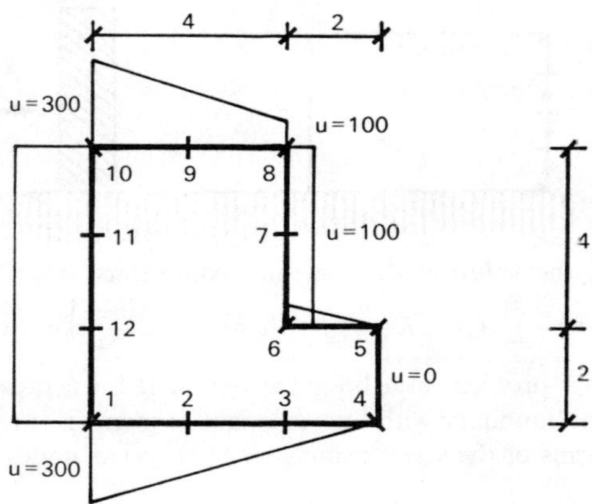

2.4. Solve again the example of 2.3 prescribing values of the potential on the whole boundary and introducing six new very small elements at the corners in such a way that the domain has blunt corners. Try several sizes of the corner elements to check the accuracy of the approximation.

2.5. Using program POLINBE, solve again the example of 2.3 with 12 linear elements and assuming that the values of the potential are known on the whole boundary. Since the potential is known along two lines merging into each corner, determine the value of one potential derivative by differences along one of these lines and leave the other as unknown.

2.6. Derive, using the method of images, the fundamental solution for a semi-finite region with the condition that the potential on the free surface is zero.

2.7. The same as 2.6 but with the condition that the flux on the free surface is zero.

2.8. Derive expressions for the derivatives of the fundamental solution $\left(\frac{\partial u^*}{\partial x_l}\right)^i$ and $\left(\frac{\partial q^*}{\partial x_l}\right)^i$ which should be used with equation (2.33) to compute internal fluxes.

2.9. Write a subroutine (DERINPC) which given an internal point XP, YP, computes the integrals $\int_{\Gamma_j}\left(\frac{\partial q^*}{\partial x_l}\right)^i d\Gamma$ and $\int_{\Gamma_j}\left(\frac{\partial u^*}{\partial x_l}\right)^i d\Gamma$ along a constant element defined by its extreme points.

2.10. Write a subroutine (IFLUXPC) that using DERINPC and the boundary potentials and fluxes of (FI and DFI) of the program POCONBE computes fluxes at internal points.

2.11. Assume an open channel section that extends from $x=0$ to $x=-\infty$. At the extreme $x=0$ there is a piston that moves harmonically with frequency ω. When the perturbations are small and the fluid is considered to be incompressible with zero viscosity, the equation that governs the motion is: $\nabla^2 p=0$, p being the pressure, and $\frac{\partial p}{\partial n}=\rho\omega^2 u_n$, where u_n is the displacement along the n direction and ρ the density.

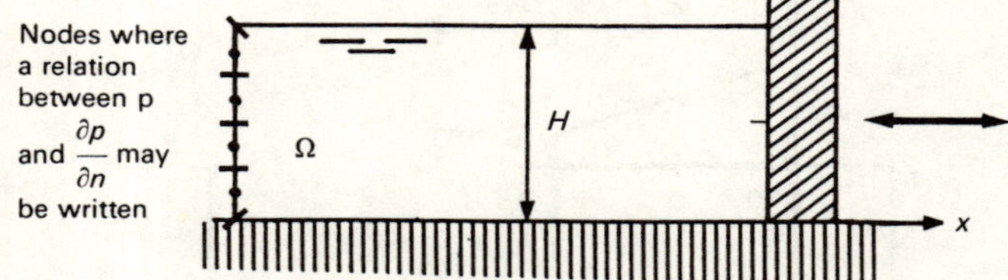

It is known that the solution that satisfies conditions at $z=H$ $(p=0)$ and $z=0$ $\left(\frac{\partial p}{\partial z}=0\right)$, is $P(z,x)=\sum_1^\infty A_n \cos K_n z\, e^{K_n x}$ where $K_n=\frac{2n-1}{2H}\pi$ $(n=1,2,\ldots)$. In order to solve a potential problem in a bounded region Ω for a prescribed motion of the piston, an artificial boundary with four constant elements is introduced at $X=-X_0$. Using as many terms of the series expansion of $p(z,x)$ as nodes exist on the artificial boundary, determine a matrix that relates the nodal values of p and $\frac{\partial p}{\partial n}$ along that boundary (Robin boundary condition).

2.12. Compute, using program POCONBE, the warping function of Saint Venant torsion of a rectangular cross section bar. Use several discretizations to verify the convergency towards the exact solution. Use symmetry to discretize only one quarter of the section.

Exact:

$$u=XY-\frac{32a^2}{\pi^3}\sum_{n=0}^{\infty}\frac{(-1)^n}{(2n+1)^3}\frac{1}{\cosh K_n b}\sin K_n X \sinh K_n Y$$

$$K_n=\frac{2n+1}{2a}\pi$$

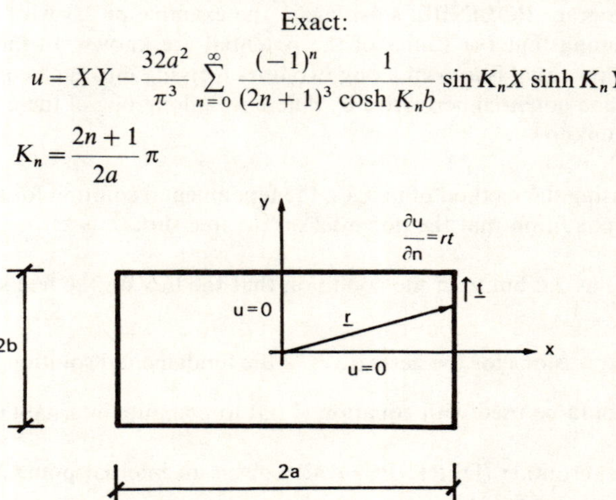

2.13. Solve the problem of exercise 2.12 using linear (POLINBE) and quadratic (POQUABE) elements. Compare the solutions.

Chapter 3

Elastostatics

3.1 Introduction

In this chapter the application of boundary elements to study linear elastostatics problems is presented. The presentation is based on the direct approach and follows the notation initiated by Alarcon, Brebbia and Dominguez in references [1] and [6], and which is consistent with the one used in previous chapters. Reference [1] gave a comprehensive treatment of higher order elements.

The direct formulation of boundary elements for elastostatics was presented by Rizzo in 1967 [2] following the work of Jaswon (see Chapter 2). The basic integral representation known as Somigliana's identity [3] was taken to the boundary which was then discretized into constant elements in a way similar to that previously used for potential problems in some of the papers presented in Chapter 2. Cruse and Rizzo [4] and Cruse [5] extended the formulation to elastodynamics.

The chapter starts with a review of the basic equations of linear elastostatics which are then used to generate the required boundary integrals. Another section deals with the fundamental solutions and explains how they can be obtained by integration of the basic governing equations. This section formulates these solutions starting with the Galerkin vector, and in this way provides a tool for obtaining fundamental solutions in many other problems.

Section 5 discusses the discretization of the body into elements and sets up the methodology to create the boundary element equations. Topics such as integration, rigid body motion and boundary conditions are contained in this section.

An important aspect of boundary elements is the correct consideration of body forces, which whenever possible should be taken to the boundary. A special section is dedicated to the treatment of these forces and how the original domain integrals can be transformed into boundary integrals, using the Galerkin vectors already defined in the part on fundamental solution.

Many engineering problems are inhomogeneous and it is then necessary to divide the body into subregions with different properties. Sometimes the subdivision is preferred for simple computational or modelling reasons. The idea is important in boundary element applications and it is discussed in detail in Section 3.7.

Although the direct formulation is usually associated with boundary elements, it may be convenient in certain cases to apply the so called indirect formulation. Section 3.8 demonstrates how these formulations can be obtained as special cases of the more general direct approach.

Section 3.9 shows how under certain conditions of geometrical and boundary conditions symmetry, a general three dimensional body can be transformed into an axisymmetric problem, with considerable savings in computer and modelling time.

The final section in this chapter deals with the case of orthotropic and anisotropic bodies for which the fundamental solution is more difficult to formulate than for isotropic elastostatics. The section shows how these solutions can be found and sets the basis for changing isotropic boundary element codes into more general anisotropic programs.

3.2 Basic Equations of Linear Elastostatics

In what follows we will restrict our discussions to linear elasticity, i.e. problems for which one assumes that the material behaves linearly and the changes in orientation of the body in the deformed state are negligible. The latter assumption leads to linear strain displacement relations and also allows the equilibrium equations to be referred to the undeformed geometry.

We will use the indicial notation throughout in addition to the matrix notation as otherwise the formulae will become difficult to write.

In solid mechanics one needs to consider the forces or state of stress in the body and the deformations or state of strain. Both states are interrelated by applying the material behaviour or constitutive equations, which establish the relationship between stresses and strains.

State of Stress

Let us define the state of stress at a point in terms of stress components (figure 3.1). In principle one has 9 different components which can be grouped together in a stress tensor, i.e.

$$\begin{bmatrix} \sigma_{11} & \sigma_{12} & \sigma_{13} \\ \sigma_{21} & \sigma_{22} & \sigma_{23} \\ \sigma_{31} & \sigma_{32} & \sigma_{33} \end{bmatrix} \tag{3.1}$$

These components are not all independent but are related through the equilibrium equations, which are of two types, (i) moment equations and (ii) direct components equations.

The moment equilibrium equation can be written by taking moments of the stress components with respect to a point in the differential element and in the limit produce the so called complementary shear relationships, i.e.

$$\sigma_{21} = \sigma_{12}; \qquad \sigma_{31} = \sigma_{13}; \qquad \sigma_{32} = \sigma_{23} \tag{3.2}$$

Equilibrium of the forces in the x_1, x_2 and x_3 directions produced the well known force equilibrium equations which need to be satisfied throughout the

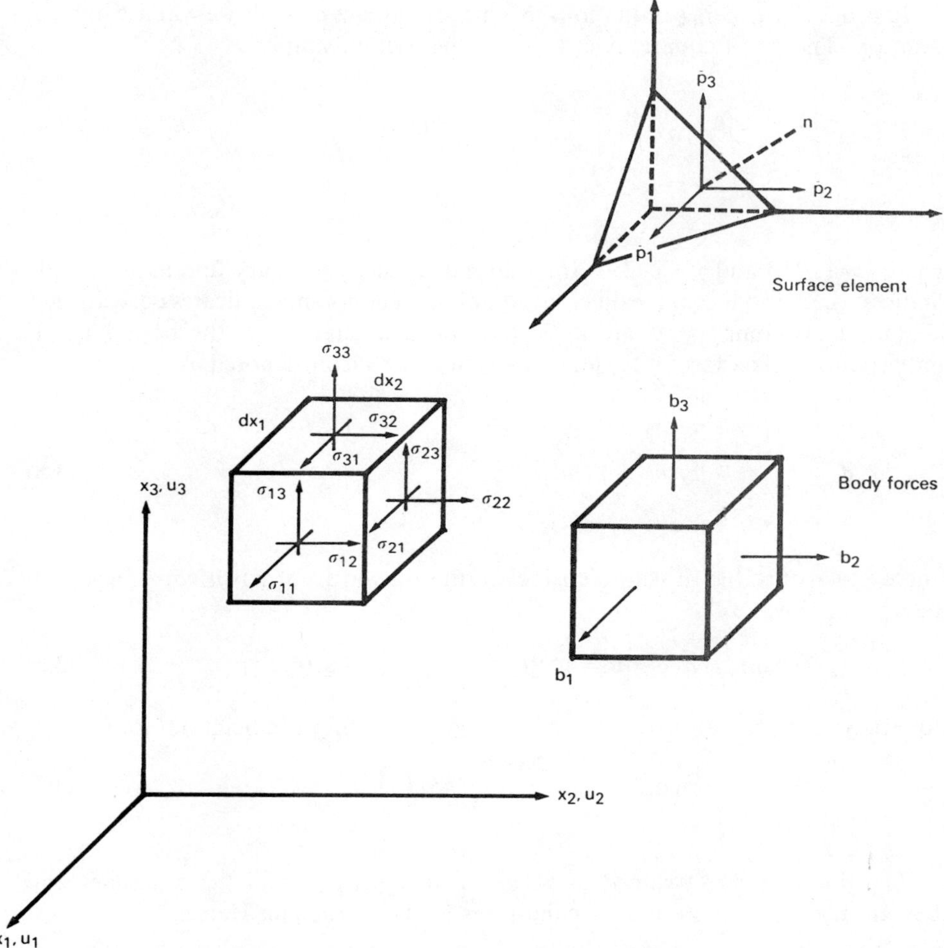

Figure 3.1 Notation for surface forces, stresses and displacements

domain (i.e. interior of the body), i.e.

$$
\begin{aligned}
&\frac{\partial \sigma_{11}}{\partial x_1}+\frac{\partial \sigma_{12}}{\partial x_2}+\frac{\partial \sigma_{13}}{\partial x_3}+b_1=0 \\
&\frac{\partial \sigma_{21}}{\partial x_1}+\frac{\partial \sigma_{22}}{\partial x_2}+\frac{\partial \sigma_{23}}{\partial x_3}+b_2=0 \\
&\frac{\partial \sigma_{31}}{\partial x_1}+\frac{\partial \sigma_{32}}{\partial x_2}+\frac{\partial \sigma_{33}}{\partial x_3}+b_3=0
\end{aligned}
\tag{3.3}
$$

where b_i are the body forces components.

In order to write these equations in a more compact manner we will use indicial notation. The set of equations in (3.3) can be written simply as

$$\frac{\partial \sigma_{ij}}{\partial x_j} + b_i = 0 \qquad \text{in } \Omega \tag{3.4}$$

or

$$\sigma_{ij,j} + b_i = 0$$

where $i = 1, 2, 3$ and $j = 1, 2, 3$. Internal indices such as j vary first and then the i indices which produce three different equations. The comma indicates derivative.

The stress components are projected into a differential of the boundary $d\Gamma$ and produce surface force intensities or tractions which are denoted by p_i such that

$$\begin{aligned} p_1 &= \sigma_{11}n_1 + \sigma_{12}n_2 + \sigma_{13}n_3 \\ p_2 &= \sigma_{21}n_1 + \sigma_{22}n_2 + \sigma_{23}n_3 \\ p_3 &= \sigma_{31}n_1 + \sigma_{32}n_2 + \sigma_{33}n_3 \end{aligned} \tag{3.5}$$

where $n_1n_2n_3$ are the direction cosines of the outward normal n with respect to the $x_1x_2x_3$ axis, i.e.

$$n_1 = \cos(n, x_1); \qquad n_2 = \cos(n, x_2); \qquad n_3 = \cos(n, x_3) \tag{3.6}$$

Equations (3.5) can also be written in a compact form in indicial notation,

$$p_i = \sigma_{ij}n_j \qquad \text{on } \Gamma \tag{3.7}$$

where $i = 1, 2, 3$ and $j = 1, 2, 3$.

The tractions are assumed to be given on the Γ_2 part of the boundary and they are the 'natural' boundary conditions for this problem. Hence,

$$\left.\begin{aligned} p_1 &= \bar{p}_1 \\ p_2 &= \bar{p}_2 \\ p_3 &= \bar{p}_3 \end{aligned}\right\} \qquad \text{on } \Gamma_2 \tag{3.8}$$

These conditions imply that the applied tractions $\bar{p}_i$ have to be in equilibrium with the traction components obtained from the internal stresses at the boundary, i.e.

$$p_i = \sigma_{ij}n_j = \bar{p}_i \qquad \text{on } \Gamma_2 \tag{3.9}$$

State of Strain

The directions or deformations of the boundary are functions of the displacements, which have the components, $u_1u_2u_3$ at every point. They produce strains which for linear cases are

Direct Strains

$$\varepsilon_{11}=\frac{\partial u_1}{\partial x_1}; \qquad \varepsilon_{22}=\frac{\partial u_2}{\partial x_2}; \qquad \varepsilon_{33}=\frac{\partial u_3}{\partial x_3} \tag{3.10}$$

Shear Strains

$$\varepsilon_{12}=\frac{1}{2}\left(\frac{\partial u_1}{\partial x_2}+\frac{\partial u_2}{\partial x_1}\right); \qquad \varepsilon_{13}=\frac{1}{2}\left(\frac{\partial u_1}{\partial x_3}+\frac{\partial u_3}{\partial x_1}\right); \qquad \varepsilon_{23}=\frac{1}{2}\left(\frac{\partial u_2}{\partial x_3}+\frac{\partial u_3}{\partial x_2}\right)$$

These expressions can also be written in indicial notation

$$\varepsilon_{ij}=\frac{1}{2}\left(\frac{\partial u_i}{\partial x_j}+\frac{\partial u_j}{\partial x_i}\right) \tag{3.11}$$

where $i=1, 2, 3$ and $j=1, 2, 3$. Another way of expressing (3.11) is as follows,

$$\varepsilon_{ij}=\tfrac{1}{2}(u_{i,j}+u_{j,i})$$

where the comma indicates derivatives.

Sometimes the state of strain is defined using the strain components arranged in a tensor (the strain tensor), i.e.

$$\begin{bmatrix} \varepsilon_{11} & \varepsilon_{12} & \varepsilon_{13} \\ \varepsilon_{21} & \varepsilon_{22} & \varepsilon_{23} \\ \varepsilon_{31} & \varepsilon_{32} & \varepsilon_{33} \end{bmatrix} \tag{3.12}$$

where $\varepsilon_{21}=\varepsilon_{12}$; $\varepsilon_{31}=\varepsilon_{13}$; $\varepsilon_{32}=\varepsilon_{23}$.

It is simpler to apply boundary conditions in terms of displacements rather than in function of strains. Hence on Γ_1 the following 'essentials' conditions can be defined

$$u_1=\bar{u}_1; \qquad u_2=\bar{u}_2; \qquad u_3=\bar{u}_3 \qquad \text{on } \Gamma_1 \tag{3.13}$$

or

$$u_j=\bar{u}_j \qquad j=1, 2, 3 \qquad \text{on } \Gamma_1$$

where $\bar{u}_j$ are the prescribed values. Note that the total Γ surface of the boundary is equal to $\Gamma_1+\Gamma_2$.

Constitutive Relationships

The states of stress and strains in a body are related throughout the strain-stress or constitutive equations for the material. For a linearly elastic material one can define two constants, called Lamé's constants, λ and μ which are associated with the volumetric and shear components. Then the stress-strain relationship can be expressed as

$$\sigma_{ij} = \lambda\delta_{ij}\varepsilon_{kk} + 2\mu\varepsilon_{ij} \tag{3.14}$$

where δ_{ij} is the Kronecker delta ($\equiv 1$ for $i = j$ and $\equiv 0$ for $i \neq j$). Notice that ε_{kk} has only internal indexes and hence it implies a ratio of the three direct strain components, and because of that is called the volumetric strain, i.e.

$$\varepsilon_{kk} = \varepsilon_{11} + \varepsilon_{22} + \varepsilon_{33} \tag{3.15}$$

The inverse of (3.14) can be written as

$$\varepsilon_{ij} = -\frac{\lambda\delta_{ij}}{2\mu(3\lambda + 2\mu)}\sigma_{kk} + \frac{1}{2\mu}\sigma_{ij} \tag{3.16}$$

where $\sigma_{kk} = \sigma_{11} + \sigma_{22} + \sigma_{33}$.

The Lamé's constant can be expressed in terms of the more familiar shear modulus G, Modulus of Elasticity E and Poisson's ratio v by the following formulae,

$$\mu = G = \frac{E}{2(1 + v)}; \qquad \lambda = \frac{vE}{(1 + v)(1 - 2v)} \tag{3.17}$$

The strain and stress in terms of E and v can be written as

$$\varepsilon_{ij} = -\frac{v}{E}\sigma_{kk}\delta_{ij} + \frac{1 + v}{E}\sigma_{ij} \tag{3.18}$$

and

$$\sigma_{ij} = \frac{E}{(1 + v)}\left[\frac{v}{(1 - 2v)}\delta_{ij}\varepsilon_{kk} + \varepsilon_{ij}\right] \tag{3.19}$$

For some particular problems (specially in soil mechanics) one may prefer to use the bulk modulus K.

In these cases one defines the *deviatoric* stress and strain components

$$\sigma'_{ij} = \sigma_{ij} - \tfrac{1}{3}\sigma_{kk}\delta_{ij} \tag{3.20}$$

$$\varepsilon'_{ij} = \varepsilon_{ij} - \tfrac{1}{3}\varepsilon_{kk}\delta_{ij} \tag{3.21}$$

Thus the constitutive equations are expressed as

$$\sigma'_{ij} = 2G\varepsilon'_{ij}, \qquad p = -K\varepsilon_{kk} \tag{3.22}$$

p is the mean pressure.

$$p = -\frac{\sigma_{kk}}{3} \tag{3.23}$$

and

$$K = \lambda + \tfrac{2}{3}G = E/(3(1-2\nu)) \tag{3.24}$$

In general for isotropic elastic materials all material constants can be expressed in function of two independent constants.

The equations of equilibrium (3), strain-displacement relations (6) and constitutive equations (6) give a complete system of equations from which one can determine the components of stress (6), displacements (3) and strains (6).

Initial Stresses or Strains

In many problems one can have initial state of stress or strain due to temperature or other causes. Consider an initial strain for instance. In this case the *elastic* strain components are obtained by subtracting from the total strain those initial strains, i.e.

$$\varepsilon_{ij}^{\mathrm{e}} = \varepsilon_{ij}^{\mathrm{t}} - \varepsilon_{ij}^{\mathrm{o}} \tag{3.25}$$

where $\varepsilon_{ij}^{\mathrm{e}}$ indicates the elastic components, $\varepsilon_{ij}^{\mathrm{t}}$ the total and $\varepsilon_{ij}^{\mathrm{o}}$ the initial strains.

One can now define the stresses using the elastic strains, i.e.

$$\sigma_{ij} = \lambda\delta_{ij}(\varepsilon_{kk}^{\mathrm{t}} - \varepsilon_{kk}^{\mathrm{o}}) + 2\mu(\varepsilon_{ij}^{\mathrm{t}} - \varepsilon_{ij}^{\mathrm{o}}) \tag{3.26}$$

or

$$\begin{aligned}\sigma_{ij} &= (\lambda\delta_{ij}\varepsilon_{kk}^{\mathrm{t}} + 2\mu\varepsilon_{ij}^{\mathrm{t}}) - (\lambda\delta_{ij}\varepsilon_{kk}^{\mathrm{o}} + 2\mu\varepsilon_{ij}^{\mathrm{o}})\\ &= \sigma_{ij}^{\mathrm{t}} + \sigma_{ij}^{\mathrm{o}}\end{aligned} \tag{3.27}$$

The σ_{ij}^{o} components are called initial stresses and are defined as,

$$\sigma_{ij}^{\mathrm{o}} = -(\lambda\delta_{ij}\varepsilon_{kk}^{\mathrm{o}} + 2\mu\varepsilon_{ij}^{\mathrm{o}}) \tag{3.28}$$

It will be seen later on how the initial stress or strain components can be analysed and included in the boundary integral formulation.

Notice that if the initial strains are due to temperature and the material is thermally isotropic, one can write

$$\varepsilon_{ii}^{\mathrm{o}} = \alpha\theta \tag{3.29}$$

where α is the dilatation coefficient and θ the difference in temperature. The values of σ_{ij}^{o} are given by

$$\sigma_{ij}^{\mathrm{o}} = -2\mu\left(\frac{1+\nu}{1-2\nu}\right)\alpha\theta\delta_{ij} \tag{3.30}$$

3.3 Fundamental Solutions

The formulation of the boundary integral equations for elastostatics to be described in section 3.4 require the knowledge of the solution of the elastic problems with the same material properties as the body under consideration but corresponding to an infinite domain loaded with a concentrated unit point load. This is the fundamental solution of elastostatics and is due to Kelvin.

If the Equilibrium equations (3.4) are expressed in terms of displacements components one obtains Navier's equations, i.e. consider (3.4)

$$\sigma_{lj,j} + b_l = 0 \tag{3.31}$$

Substitute above the stress strain relationships (3.19), i.e.

$$\sigma_{lj} = 2\mu\left[\frac{\nu}{1-2\nu}\,\delta_{lj}\varepsilon_{mm} + \varepsilon_{lj}\right] \tag{3.32}$$

and the strain displacement equation (3.11)

$$\varepsilon_{lj} = \tfrac{1}{2}(u_{l,j} + u_{j,l}) \tag{3.33}$$

The results are the Navier equations or equilibrium equations in terms of displacements, that is

$$\left(\frac{1}{1-2\nu}\right)u_{j,jl} + u_{l,jj} + \frac{1}{\mu}b_l = 0 \tag{3.34}$$

Kelvin solution is obtained from equation (3.34) when a unit contributed load applied at a point 'i' in the direction of the unit vector e_l, i.e.

$$b_l = \Delta^i e_l \tag{3.35}$$

An easy way of computing the fundamental solution is using the representation of the displacement in terms of Galerkin's vector. One assumes a vector **G** from which the displacement components may be obtained as

$$u_j = G_{j,mm} - \frac{1}{2(1-\nu)}G_{m,jm} \tag{3.36}$$

Substitution of equations (3.35) and (3.36) into equation (3.34) gives

$$G_{l,mmjj} + \frac{1}{\mu}\Delta^i e_l = 0 \tag{3.37}$$

or

$$\nabla^2(\nabla^2 G_l) + \frac{1}{\mu}\Delta^i e_l = 0 \tag{3.38}$$

This equation may be written for three-dimensional or two-dimensional plane strain problems as

$$\nabla^2(F_l) + \frac{1}{\mu}\Delta^i e_l = 0 \tag{3.39}$$

where

$$F_l = \nabla^2 G_l \tag{3.40}$$

Notice that equation (3.39) is similar to (2.9) from which the fundamental solution for potential problems was obtained. Solution of equation (3.39) gives

$$F_l = \frac{1}{4\pi r\mu} e_l \tag{3.41}$$

for three-dimensions, and

$$F_l = \frac{1}{2\pi\mu} \ln\left(\frac{1}{r}\right) e_l \tag{3.42}$$

for two-dimensions. Substitution of equations (3.41) or (3.42) into (3.40) gives

$$\nabla^2 G_l = \frac{1}{4\pi r\mu} e_l \tag{3.43}$$

for three-dimensions and

$$\nabla^2 G_l = \frac{1}{2\pi\mu} \ln\left(\frac{1}{r}\right) e_l \tag{3.44}$$

for two-dimensions, which solutions are

$$G_l = G \cdot e_l \tag{3.45}$$

where

$$G = \frac{1}{8\pi\mu} r \tag{3.46}$$

for three-dimensions and

$$G = \frac{1}{8\pi\mu} r^2 \ln\left(\frac{1}{r}\right) \tag{3.47}$$

for two dimensions. Taking each load as independent, one can write,

$$G_{lk} = G\delta_{lk} \tag{3.48}$$

where G_{lk} is the k component of the Galerkin's vector at any point when a unit load is applied at 'i' in the l direction. The displacement at any point in the domain for the point load considering each direction as independent is written

$$u_k^* = u_{lk}^* e_l \tag{3.49}$$

where u_{lk}^* represents the displacement at any point in the k direction when a unit load is applied at 'i' in the l direction.

In accordance with the definition of equation (3.36) one can now write,

$$u_{lk}^* = G_{lk,mm} - \frac{1}{2(1-\nu)} G_{lm,km} \tag{3.50}$$

Substituting equations (3.48) and (3.46) into (3.50) one obtains

$$u_{lk}^* = \frac{1}{16\pi\mu(1-\nu)r}[(3-4\nu)\delta_{lk} + r_{,l}r_{,k}] \tag{3.51}$$

for three dimensional problems. Notice that

$$r_{,l} = \frac{\partial r}{\partial x_l}; \qquad r_{,k} = \frac{\partial r}{\partial x_k} \tag{3.52}$$

These are the ratios between the projection of $\mathbf{r}$ in the $x_1x_2x_3$ directions, which we can call r_1, r_2 and r_3 and the length of $\mathbf{r}$ (see figure 3.3); i.e.

$$r_{,l} = \frac{r_l}{r} \tag{3.53}$$

For the two dimensional plane strain problem the fundamental solution in terms of displacements is obtained by substituting equations (3.48) and (3.47) into (3.50), which gives

$$u_{lk}^* = \frac{1}{8\pi\mu(1-\nu)}\left[(3-4\nu)\ln\frac{1}{r}\,\delta_{lk} + r_{,l}r_{,k}\right] \tag{3.54}$$

Notice that when Laplace's operator is applied to the Galerkin's vector given by equation (3.47) the computed value of F_l differs from that of equation (3.42) in a constant, the resulting value being also the solution of (3.39). Any value of G different from that given by (3.47) in r^2 terms may also be used and in fact one of those fundamental solutions of the biharmonic equation for two dimensions was used in Chapter 2 (equation (2.110)). These values produce fundamental solution displacements u^* identical to equation (3.54) except for a rigid body motion that is neglected because it does not change the solution of the problem as can be seen below.

Stresses at any internal point can be written using the strain-displacement relations (3.11) and the strain-stress equation (3.19). They can be expressed as,

$$\sigma^*_{kj} = S^*_{lkj} e_l \tag{3.55}$$

where the kernel S_{lkj} has been obtained from u^*_{lk} and will be written in full later on.

The tractions or surface forces on the Γ boundary with normal $\vec{n}$ can be written through (3.55) and equation (3.9) as,

$$p^*_k = p^*_{lk} e_l \tag{3.56}$$

where the traction components for three dimensional case are

$$p^*_{lk} = -\frac{1}{8\pi(1-v)r^2}\left[\frac{\partial r}{\partial n}[(1-2v)\delta_{lk} + 3r_{,l}r_{,k}] + (1-2v)(n_l r_{,k} - n_k r_{,l})\right] \tag{3.57}$$

n_l and n_k are the direction cosines of the normal with respect to x_l and x_k. $\partial r/\partial n$ is the derivative of the distance vector r with respect to the normal.

For 2 dimensional plane strain problems one has

$$p^*_{lk} = -\frac{1}{4\pi(1-v)r}\left[\frac{\partial r}{\partial n}[(1-2v)\delta_{lk} + 2r_{,l}r_{,k}] + (1-2v)(n_l r_{,k} - n_k r_{,l}]\right] \tag{3.58}$$

Example 3.1

Consider for instance one component of the Galerkin vector, say G_1, for the case of two dimensional elasticity. Substituting it into equation (3.36) will produce the following two displacement components (which correspond to a force acting in the direction $l = 1$), i.e.

$$\begin{aligned} u_1 &= \nabla^2 G_1 - \frac{1}{2(1-v)}\left\{\frac{\partial^2 G_1}{\partial x_1^2}\right\} \\ u_2 &= -\frac{1}{2(1-v)}\frac{\partial^2 G_1}{\partial x_2 \partial x_1} \end{aligned} \tag{a}$$

One can now substitute these components into the Navier equations (3.34) which are expanded into

$$\begin{aligned} \nabla^2 u_1 + \left(\frac{1}{1-2v}\right)\left\{\frac{\partial^2 u_1}{\partial x_1^2} + \frac{\partial^2 u_2}{\partial x_2 \partial x_1}\right\} + \frac{\Delta^i}{\mu} e_l = 0 \\ \nabla^2 u_2 + \left(\frac{1}{1-2v}\right)\left\{\frac{\partial^2 u_1}{\partial x_1 \partial x_2} + \frac{\partial^2 u_2}{\partial x_2^2}\right\} = 0 \end{aligned} \tag{b}$$

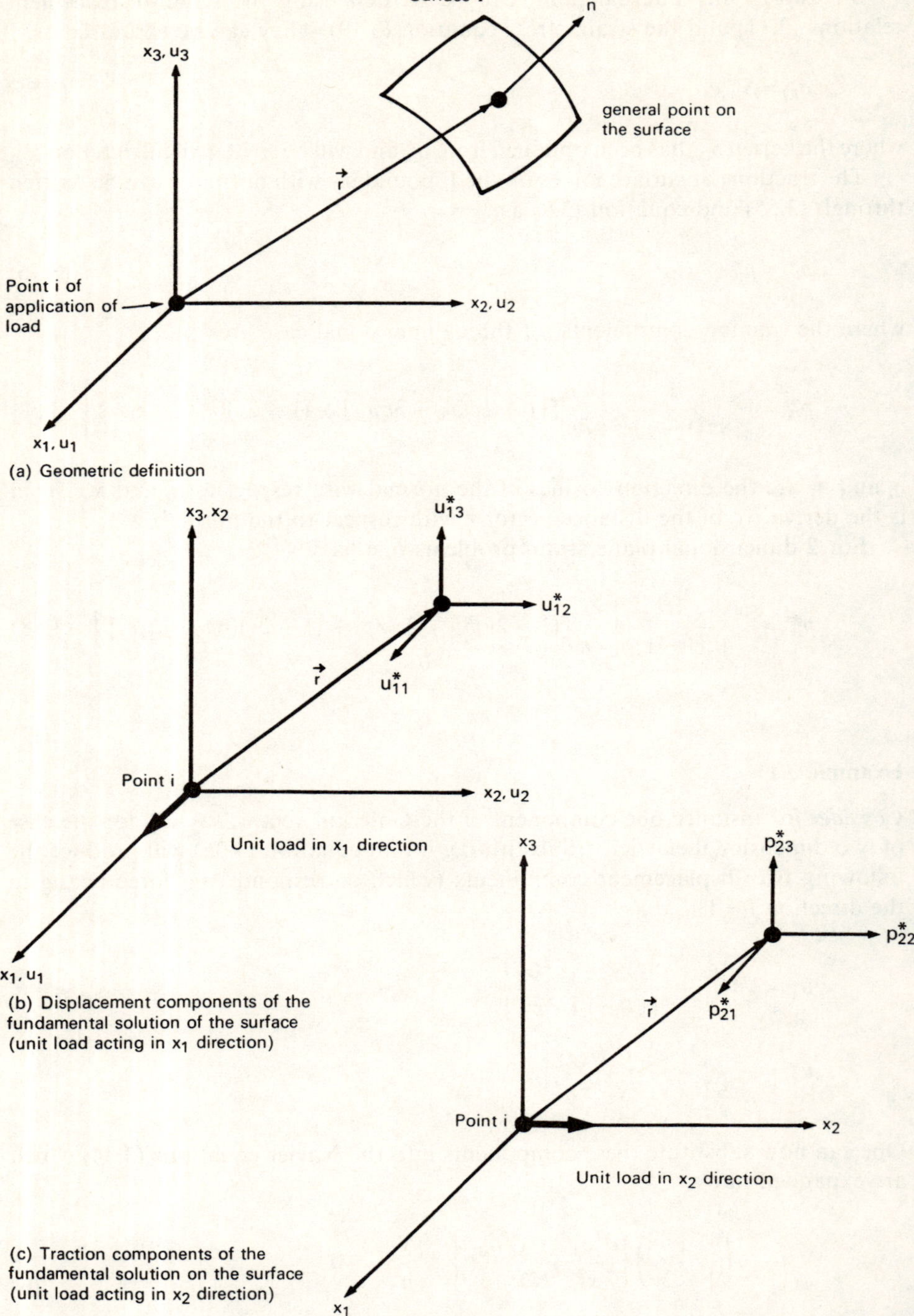

Figure 3.2 Geometrical interpretation of the components of the fundamental solution

This gives for the first equation

$$\nabla^2(\nabla^2 G_1) + \frac{\Delta^i}{\mu} e_1 = 0 \tag{c}$$

while the second is identically satisfied.

The reader can verify that the G_2 component will behave in a similar way.

3.4 Boundary Integral Formulation

The governing integral equations for elastostatics will be deduced using considerations of weighted residual. The fact that some terms which are assumed to be approximate may not be so does not detract from the use of these concepts as a general way of producing the required statement. The concepts are very similar to those used in virtual work.

Consider first that one desires to minimize the errors involved in the numerical approximation of the governing equations of elastostatics, i.e.

$$\sigma_{kj,j} + b_k = 0 \qquad \text{in } \Omega \tag{3.59}$$

which usually have to satisfy the following conditions

(i) Essential or displacement conditions

$$u_k = \bar{u}_k \qquad \text{on } \Gamma_1 \tag{3.60}$$

(ii) Natural or traction conditions

$$p_k = \sigma_{kj} n_j = \bar{p}_k \qquad \text{in } \Gamma_2 \tag{3.61}$$

Consider first that we are only interested in minimizing (3.59). To this end one can weight each of these equations by displacement type functions u_k^* and orthogonalize the product, i.e.

$$\int_\Omega (\sigma_{kj,j} + b_k) u_k^* \, d\Omega = 0 \tag{3.62}$$

If we carry out the integration by parts on the first term of this equation and group the corresponding terms together, one finds the following expression

$$-\int_\Omega \sigma_{kj} \varepsilon_{kj}^* \, d\Omega + \int_\Omega b_k u_k^* \, d\Omega = -\int_\Gamma p_k u_k^* \, d\Gamma \tag{3.63}$$

Integrating by parts one finds the adjoint of the equation (3.59), i.e.

$$\int_\Omega \sigma_{kj,j}^* u_k \, d\Omega + \int_\Omega b_k u_k^* \, d\Omega = -\int_\Gamma p_k u_k^* \, d\Gamma + \int_\Gamma p_k^* u_k \, d\Gamma \tag{3.64}$$

This expression corresponds to Betti's reciprocal theorem (notice that $\sigma^*_{kj,j} = -b^*_k$) which is sometimes used as the starting point for the boundary integral formulation.

Notice that the two terms on the right hand side are integrals on the Γ surface of the body. Let us now consider that the boundary is divided into two parts Γ_1 and Γ_2 and on each of them the boundary conditions (3.60) and (3.61) apply. Hence on can write (3.64)

$$\int_\Omega \sigma^*_{kj,j} u_k \, d\Omega + \int_\Omega b_k u^*_k \, d\Omega = -\int_{\Gamma_1} p_k u^*_k \, d\Gamma - \int_{\Gamma_2} \bar{p}_k u^*_k \, d\Gamma + \int_{\Gamma_1} \bar{u}_k p^*_k \, d\Gamma + \int_{\Gamma_2} u_k p^*_k \, d\Gamma \tag{3.65}$$

The bars represent known values of displacements u_k and tractions p_k components. One can now integrate again by parts trying to return to equation (3.62) but we will find that the resulting expression is slightly different as we have now imposed boundary conditions in Γ_1 andΓ_2. Integrating by parts twice the first integral in (3.65) one obtains,

$$\int_\Omega (\sigma_{kj,j} + b_k) u^*_k \, d\Omega = \int_{\Gamma_2} (p_k - \bar{p}_k) u^*_k \, d\Gamma + \int_{\Gamma_1} (\bar{u}_k - u_k) p^*_k \, d\Gamma \tag{3.66}$$

This expression is a generalized statement that can be used to obtain the boundary integral equations. Having established this starting principle one can now return to expression (3.65) and use as weighting functions the fundamental solution presented in 3.3, which was obtained for a point load $b_l = \Delta^i$ along the direction of the unit vector e_l, i.e.

$$\sigma^*_{lj,j} + \Delta^i e_l = 0 \tag{3.67}$$

The fundamental solution may be written as before, i.e.

$$\begin{aligned} u^*_k &= u^*_{lk} e_l \\ p^*_k &= p^*_{lk} e_l \end{aligned} \tag{3.68}$$

where u^*_{lk}, p^*_{lk} are k components of displacements and tractions due to a unit point load in the l direction. The first integral in (3.65) for a particular direction e_l of the unit load becomes

$$\int_\Omega \sigma^*_{kj,j} u_k \, d\Omega = \int_\Omega \sigma^*_{lj,j} u_l \, d\Omega = -\int_\Omega \Delta^i u_l e_l \, d\Omega = -u^i_l e_l \tag{3.69}$$

where u^i_l represents the l component of the displacement at the point i of application of the load.

Equation (3.65) can now be written to represent the three separate components of the displacement at i by taking the three directions of the point load at 'i'

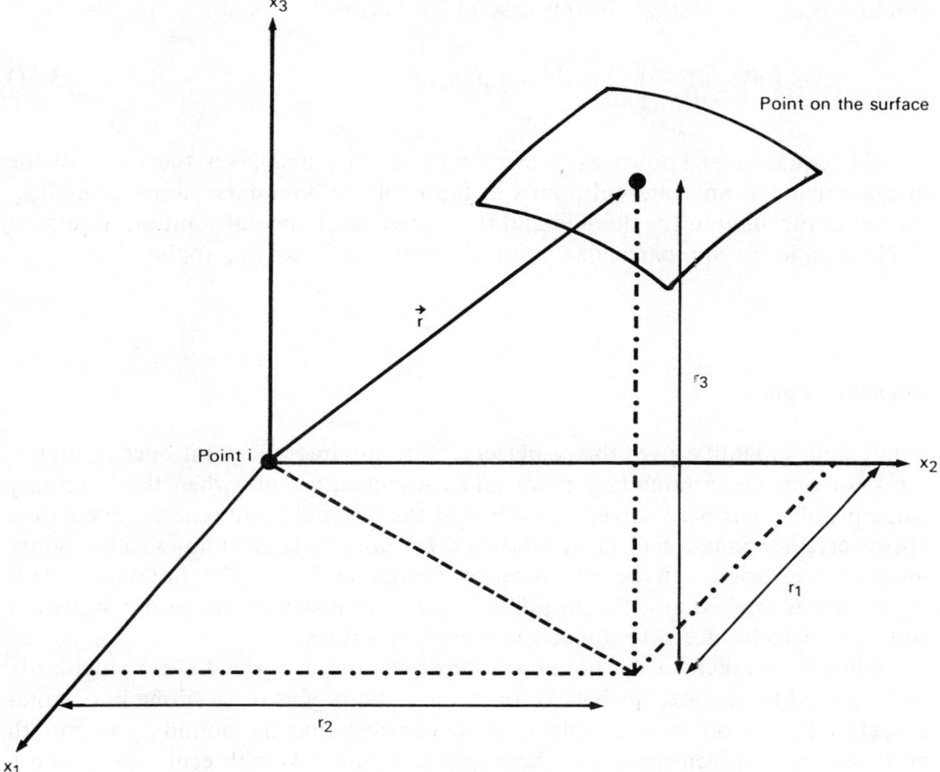

Figure 3.3 Interpretation of the components of the distance vector $\vec{r}$

independently, i.e.

$$u_l^i + \int_{\Gamma_1} p_{lk}^* \bar{u}_k \, d\Gamma + \int_{\Gamma_2} p_{lk}^* u_k \, d\Gamma$$

$$= \int_{\Gamma_1} u_{lk}^* p_k \, d\Gamma + \int_{\Gamma_2} u_{lk}^* \bar{p}_k \, d\Gamma + \int_{\Omega} u_{lk}^* b_k \, d\Omega \tag{3.70}$$

Notice that when one applies a unit point load along a particular direction 'l', the tractions and displacements at any point in the domain have components along the three or two directions (equations (3.68), and (3.51) to (3.58)) while terms of the type $\sigma_{lj,j}$ only are different from zero along the direction of the load. This leads to the fact that for a given direction 'l' at the point the first term of (3.65) only produces displacements along the direction of the load (first term of (3.70)). The rest of the terms however include products for all the components.

Equation (3.70) can be written in a more compact way if one considers the two parts of the boundary together (i.e. $\Gamma = \Gamma_1 + \Gamma_2$) and applies the boundary conditions at a later stage. In this case (3.70) becomes

$$u_l^i + \int_\Gamma p_{lk}^* u_k \, d\Gamma = \int_\Gamma u_{lk}^* p_k \, d\Gamma + \int_\Omega u_{lk}^* b_k \, d\Omega \tag{3.71}$$

This equation is known as Somigliana's identity and gives the value of the displacements at any internal points in terms of the boundary values u_k and p_k, the forces throughout the domain and the known fundamental solution. Equation (3.71) is valid for any particular point 'i' where the forces are applied.

Boundary Points

Somigliana's identity gives the displacement at any internal point once u_k and p_k are known at every boundary point and consequently only when the boundary value problem has been solved the values at the internal points can be calculated. However, since equation (3.71) is valid for every point in Ω including Γ, a boundary integral expression can be obtained by taking (3.71) to the boundary. This expression is applied at different points on the boundary to produce a system of equations which once solved give the boundary values.

When 'i' is taken to the boundary, however, the integrals have a singularity and we need to analyse this behaviour in the same way as it was done in Chapter 2, section 2.2 for potential problems. If we consider that the boundary is smooth at 'i' one can supplement it by a hemisphere (figure 3.4) with centre at 'i' and a small radius ε which will afterwards be taken to the limit, i.e.$\varepsilon \to 0$.

There are two types of boundary integrals in equation (3.71). Consider first the one on the right hand side and write it in function of Γ_ε surface of the hemisphere, i.e.

$$\int_\Gamma u_{lk}^* p_k \, d\Gamma = \lim_{\varepsilon \to 0} \left\{ \int_{\Gamma - \Gamma_\varepsilon} u_{lk}^* p_k \, d\Gamma \right\} + \lim_{\varepsilon \to 0} \left\{ \int_{\Gamma_\varepsilon} u_{lk}^* p_k \, d\Gamma \right\} \tag{3.72}$$

The first integral on the right hand side of (3.72) will simply become an integral on the whole boundary Γ when $\varepsilon \to 0$. The second integral can be written as,

$$p_k^i \lim_{\varepsilon \to 0} \left\{ \int_{\Gamma_\varepsilon} u_{lk}^* \, d\Gamma \right\} \tag{3.73}$$

Noticing that the fundamental solution is of order $1/\varepsilon$ and the surface integral in (3.73) will produce a ε^2, one can conclude that (3.73) will tend to zero as $\varepsilon \to 0$, i.e.

$$\lim_{\varepsilon \to 0} \left\{ \int_{\Gamma_\varepsilon} u_{lk}^* \, d\Gamma \right\} \equiv 0 \tag{3.74}$$

In other words the integral investigated is not affected by the singularity at 'i'.

The left hand side integral in (3.71) however behaves differently. If one writes it as

$$\int_\Gamma p_{lk}^* u_k \, d\Gamma = \lim_{\varepsilon \to 0} \left\{ \int_{\Gamma - \Gamma_\varepsilon} p_{lk}^* u_k \, d\Gamma \right\} + \lim_{\varepsilon \to 0} \left\{ \int_{\Gamma_\varepsilon} p_{lk}^* u_k \, d\Gamma \right\} \tag{3.75}$$

one can see that the limit of the last integral can be written as

$$\lim_{\varepsilon \to 0} \left\{ \int_{\Gamma_\varepsilon} p_{lk}^* u_k \, d\Gamma \right\} = u_k^i \lim_{\varepsilon \to 0} \left\{ \int_{\Gamma_\varepsilon} p_{lk}^* \, d\Gamma \right\} \tag{3.76}$$

The p_{lk}^* values are now of $1/\varepsilon^2$ order while the terms resulting from integration over the surface of the hemisphere are of order ε^2. Hence the integral (3.76) does not vanish when $\varepsilon \to 0$ but produces a free term. By substituting the values of p_{lk}^* as given in equation (3.57) and integrating over Γ_ε one finds

$$\lim_{\varepsilon \to 0} \left\{ \int_{\Gamma_\varepsilon} p_{lk}^* \, d\Gamma \right\} = -\tfrac{1}{2}\delta_{lk} \tag{3.77}$$

Hence the left hand side integral (3.75) can be written in the limit as

$$\int_\Gamma p_{lk}^* u_k \, d\Gamma - \tfrac{1}{2}\delta_{lk} u_k^i = \int_\Gamma p_{lk}^* u_k \, d\Gamma - \tfrac{1}{2} u_l^i \tag{3.78}$$

where the integral on Γ is defined in the sense of the Cauchy Principal value. This is demonstrated in detail in example 3.2 where the different terms of the fundamental solution are integrated one by one.

Therefore for boundary points equation (3.71) transforms into

$$c_{lk}^i u_k^i + \int_\Gamma p_{lk}^* u_k \, d\Gamma = \int_\Gamma u_{lk}^* p_k \, d\Gamma + \int_\Omega u_{lk}^* b_k \, d\Omega \tag{3.79}$$

where the integrals are in the sense of Cauchy principal value and where Γ is smooth at 'i', $c_{lk}^i = \tfrac{1}{2}\delta_{lk}$. When '$i$' is at a point where the boundary is not smooth, the value of the integrals in equation (3.78) give different results and it is generally difficult to obtain a general expression in three dimensions.

Fortunately, explicit calculations of this value is not usually necessary as it can be obtained using rigid body motions as will be shown in section 3.5.

Boundary equation (3.79) permits to solve the general boundary value problem of elastostatics. If displacements are known over the whole domain, equation (3.79) produces an integral equation of the first kind, if tractions are known over all the boundary an integral equation of the second kind is obtained and finally a combination of both types of boundary conditions results in a mixed integral equation.

Example 3.2

Consider the behaviour of the two types of integrals in (3.71) for the case of the point 'i' being on the smooth boundary Γ surrounded by a hemisphere as shown in figure 3.4. In the limit the radius of the hemisphere will tend to zero.

The first type of integral is as follows.

$$\int_{\Gamma} u_k p_{lk}^* \, d\Gamma = \lim_{\varepsilon \to 0} \left\{ \int_{\Gamma - \Gamma_\varepsilon} u_k p_{lk}^* \, d\Gamma \right\} + \lim_{\varepsilon \to 0} \left\{ \int_{\Gamma_\varepsilon} u_k p_{lk}^* \, d\Gamma \right\} \tag{a}$$

Consider the Γ_ε integral only, i.e.

$$\begin{aligned} I &= \lim_{\varepsilon \to 0} \left\{ \int_{\Gamma_\varepsilon} u_k \, p_{lk}^* \, d\Gamma \right\} \\ &= \lim_{\varepsilon \to 0} \left\{ - \int_{\Gamma_\varepsilon} u_k \left[\frac{\partial r}{\partial n} \left\{ (1 - 2\nu)\delta_{lk} + 3 \frac{\partial r}{\partial x_l} \frac{\partial r}{\partial x_k} \right\} \right. \right. \\ &\quad \left. \left. - (1 - 2\nu) \left\{ \frac{\partial r}{\partial x_l} n_k - \frac{\partial r}{\partial x_k} n_l \right\} \right] \frac{1}{8\pi(1 - \nu) r^2} \right\} d\Gamma \end{aligned} \tag{b}$$

Note that $\varepsilon \equiv r$. Consider figure 3.5 where for simplicity a spherical system of coordinates is used. For this particular case the second term in equation (3.57) will disappear as,

$$\frac{\partial r}{\partial x_l} n_k - \frac{\partial r}{\partial x_k} n_l = \frac{\partial r}{\partial x_l} \frac{\partial r}{\partial x_k} - \frac{\partial r}{\partial x_k} \frac{\partial r}{\partial x_l} \equiv 0 \tag{c}$$

Hence one only needs to consider the first term in the integral, that is,

$$I = \lim_{\varepsilon \to 0} \left\{ - \int_{\Gamma_\varepsilon} u_k \frac{\partial r}{\partial n} \left\{ (1 - 2\nu)\delta_{lk} + 3 \frac{\partial r}{\partial x_l} \frac{\partial r}{\partial x_k} \right\} \frac{d\Gamma}{8\pi(1 - \nu) r^2} \right\} \tag{d}$$

Note that $\partial r / \partial n = 1$.

This can be expanded taking into account the geometric relationships shown in figure 3.5, for instance when $l = 1$

$$\begin{aligned} I = \lim_{\varepsilon \to 0} \left\{ - \int_{\Gamma_\varepsilon} \{ u_1^i (1 - 2\nu) + 3u_1^i e_1 e_1 + 3u_2^i e_1 e_2 \right. \\ \left. + 3u_3^i e_1 e_3 \} \cdot \frac{\sin\theta \, d\theta \, d\phi}{8\pi(1 - \nu)} \right\} \end{aligned} \tag{e}$$

The e_i are unit vectors in the x_i direction (see figure 3.5) such that

$$e_i = n_i = \frac{\partial r}{\partial x_i} = \frac{r_i}{r}$$

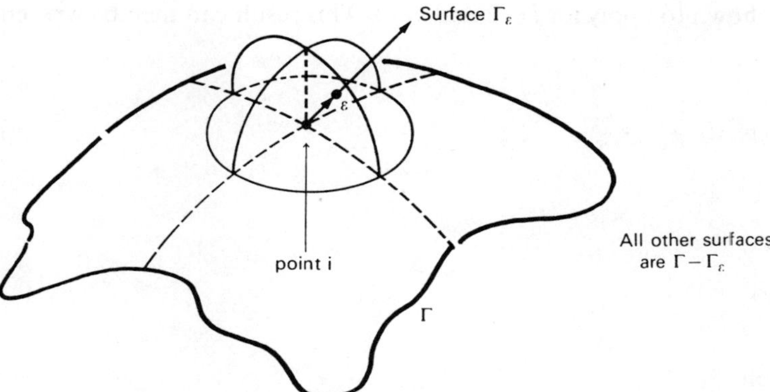

Figure 3.4 Full boundary surface Γ_ε assumed hemispherical for integration purposes

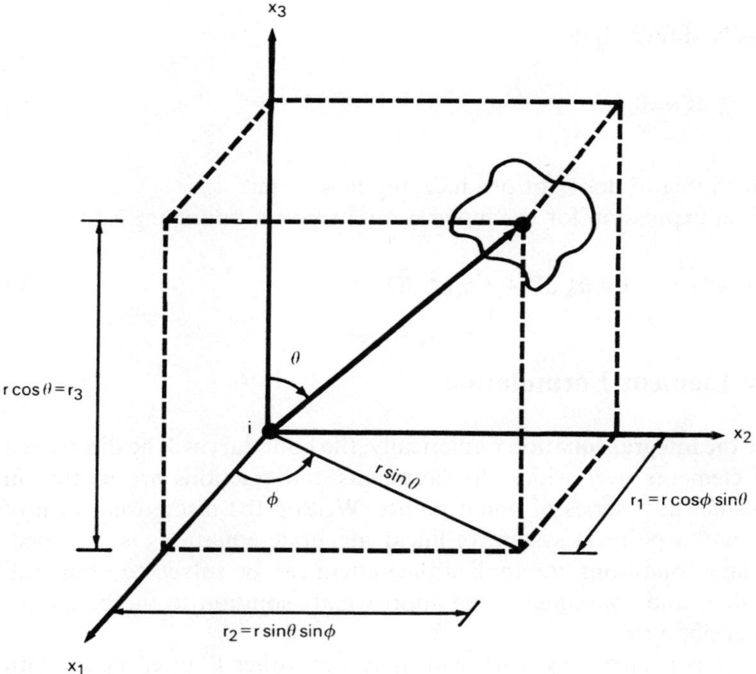

Figure 3.5 Geometry definitions

After integration we find that some of the integrals are zero and the final result is,

$$I = -\frac{1}{8\pi(1-\nu)}[(1-2\nu)2\pi + 2\pi]u_1^i = -\frac{4(1-\nu)}{8(1-\nu)}u_1^i = -\tfrac{1}{2}u_1^i \tag{f}$$

The same can be shown to apply for $l = 2$ and $l = 3$. This result can then be written as,

$$\lim_{\varepsilon \to 0} \left\{ \int_{\Gamma_\varepsilon} u_k p_{lk}^* \, d\Gamma \right\} = -\frac{u_l^i}{2} \tag{g}$$

The integral

$$\int_\Gamma p_k u_{lk}^* \, d\Gamma \tag{h}$$

can also be written

$$\int_{\Gamma - \Gamma_\varepsilon} p_k u_{lk}^* \, d\Gamma + \int_{\Gamma_\varepsilon} p_k u_{lk}^* \, d\Gamma \tag{i}$$

but it can be easily shown that

$$\lim_{\varepsilon \to 0} \int_{\Gamma_\varepsilon} p_k u_{lk}^* \, d\Gamma = 0 \tag{j}$$

and therefore this integral does not produce any new term.

Hence the final expression for the integral on a smooth boundary is,

$$\tfrac{1}{2} u_l^i + \int_\Gamma u_k p_{lk}^* \, d\Gamma = \int_\Gamma p_k u_{lk}^* \, d\Gamma + \int_\Omega b_k u_{lk}^* \, d\Omega \tag{k}$$

3.5 Boundary Element Formulation

In order to solve the integral equation numerically, the boundary will be discretized into a series of elements over which displacements and tractions are written in terms of their values at a series of nodal points. Writing the discretized form of (3.79) for every nodal point, a system of linear algebraic equations is obtained. Once the boundary conditions are applied the system can be solved to obtain all the unknown values and consequently an approximate solution to the boundary values problem is obtained.

It is now more convenient to work with matrices rather than carry on with the indicial notation. To this effect we can start by defining the u and p functions which apply over each element 'j', i.e.

$$\mathbf{u} = \mathbf{\Phi} \mathbf{u}^j \tag{3.80}$$

and

$$\mathbf{p} = \mathbf{\Phi} \mathbf{p}^j \tag{3.81}$$

where $\mathbf{u}^j$ and $\mathbf{p}^j$ are the element nodal displacements and tractions, of dimensions $3 \times Q$ for three dimensions and $2 \times Q$ for two dimensions, Q being the number of nodes on the element. **u** and **p** are the displacements and tractions at any point on the boundary Γ_e, i.e.

$$\mathbf{p} = \begin{Bmatrix} p_1 \\ p_2 \\ p_3 \end{Bmatrix}; \qquad \mathbf{u} = \begin{Bmatrix} u_1 \\ u_2 \\ u_3 \end{Bmatrix} \tag{3.82}$$

The interpolation function matrix $\mathbf{\Phi}$ is a $3 \times 3Q$ (or $2 \times 2Q$ two dimensions) array of shape functions, i.e.

$$\mathbf{\Phi} = \begin{bmatrix} \phi_1 & 0 & 0 & \phi_2 & 0 & 0 & \dots & \phi_Q & 0 & 0 \\ 0 & \phi_1 & 0 & 0 & \phi_2 & 0 & \dots & 0 & \phi_Q & 0 \\ 0 & 0 & \phi_1 & 0 & 0 & \phi_2 & \dots & 0 & 0 & \phi_Q \end{bmatrix}$$
$$= [\boldsymbol{\phi}_1 \quad \boldsymbol{\phi}_2 \quad \dots \quad \boldsymbol{\phi}_Q] \tag{3.83}$$

These functions are the standard two dimensional finite element type function discussed in section 2.7 (see figure 3.6).

Notice that the body forces at any point on the Ω domain can also be expressed in vector form in function of the three components, i.e.

$$\mathbf{b} = \begin{Bmatrix} b_1 \\ b_2 \\ b_3 \end{Bmatrix} \tag{3.84}$$

The fundamental solution coefficients can be expressed as,

$$\mathbf{p}^* = \begin{bmatrix} p^*_{11} & p^*_{12} & p^*_{13} \\ p^*_{21} & p^*_{22} & p^*_{23} \\ p^*_{31} & p^*_{32} & p^*_{33} \end{bmatrix} = \tag{3.85}$$

matrix whose coefficients, p^*_{lk}, are the tractions in k direction due to a unit force at 'i' acting in the 'l' direction

$$\mathbf{u}^* = \begin{bmatrix} u^*_{11} & u^*_{12} & u^*_{13} \\ u^*_{21} & u^*_{22} & u^*_{23} \\ u^*_{31} & u^*_{32} & u^*_{33} \end{bmatrix} = \tag{3.86}$$

matrix whose coefficients u^*_{lk} are the displacements in the 'k' direction due to a unit force at 'i' acting on the 'l' directon

With this notation equation (3.79) valid for each i point can be rewritten as follows,

$$c^i\mathbf{u}^i + \int_\Gamma \mathbf{p}^*\mathbf{u}\, d\Gamma = \int_\Gamma \mathbf{u}^*\mathbf{p}\, d\Gamma + \int_\Omega \mathbf{u}^*\mathbf{b}\, d\Omega \tag{3.87}$$

where $c^i = \frac{1}{2}$ for smooth boundary. Otherwise it will be a 3×3 (or 2×2 in two dimensions) array.

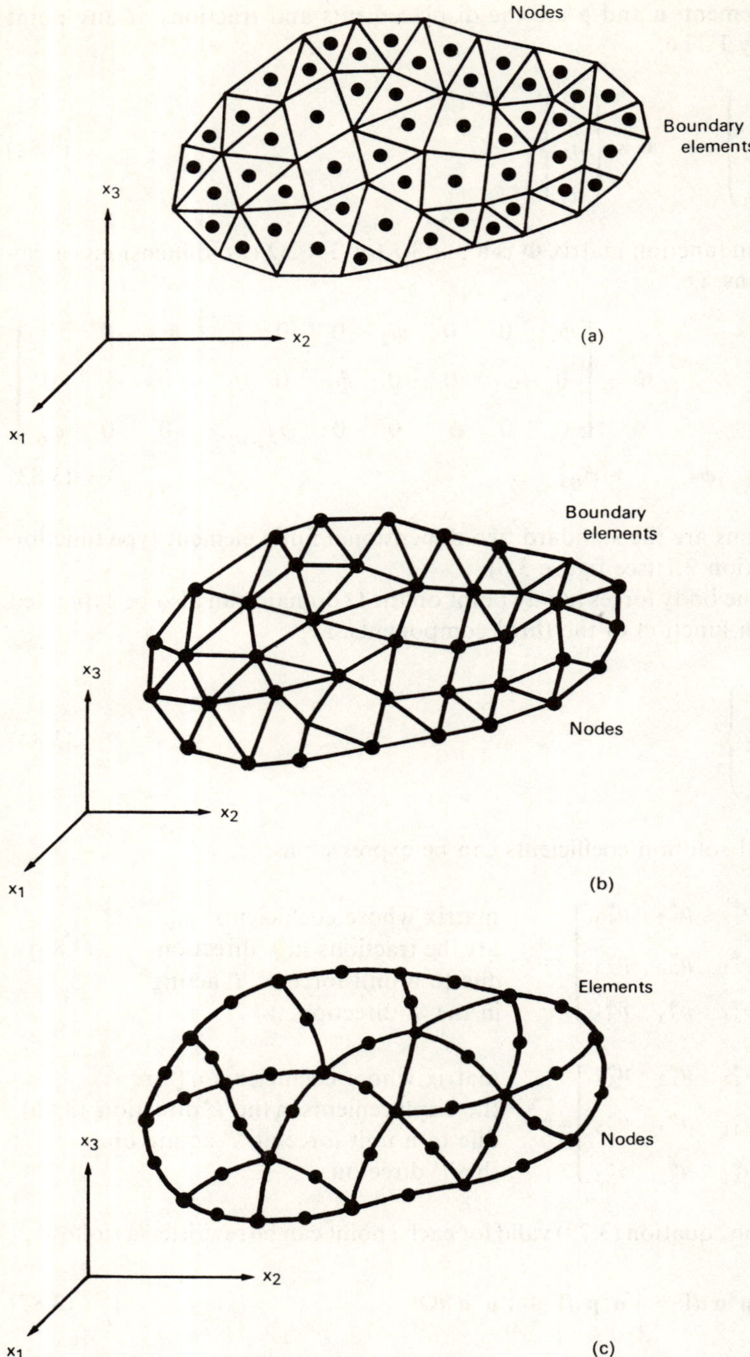

Figure 3.6 Three-dimensional body divided into (a) constant boundary elements, (b) linear boundary elements and (c) quadratic boundary elements

Notice that the Cartesian coordinates of the boundary may also be written in terms of nodal coordinates to define curved elements. If this is the case we will need to transform from one to another system and this transformation will involve introducing a Jacobian as it was shown in section 2.7. We will discuss this in what follows.

Consider now that we substitute the above functions into equation (3.87) and discretize the boundary obtaining the following equation for a nodal point.

$$\mathbf{c}^i\mathbf{u}^i + \sum_{j=1}^{NE}\left\{\int_{\Gamma_j}\mathbf{p}^*\boldsymbol{\Phi}\,d\Gamma\right\}\mathbf{u}^j = \sum_{j=1}^{NE}\left\{\int_{\Gamma_j}\mathbf{u}^*\boldsymbol{\Phi}\,d\Gamma\right\}\mathbf{p}^j + \sum_{s=1}^{M}\left\{\int_{\Omega_s}\mathbf{u}^*\mathbf{b}\,d\Omega\right\} \tag{3.88}$$

Note that summation for $j = 1$ to NE indicates summation over all the NE elements on the surface and Γ_j is the surface of a 'j' element. $\mathbf{u}^j$ and $\mathbf{p}^j$ are the nodal displacement and tractions in the element 'j'.

Notice also that we have considered that the domain is divided into M internal cells over which the body forces integrals are to be computed. These domain regions over which generally a numerical integration is carried out can in some cases be avoided by taking the body force integrals to the boundary as will be seen later.

The integrals in (3.88) are usually solved numerically, particularly if the elements are curved, as it is then difficult to integrate analytically. The interpolation functions $\boldsymbol{\Phi}$ tend to be expressed in a homogeneous system of coordinates such as those described in section 2.7 and of the type drawn in figure 3.7. The coordinates need then to be transferred from the ξ_i system to the global x_i system.

Transformation of Coordinates

The transformation of coordinates is identical to the one described in section 2.10 where two types of Jacobian were found.

(i) *Volume to Volume Transformation* relating derivatives in x_i system to those in the $\xi_1\xi_2\eta$ (figure 3.7)

$$d\Omega = |\mathbf{J}|\,d\xi_1\,d\xi_2\,d\eta \tag{3.89}$$

(ii) *Surface to Volume Transformation* which produces

$$d\Gamma = |\mathbf{G}|\,d\xi_1\,d\xi_2 \tag{3.90}$$

To compute the values of these Jacobians one needs to know the variation of x_i coordinates in function of the homogeneous system $\xi_1\xi_2\eta$, which are given in function of the same interpolation functions as used for displacements and tractions (equations (3.80) and (3.81)), i.e.

$$\mathbf{x} = \boldsymbol{\Phi}\mathbf{x}^j \tag{3.91}$$

$\mathbf{x}^j$ are the nodal values of the coordinates over the element under consideration,

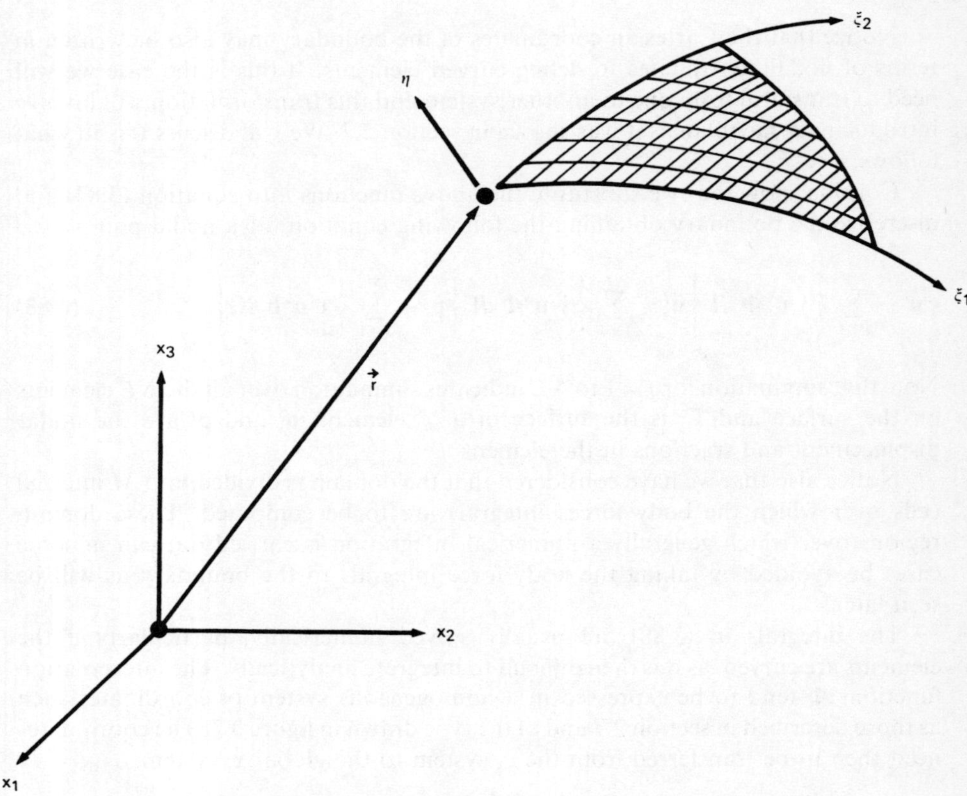

Figure 3.7 Coordinate systems for an element on a curved surface

and $\boldsymbol{\phi}$ is the same interpolation function used for displacements and tractions as given in equation (3.83).

Equation (3.88) can now be written as

$$\mathbf{c}^i\mathbf{u}^i + \sum_{j=1}^{NE} \left\{ \int_{\Gamma_j} \mathbf{p}^*\boldsymbol{\Phi}|\mathbf{G}|\, d\xi_1\, d\xi_2 \right\} \mathbf{u}^j$$

$$= \sum_{j=1}^{NE} \left\{ \int_{\Gamma_j} \mathbf{u}^*\boldsymbol{\Phi}|\mathbf{G}|\, d\xi_1\, d\xi_2 \right\} \mathbf{p}^j + \sum_{s=1}^{M} \left\{ \int_{\Omega_s} \mathbf{u}^*\mathbf{b}|\mathbf{J}|\, d\xi_1\, d\xi_2\, d\eta \right\} \quad (3.92)$$

Applying numerical integration to the above formula (see Appendix A) one obtains,

$$\mathbf{c}^i\mathbf{u}^i + \sum_{j=1}^{NE} \left\{ \sum_{k=1}^{l} w_k(\mathbf{p}^*\boldsymbol{\Phi})_k|\mathbf{G}| \right\} \mathbf{u}^j$$

$$= \sum_{j=1}^{NE} \left\{ \sum_{k=1}^{l} w_k(\mathbf{u}^*\boldsymbol{\Phi})_k|\mathbf{G}| \right\} \mathbf{p}^j$$

$$+ \sum_{s=1}^{M} \left\{ \sum_{p=1}^{r} w_p(\mathbf{u}^*\mathbf{b}^*)_p|\mathbf{J}| \right\} \quad (3.93)$$

where l is the number of integration points on the surface elements and w_k the weight at those points. r is the number of integration points on the cells. Functions such as $(\mathbf{p}^*\mathbf{\Phi})$, $(\mathbf{u}^*\mathbf{\Phi})$ and $(\mathbf{u}^*\mathbf{b})$ have to be evaluated at the integration points.

System of Equations

Equation (3.88) or its numerical integral form (3.93) correspond to a particular node 'i' and once integrated can be written as,

$$\mathbf{c}^i\mathbf{u}^i + \sum_{j=1}^{N} \hat{\mathbf{H}}^{ij}\mathbf{u}^j = \sum_{j=1}^{N} \mathbf{G}^{ij}\mathbf{p}^j + \sum_{s=1}^{M} \mathbf{B}^{is} \tag{3.94}$$

where N is the number of nodes, $\mathbf{u}^j$ and $\mathbf{p}^j$ are the displacements and tractions at node 'j'. The influence matrices $\mathbf{H}$ and $\mathbf{G}$ (now 3×3 for three dimensions and 2×2 in two dimensional cases) are

$$\begin{aligned} \hat{\mathbf{H}}^{ij} &= \sum_t \int_{\Gamma_t} \mathbf{p}^*\mathbf{\Phi}_q \, d\Gamma \\ \mathbf{G}^{ij} &= \sum_t \int_{\Gamma_t} \mathbf{u}^*\mathbf{\Phi}_q \, d\Gamma \end{aligned} \tag{3.95}$$

where the summation extends to all the elements to which node 'j' belongs and q is number of order of the node 'j' within element t. For constant elements, the summation extends only to one element, $t \equiv j$ and $\mathbf{\Phi}_q$ is the identity matrix.

$$\mathbf{B}^{is} = \int_{\Omega_s} \mathbf{u}^*\mathbf{b} \, d\Omega$$

Calling

$$\begin{aligned} \mathbf{H}^{ij} &= \hat{\mathbf{H}}^{ij} & \text{if } i \neq j \\ \mathbf{H}^{ij} &= \hat{\mathbf{H}}^{ij} + \mathbf{c}^i & \text{if } i = j \end{aligned} \tag{3.96}$$

equation (3.94) for node 'i' becomes,

$$\sum_{j=1}^{N} \mathbf{H}^{ij}\mathbf{u}^j = \sum_{j=1}^{N} \mathbf{G}^{ij}\mathbf{p}^j + \sum_{s=1}^{M} \mathbf{B}^{is} \tag{3.97}$$

The contribution for all 'i' nodes can be written together in matrix form to give the global system equations, i.e.

$$\mathbf{HU} = \mathbf{GP} + \mathbf{B} \tag{3.98}$$

Notice that the elements $\mathbf{c}^i$ will be a series of 3×3 submatrices on the diagonal $\mathbf{H}$ (or 2×2 in two dimensional cases). The elements of these submatrices are not simply given by the solid angle but can become very cumbersome to compute analytically. Fortunately this is not required as they can be found by consideration of rigid body movement as we will see shortly.

The vectors **U** and **P** represent all the values of displacements and tractions before applying boundary conditions. These conditions can be introduced by rearranging the columns in **H** and **G**, passing all unknowns to a vector **X** on the left hand side. This gives the final system of equations, i.e.

$$\mathbf{AX} = \mathbf{F} \tag{3.99}$$

Notice that the **B** vector has been incorporated in **F**. Solving the above system all boundary values are fully determined.

Rigid Body Considerations

As it was pointed out the diagonal submatrices $\mathbf{H}^{ii}$ in **H** include terms in $\hat{\mathbf{H}}^{ii}$ and $\mathbf{c}^i$. Difficulties appear when trying to compute explicitly these terms particularly at corners due to the singularity of the fundamental solution. Assuming a rigid body displacement in the direction of one of the cartesian coordinates the traction and body force vector must be zero and hence from (3.98)

$$\mathbf{HI}^q = \mathbf{0} \tag{3.100}$$

where $\mathbf{I}^q$ is a vector that for all nodes has unit displacement along the 'q' direction ($q = 1$, 2 or 3) and zero displacement in any other direction. Since (3.100) has to be satisfied for any rigid body displacement one can write,

$$\mathbf{H}^{ii} = -\sum_{j=1}^{N} \mathbf{H}^{ij} \qquad (\text{for } j \neq i) \tag{3.101}$$

which gives the diagonal submatrices in terms of the rest of the terms of the **H** matrix.

The above considerations are strictly valid for closed domains. When dealing with infinite or semi-infinite regions equation (3.101) must be modified. If the rigid body displacement is prescribed for a boundless domain the integral

$$\int_{\Gamma_\infty} \mathbf{p}^* \mathbf{I}^q \, d\Gamma = \left\{ \int_{\Gamma_\infty} \mathbf{p}^* \, d\Gamma \right\} \mathbf{I}^q \tag{3.102}$$

over the external boundary Γ_∞ at infinity will not be zero and since the tractions $\mathbf{p}^*$ are due to a point load, this integral must be,

$$\int_{\Gamma_\infty} \mathbf{p}^* \, d\Gamma = -\mathbf{I} \tag{3.103}$$

where $\mathbf{I}$ is the 3×3 (or 2×2 in two dimensional cases) identity matrix. The diagonal submatrices for this case are,

$$\mathbf{H}^{ii} = \mathbf{I} - \sum_{j=1}^{N} \mathbf{H}^{ij} \qquad (\text{for } j \neq i) \tag{3.104}$$

Internal Points

Somigliana's identity (3.71) gives the displacement at any internal point in terms of the boundary displacements and tractions. Considering again its integral representation as in (3.88) one has,

$$\mathbf{u}^i = \sum_{j=1}^{NE} \left\{ \int_{\Gamma_j} \mathbf{u}^* \mathbf{\Phi} \, d\Gamma \right\} \mathbf{p}^j - \sum_{j=1}^{NE} \left\{ \int_{\Gamma_j} \mathbf{p}^* \mathbf{\Phi} \, d\Gamma \right\} \mathbf{u}^j + \sum_{s=1}^{M} \left\{ \int_{\Omega_s} \mathbf{u}^* \mathbf{b} \, d\Omega \right\} \tag{3.105}$$

where Γ_j is once more the surface corresponding to element j and 'i' is now an internal point. The internal point displacements in terms of the nodal displacements and tractions can be written in the same way as (3.94), i.e.

$$\mathbf{u}^i = \sum_{j=1}^{N} \mathbf{G}^{ij} \mathbf{p}^j - \sum_{j=1}^{N} \hat{\mathbf{H}}^{ij} \mathbf{u}^j + \sum_{s=1}^{M} \mathbf{B}^{is} \tag{3.106}$$

The terms $\mathbf{G}^{ij}$ and $\mathbf{H}^{ij}$ consist of integrals over the elements to which node j belongs. Those integrals do not contain any singularity and can be easily computed using numerical integration. Terms like $\mathbf{B}^{is}$ however will contain a singularity (notice that they are domain terms and the point i is now in the domain) and special care should be taken when computing them numerically. Being domain integrals however their order of singularity is one less than the integrals on the boundary and consequently can be more accurately computed using numerical integration formulae.

For an isotropic medium the internal stresses can be computed by differentiating the displacements at internal points and introducing the corresponding strains into the stress-strain relationships, i.e.

$$\sigma_{ij} = \frac{2\mu\nu}{1 - 2\nu} \delta_{ij} \frac{\partial u_l}{\partial x_l} + \mu \left(\frac{\partial u_i}{\partial x_j} + \frac{\partial u_j}{\partial x_i} \right) \tag{3.107}$$

After carrying out the derivatives inside the integral equations one obtains,

$$\sigma_{ij} = \int_\Gamma \left\{ \frac{2\mu\nu}{1-2\nu} \delta_{ij} \frac{\partial u^*_{lk}}{\partial x_l} + \mu \left(\frac{\partial u^*_{ik}}{\partial x_j} + \frac{\partial u^*_{jk}}{\partial x_i} \right) \right\} p_k \, d\Gamma$$

$$+ \int_\Omega \left\{ \frac{2\mu\nu}{1-2\nu} \delta_{ij} \frac{\partial u^*_{lk}}{\partial x_l} + \mu \left(\frac{\partial u^*_{ik}}{\partial x_j} + \frac{\partial u^*_{jk}}{\partial x_i} \right) \right\} b_k \, d\Omega$$

$$- \int_\Gamma \left\{ \frac{2\mu\nu}{1-2\nu} \delta_{ij} \frac{\partial p^*_{lk}}{\partial x_l} + \mu \left(\frac{\partial p^*_{ik}}{\partial x_j} + \frac{\partial p^*_{jk}}{\partial x_i} \right) \right\} u_k \, d\Gamma \tag{3.108}$$

All derivatives are taken at the internal point under consideration, which is the point of application of the fundamental solution. Taking the corresponding derivatives of the fundamental solution the above equation can be written, in a compact form, as,

$$\sigma_{ij} = \int_\Gamma D_{kij} p_k \, d\Gamma - \int_\Gamma S_{kij} u_k \, d\Gamma + \int_\Omega D_{kij} b_k \, d\Omega \tag{3.109}$$

where the third order tensor components D_{kij} and S_{kij} are

$$D_{kij} = \frac{1}{r^\alpha} \{(1-2\nu)\{\delta_{ki} r_{,j} + \delta_{kj} r_{,i} - \delta_{ij} r_{,k}\} + \beta r_{,i} r_{,j} r_{,k}\} \frac{1}{4\alpha\pi(1-\nu)} \tag{3.110}$$

$$S_{kij} = \frac{2\mu}{r^\beta} \left\{ \beta \frac{\partial r}{\partial n} [(1-2\nu)\delta_{ij} r_{,k} + \nu(\delta_{ik} r_{,j} + \delta_{jk} r_{,i}) - \gamma r_{,i} r_{,j} r_{,k}] + \beta\nu(n_i r_{,j} r_{,k} + n_j r_{,i} r_{,k}) + (1-2\nu)(\beta n_k r_{,i} r_{,j} + n_j \delta_{ik} + n_i \delta_{jk}) - (1-4\nu) n_k \delta_{ij} \right\} \frac{1}{4\alpha\pi(1-\nu)} \tag{3.111}$$

The above formulae are applicable for 2 or 3 dimensions. For the former case $\alpha = 1$, $\beta = 2$ and $\gamma = 4$ and for the latter $\alpha = 2$, $\beta = 3$ and $\gamma = 5$.

All the derivatives indicated by commas are taken at the boundary point x_i^B of figure 3.8, i.e.

$$r_{,i} = \frac{\partial r}{\partial x_i} = \frac{r_i^B}{r^B} \tag{3.112}$$

This derivative is equal and opposite in sign to those taken at an internal point.

Equation (3.109) is discretized by dividing the Γ boundary into a summation over all the element surfaces and assuming the corresponding interpolation functions for u_k and p_k.

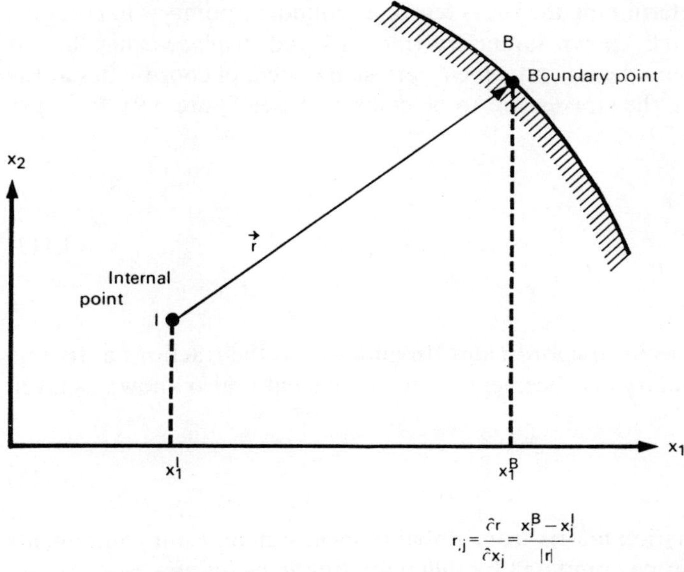

$$r_{,j} = \frac{\partial r}{\partial x_j} = \frac{x_j^B - x_j^I}{|r|}$$

Figure 3.8 Definition of derivatives required in the internal stress formulae (3.109)

The values obtained for the internal stresses using the above formulae are in general more accurate than those computed using other numerical methods and similar discretization. The same can be said of the internal displacements computed through (3.105). However when the internal point is very close to the boundary (say less than 1/4 of the smallest length of the nearest element) because of the peak in the fundamental solution, special numerical integration schemes have to be used to obtain accurate stresses and displacements.

Notice also that the values of displacements at the boundary are known from the integral equation solution. The same can not be said of the stresses since the stress vector has more independent terms than the traction vector. The problem will be studied in the next section.

Stresses on the Boundary

The numerical solution of the governing equations produces all the boundary displacements and tractions. In many practical applications however the boundary stresses rather than the tractions are required. One possibility would be to take (3.109) to the boundary but this produces higher order singularity of a type that has not yet been effectively resolved. For three dimensional problems, the D_{kij} and S_{kij} terms contain singularities of order $1/r^2$ and $1/r^3$ respectively and taking equation (3.109) to the boundary requires special consideration of how the principal values should be computed.

The simplest way of determining the stress tensor at boundary points is to compute its components from the known boundary tractions and displacements. Let us assume a three dimensional case and a local cartesian system of coordinates at the boundary points where the stresses are to be computed (see figure 3.9). It is easy to see that,

$$\begin{aligned} \sigma'_{13} &= \sigma'_{31} = p'_1 \\ \sigma'_{23} &= \sigma'_{32} = p'_2 \\ \sigma'_{33} &= p'_3 \end{aligned} \tag{3.113}$$

where the dash indicates local coordinates. In addition to the tractions a discrete expansion for the boundary displacements over the element is also known as given by (3.80), i.e.

$$\mathbf{u}' = \mathbf{R}^T \mathbf{\Phi} \mathbf{u}^j \tag{3.114}$$

where $\mathbf{R}$ is a transformation matrix from global to local system. Four components of the strain tensor can be computed by differentiating u' as follows

$$\varepsilon'_{ij} = \frac{1}{2}\left(\frac{\partial u'_j}{\partial x'_i} + \frac{\partial u'_i}{\partial x'_j}\right) \qquad i, j = 1, 2 \tag{3.115}$$

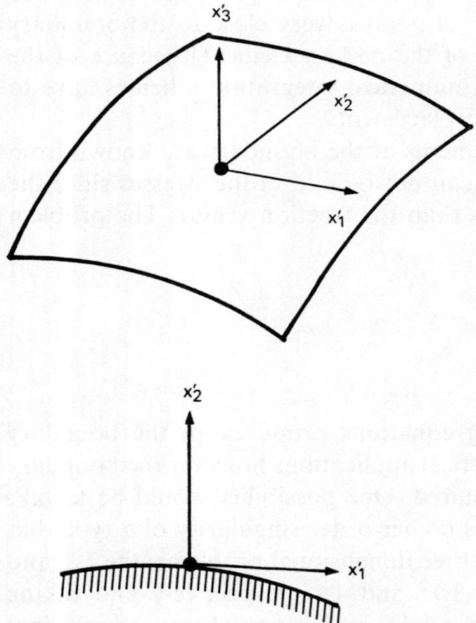

Figure 3.9 Local system of coordinates over the element

Notice that the ε'_{ij} will depend on the derivatives of the shape functions and the nodal displacements. If constant elements are used, the displacement derivatives can be computed using a finite difference approximation between adjacent nodes.

The rest of the terms of the stress tensor can now be computed from the constitutive equations as follows,

$$\begin{aligned}
\sigma'_{12} &= \sigma'_{21} = 2\mu\varepsilon'_{12} \\
\sigma'_{11} &= \frac{1}{1-\nu}\left[\nu\sigma'_{33} + 2\mu(\varepsilon'_{11} + \nu\varepsilon'_{22})\right] \\
\sigma'_{22} &= \frac{1}{1-\nu}\left[\nu\sigma'_{33} + 2\mu(\varepsilon'_{22} + \nu\varepsilon'_{11})\right]
\end{aligned} \tag{3.116}$$

For two dimensional problems the procedure is analogous. Three components of the stress tensor are obtained from the tractions, i.e.

$$\begin{aligned}
\sigma'_{12} &= \sigma'_{21} = p'_1 \\
\sigma'_{22} &= p'_2
\end{aligned} \tag{3.117}$$

and one component of the stress tensor is computed from the surface displacements as follows,

$$\varepsilon'_{11} = \frac{\partial u'_1}{\partial x'_1} \tag{3.118}$$

and the last stress component is computed from

$$\sigma'_{11} = \frac{1}{1-\nu}\left(\nu\sigma'_{22} + 2\mu\varepsilon'_{11}\right) \tag{3.119}$$

using the stress-strain relationships corresponding to plane strain.

Traction Discontinuities at Corner Points

When a node is located at a point where the boundary is not smooth, i.e. has a corner for two dimensional problems or corners and edges in three dimensional cases, a discontinuity in the traction will occur at that node. This implies that if the nodal tractions are unknown the number of equations at that node is smaller than the number of unknowns.

In order to explain what occurs, let us consider a two dimensional corner for simplicity (figure 3.10). When the tractions are known at both sides of the corner node, only the two components of the nodal displacements are unknown and no special treatment of the corner node is required. It may also happen, for any of the two components, that the displacement and one of the tractions, either 'before'

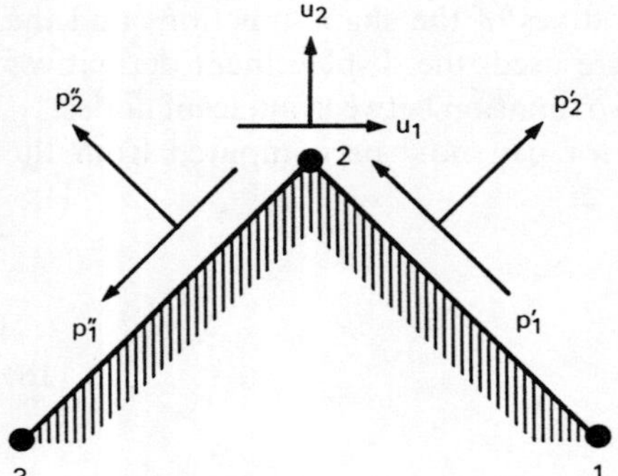

Figure 3.10 Continuous corner node

or 'after' the node, is known; then the other traction, 'after' or 'before' the node will be the unknown and the problem is solved without difficulties as will be seen in sections 4.7 and 4.8. However, when two different values ('before' and 'after' the node) of any component of the traction are unknown and only the displacement is known a special treatment of the corner is required.

The easy way then of solving the problem is by duplicating the corner node. The geometry of the problem is slightly modified and only two traction components are assigned to each node (figure 3.11(a)). The problem may now be solved by the standard procedure. The distance between the two corner nodes must be very small and it is limited by the numerical problems that may be originated by the existence of two sets of equations whose coefficients are very close to each other. In practice excellent results are obtained if the distances are not too small. When the corner node is duplicated a small gap may be left between the two nodes (figure 3.11(a)) or a small element may be assumed between the two (figure 3.11(b)). In the latter case, tractions over the small elements are assumed to be p'_1, p'_2 and p''_1, p''_2 for nodes $2'$ and $2''$ respectively.

Another version of the double node approach is the use of discontinuous elements already mentioned in section 2.5. This consists of displacing inside the element the nodes that meet or that would meet at corners or edges (figure 3.11(c)). The approach is very simple and effective and has the added advantage that it can model better corners with high stress concentrations. When used to model singularities – such as in fracture mechanics problems – the results using discontinuous elements have converged well to the correct solution.

The problem of having more than two unknowns at a corner can also be solved as follows. If the tractions at both sides of a corner are unknown the displacements along the two elements to which the node belongs will be known. In such a case the displacement derivatives along those elements can be obtained by derivation of the shape functions. Since the two elements follow different directions all the components of the strain tensor at the corner can be written in terms of nodal displacements of those elements and the stress tensor and any traction at the corner may be computed. This procedure can not be applied for corners where a singularity of the stress tensor occurs.

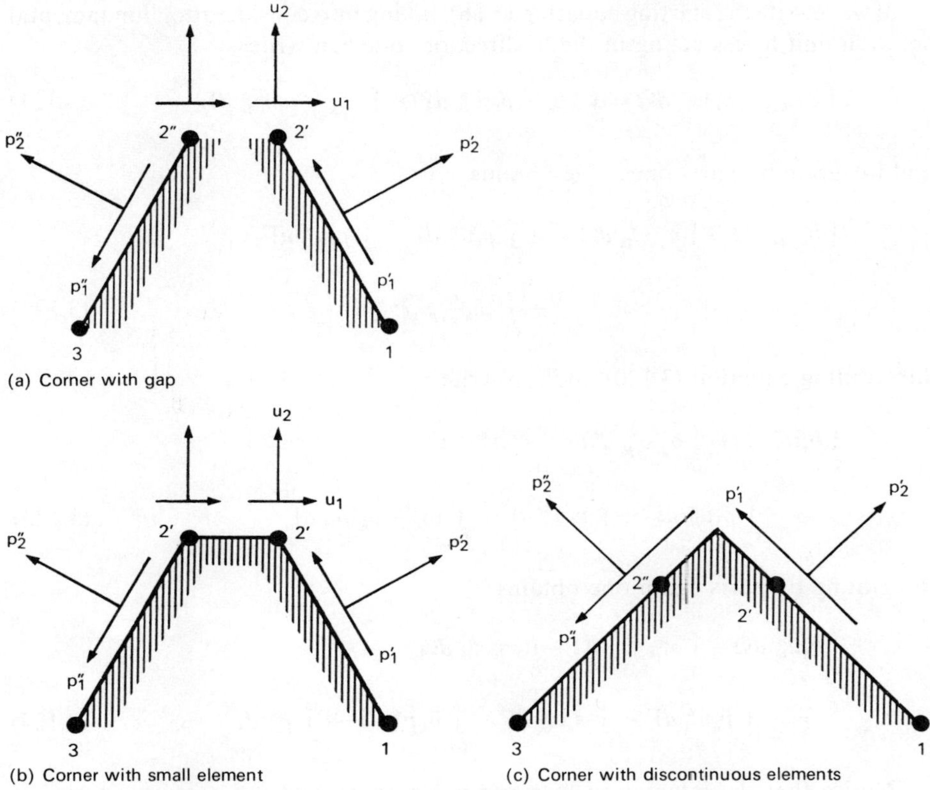

Figure 3.11 Corner tractions and displacements modelling

3.6 Treatment of Domain Integrals and Body Force Terms

Domain integrals are important in boundary elements as they can not only represent body or thermal forces but also a wide range of non-linear effects.

Initial Stress

Many problems produce an initial stress or strain field and in this case one needs to differentiate between the elastic, initial and total components of those variables. If an initial stress state exists, the elastic stresses are given by (see section 3.2)

$$\sigma_{ij} = \sigma_{ij}^{t} + \sigma_{ij}^{o} \qquad (3.120)$$

σ_{ij}^{t} represents the 'total' stress and σ_{ij}^{o} the initial one.

If we rewrite the starting equation (3.66), taking into consideration fundamental solution unit forces acting in the 'l' direction, one can write

$$\int_{\Omega} (\sigma_{jk,j} + b_k) u^*_{lk} \, d\Omega = \int_{\Gamma_2} (p_k - \bar{p}_k) u^*_{lk} \, d\Gamma + \int_{\Gamma_1} (\bar{u}_k - u_k) p^*_{lk} \, d\Gamma \tag{3.121}$$

and integrate by parts once, one obtains

$$\int_{\Omega} b_k u^*_{lk} \, d\Omega - \int_{\Omega} \sigma_{jk} \varepsilon^*_{ljk} \, d\Omega = -\int_{\Gamma_2} \bar{p}_k u^*_{lk} \, d\Gamma - \int_{\Gamma_1} p_k u^*_{lk} \, d\Gamma + \int_{\Gamma_1} (\bar{u}_k - u_k) p^*_{lk} \, d\Gamma \tag{3.122}$$

Substituting equation (3.120) one can write,

$$\int_{\Omega} b_k u^*_{lk} \, d\Omega - \int_{\Omega} \sigma^{\mathrm{t}}_{jk} \varepsilon^*_{ljk} \, d\Omega - \int_{\Omega} \sigma^{\mathrm{o}}_{jk} \varepsilon^*_{ljk} \, d\Omega = -\int_{\Gamma_2} \bar{p}_k u^*_{lk} \, d\Gamma - \int_{\Gamma_1} p_k u^*_{lk} \, d\Gamma + \int_{\Gamma_1} (\bar{u}_k - u_k) p^*_{lk} \, d\Gamma \tag{3.123}$$

Integrating by parts again one obtains

$$\int_{\Omega} b_k u^*_{lk} \, d\Omega + \int_{\Omega} \sigma^*_{ljk,j} u_k \, d\Omega - \int_{\Omega} \sigma^{\mathrm{o}}_{jk} \varepsilon^*_{ljk} \, d\Omega = -\int_{\Gamma_2} \bar{p}_k u^*_{lk} \, d\Gamma - \int_{\Gamma_1} p_k u^*_{lk} \, d\Gamma + \int_{\Gamma_1} \bar{u}_k p^*_{lk} \, d\Gamma + \int_{\Gamma_2} p^*_{lk} \, d\Gamma \tag{3.124}$$

Notice that the integration by parts has been carried out on σ^{t}_{jk} and not on σ_{jk} because the former is the one related to the total displacements. Equation (3.124) gives in more compact form, and after substituting for the fundamental solution, the following expression

$$\int_{\Gamma} u^*_{lk} p_k \, d\Gamma + \int_{\Omega} b_k u^*_{lk} \, d\Omega - \int_{\Omega} \sigma^{\mathrm{o}}_{jk} \varepsilon^*_{ljk} \, d\Omega = c^i_{lk} u^i_k + \int_{\Gamma} p^*_{lk} u_k \, d\Gamma \tag{3.125}$$

This demonstrates that the initial stress (and similarly for strain) field can be treated in a similar way as a body force field, b_k, although it is generally difficult to transform the σ^{o} term into a boundary integral.

Body Forces

Let us study in this section how the body force term in (3.125) can be taken to the boundary. The integral under study is

$$\int_{\Omega} b_k u^*_{lk} \, d\Omega \tag{3.126}$$

The integral does not include any unknown values but in order to compute it numerically the whole domain has to be discretized into cells which represents a considerable difficulty and diminishes the elegance and computational efficiency of the method. Fortunately domain integrals can be avoided in many practical cases by reducing them to the boundary. Consider in this case that the body force components b_k can be obtained from a potential function ψ such that,

$$b_k = \frac{\partial \psi}{\partial x_k} \tag{3.127}$$

where the potential ψ is assumed to satisfy the following harmonic relationship

$$\nabla^2 \psi = K_0 = \text{constant} \tag{3.128}$$

Then integral (3.126) can be rewritten as

$$\begin{aligned} \int_\Omega u_{lk}^* b_k \, d\Omega &= \int_\Omega u_{lk}^* \left(\frac{\partial \psi}{\partial x_k} \right) d\Omega \\ &= \int_\Omega \frac{\partial}{\partial x_k} (u_{lk}^* \psi) \, d\Omega - \int_\Omega \frac{\partial u_{lk}^*}{\partial x_k} \psi \, d\Omega \end{aligned} \tag{3.129}$$

where derivatives by parts have been used. For the divergence theorem one can now write

$$\int_\Omega u_{lk}^* b_k \, d\Omega = \int_\Gamma u_{lk}^* \psi n_k \, d\Gamma - \int_\Omega \frac{\partial u_{lk}^*}{\partial x_k} \psi \, d\Omega \tag{3.130}$$

where n_k is the direction cosine of the normal to Γ with respect to x_k axis. In order to take the last integral to the boundary one can write the fundamental solution u_{lk}^* in terms of its Galerkin vector G_{lk} in the form

$$u_{lk}^* = G_{lk,jj} - \frac{1}{2(1-\nu)} G_{lj,kj} = \nabla^2 (G_{lk}) - \frac{1}{2(1-\nu)} G_{lj,kj} \tag{3.131}$$

where the comma indicates derivatives.

The Galerkin vector for the fundamental solution of three dimensional elastostatics problems was given in section 3.3 and is repeated below, i.e.

$$G_{kl} = \frac{1}{8\pi\mu} r \delta_{kl} \tag{3.132}$$

For two dimensional problems,

$$G_{kl} = \frac{1}{8\pi\mu} r^2 \ln\left(\frac{1}{r}\right) \delta_{kl} \tag{3.133}$$

It must be pointed out that when the two-dimensional Galerkin vector is derived, the following fundamental solution is obtained,

$$u_{lk}^* = \frac{1}{8\pi\mu(1-\nu)}\left[(3-4\nu)\ln\frac{1}{r}\,\delta_{kl} - \left(\frac{7-8\nu}{2}\right)\delta_{kl} + r_{,k}r_{,l}\right] \tag{3.134}$$

This solution differs from the previous fundamental solution (equation (3.54)) by a constant term, which was dropped before because a rigid body motion does not change the solution of the systems of equations. However, when body forces exist and the Galerkin vector given by (3.133) is used to take the domain integral to the boundary, the fundamental solution given by (3.134) must be used.

By applying (3.131) one can write the last integral in (3.130) as follows,

$$\int_\Omega u_{lk,k}^*\psi\, d\Omega = \frac{1-2\nu}{2(1-\nu)}\int_\Omega \nabla^2(G_{lk,k})\psi\, d\Omega \tag{3.135}$$

Finally, the Green's theorem can be applied between the field $G_{lk,k}$ and ψ, i.e.

$$\int_\Omega (G_{lk,k}\nabla^2\psi - \nabla^2(G_{lk,k})\psi)\, d\Omega = \int_\Gamma (G_{lk,k}\psi_{,j}n_j)\, d\Gamma - \int_\Gamma (G_{lk,kj}\psi n_j)\, d\Gamma \tag{3.136}$$

Notice that the first left hand side integral in (3.136) is simply

$$\int_\Omega G_{lk,k}\nabla^2\psi\, d\Omega = K_0\int_\Omega G_{lk,k}\, d\Omega = K_0\int_\Gamma G_{lk}n_k\, d\Gamma \tag{3.137}$$

Here, equation (3.136) can be written as,

$$-\int_\Omega \nabla^2(G_{lk,k})\psi\, d\Omega = \int_\Gamma (G_{lk,k}\psi_{,j}n_j)\, d\Gamma - \int_\Gamma (G_{lk,kj}\psi n_j)\, d\Gamma - K_0\int_\Gamma G_{lk}n_k\, d\Gamma \tag{3.138}$$

One can now substitute equation (3.138) into (3.135) and this into (3.130) to obtain the expression for the body force integral in terms of boundary integrals, i.e.

$$\int_\Omega u_{lk}^*b_k\, d\Omega = \int_\Gamma u_{lk}^*\psi n_k\, d\Gamma + \frac{1-2\nu}{2(1-\nu)} \times \left\{\int_\Gamma (G_{lk,k}\psi_{,j}n_j)\, d\Gamma - \int_\Gamma (G_{lk,kj}\psi n_j)\, d\Gamma - K_0\int_\Gamma G_{lk}n_k\, d\Gamma\right\} \tag{3.139}$$

We will now specialize equation (3.139) for different types of loads.

Gravitational Loads

As an example of body forces, the case of gravitational loads will be considered first as they are simple to express, i.e.

$$\psi = -\rho g x_3; \qquad \mathbf{b} = \begin{Bmatrix} 0 \\ 0 \\ -\rho g \end{Bmatrix}; \qquad \nabla^2 \psi = 0 \tag{3.140}$$

where ρ is the density and g the gravitational acceleration.

The domain integral for this case can be written as

$$\int_\Omega u^*_{lk} b_k \, d\Omega = -\int_\Gamma u^*_{lk} \rho g x_3 n_k \, d\Gamma - \frac{(1-2\nu)\rho g}{16\pi\mu(1-\nu)} \int_\Gamma (r_{,l} n_3 - x_3 r_{,lj} n_j) \, d\Gamma \tag{3.141}$$

Notice that in this case the internal stress σ_{ij} will also have a body force term which can also be taken to the boundary using the above method.

Thermoelastic Problems

The temperature changes θ in an elastic body are equivalent to adding a body force equal to $(-\gamma\theta_{,k})$ at each point and increasing the tractions by $(\gamma\theta n_k)$ where

$$\gamma = \frac{2\mu\alpha(1+\nu)}{(1-2\nu)} \tag{3.142}$$

α is the coefficient of thermal expansion. Hence the thermoelastic problem is a particular case of the elastostatic problem with body forces.

The boundary integral equation for a thermoelastic body without any other types of body forces is

$$c_{lk} u^i_k + \int_\Gamma p^*_{lk} u_k \, d\Gamma = \int_\Gamma u^*_{lk} p_k \, d\Gamma + \int_\Gamma u^*_{lk} (\gamma\theta) n_k \, d\Gamma - \int_\Omega u^*_{lk} \gamma \theta_{,k} \, d\Omega \tag{3.143}$$

Comparing to (3.139) we can write the potential of the equivalent body forces as,

$$\psi = -\gamma\theta \tag{3.144}$$

Taking into account that for steady state conduction one has $\nabla^2\theta \equiv 0$ the domain integral in (3.143) then becomes a boundary integral and (3.143) can be written as

$$c_{lk} u^i_k + \int_\Gamma p^*_{lk} u_k \, d\Gamma = \int_\Gamma u^*_{lk} p_k \, d\Gamma + \frac{(1-2\nu)\gamma}{2(1-\nu)} \int_\Gamma (G_{lk,kj} \theta n_j - G_{lk,k} \theta_{,j} n_j) \, d\Gamma \tag{3.145}$$

The above integral equation may be discretized in the usual way. All the functions in the last integral of equation (3.145) are known and may be easily integrated by numerical procedures over the boundary elements on the surface.

Thermoelastic problems may also be studied considering the temperature effects as initial strains, i.e

$$\varepsilon_{jk}^{\circ} = \alpha\theta\delta_{jk} \tag{3.146}$$

with the initial stresses given by

$$\sigma_{jk}^{\circ} = -(\lambda\varepsilon_{ii}^{\circ}\delta_{jk} + 2\mu\varepsilon_{jk}^{\circ}) \tag{3.147}$$

The values of σ_{jk}° are

$$\sigma_{jk}^{\circ} = -2\mu\left(\frac{1+v}{1-2v}\right)\alpha\theta\delta_{jk} = -\gamma\theta\delta_{jk} \tag{3.148}$$

Substituting equation (3.148) into the initial stress term of the elastic equation (3.125) one obtains,

$$\begin{aligned} -\int_{\Omega} \sigma_{jk}^{\circ}\varepsilon_{ljk}^{*}\, d\Omega &= \int_{\Omega} \gamma\theta\delta_{jk}\varepsilon_{ljk}^{*}\, d\Omega = \int_{\Omega} \gamma\theta u_{lk,k}^{*}\, d\Omega \\ &= -\int_{\Omega} \gamma\theta_{,jk} u_{lk}^{*}\, d\Omega + \int_{\Gamma} \gamma\theta u_{lk}^{*} n_k\, d\Gamma \end{aligned} \tag{3.149}$$

When equation (3.149) is substituted into (3.125) without any other type of body force, one obtains once more equation (3.143) that after taking the domain integral to the boundary gives (3.145). As could be expected both interpretations of thermoelastic effects give identical results.

3.7 Subregions in Elasticity

In the previous sections the boundary elements for homogeneous linear elastic isotropic media has been formulated. In many cases media that are not homogeneous but consist of several zones each one being homogeneous must be analysed. Those problems may be studied using the above formulation of the BEM and continuity conditions. All the boundaries of the body have to be discretized, including internal boundaries that separate homogeneous zones within the medium. The equations formulated for every homogeneous zone plus the displacement and traction continuity conditions over the internal boundaries produce a system that may be solved once the external boundary conditions are taken into account.

Consider the problem shown in figure 3.12 consisting of three zones of different elastic materials. A two-dimensional domain has been represented for simplicity; however, the procedure described below applies to both two and three-dimensional problems.

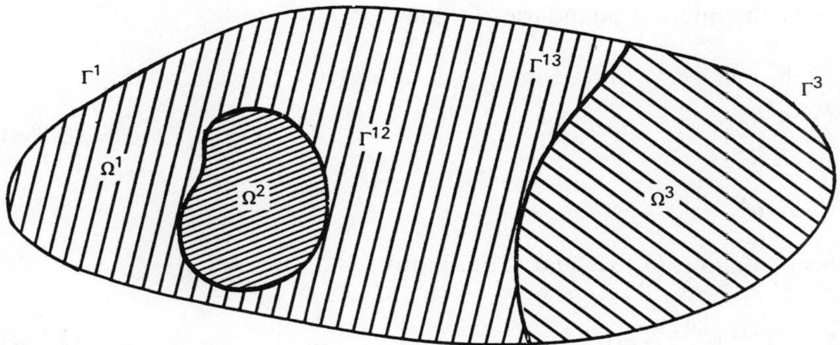

Figure 3.12 Zoned media

The following symbols are used:

Γ^i : external part of the boundary of zone Ω^i
Γ^{ij} : boundary between zones Ω^i and Ω^j
$\mathbf{u}^i$, $\mathbf{p}^i$: nodal displacements and tractions at nodes on the boundary Γ^i of zone Ω^i
$\mathbf{u}^{ij}$, $\mathbf{p}^{ij}$: nodal displacements and tractions at nodes on Γ^{ij} as part of zone Ω^i
$\mathbf{H}^i$, $\mathbf{G}^i$: parts of the H and G matrices obtained for zone Ω^i that multiply $\mathbf{u}^i$ and $\mathbf{p}^i$, respectively
$\mathbf{H}^{ij}$, $\mathbf{G}^{ij}$: parts of the H and G matrices obtained for zone Ω^i that multiply $\mathbf{u}^{ij}$ and $\mathbf{p}^{ij}$, respectively.

The BEM equations for the three homogeneous zones of figure 3.12 are

$$[\mathbf{H}^1 \quad \mathbf{H}^{12} \quad \mathbf{H}^{13}]\begin{Bmatrix} \mathbf{u}^1 \\ \mathbf{u}^{12} \\ \mathbf{u}^{13} \end{Bmatrix} = [\mathbf{G}^1 \quad \mathbf{G}^{12} \quad \mathbf{G}^{13}]\begin{Bmatrix} \mathbf{p}^1 \\ \mathbf{p}^{12} \\ \mathbf{p}^{13} \end{Bmatrix} \tag{3.150}$$

for zone Ω^1

$$\mathbf{H}^{21} \quad \mathbf{u}^{21} = \mathbf{G}^{21} \quad \mathbf{p}^{21} \tag{3.151}$$

for zone Ω^2 and

$$[\mathbf{H}^3 \quad \mathbf{H}^{31}]\begin{Bmatrix} \mathbf{u}^3 \\ \mathbf{u}^{31} \end{Bmatrix} = [\mathbf{G}^3 \quad \mathbf{G}^{31}]\begin{Bmatrix} \mathbf{p}^3 \\ \mathbf{p}^{31} \end{Bmatrix} \tag{3.152}$$

for zone Ω^3. The traction equilibrium conditions and displacement compatibility conditions over the internal boundaries Γ^{ij} are

$$\begin{aligned}\mathbf{u}^{12} &= \mathbf{u}^{21}\\ \mathbf{u}^{13} &= \mathbf{u}^{31}\\ \mathbf{p}^{12} &= -\mathbf{p}^{21}\\ \mathbf{p}^{13} &= -\mathbf{p}^{31}\end{aligned} \tag{3.153}$$

that transform equation (3.151) and (3.152) into

$$[\mathbf{H}^{21} \quad \mathbf{G}^{21}]\begin{Bmatrix}\mathbf{u}^{12}\\ \mathbf{p}^{12}\end{Bmatrix} = 0 \tag{3.154}$$

$$[\mathbf{H}^{3} \quad \mathbf{H}^{31} \quad \mathbf{G}^{31}]\begin{Bmatrix}\mathbf{u}^{3}\\ \mathbf{u}^{13}\\ \mathbf{p}^{13}\end{Bmatrix} = \mathbf{G}^{3}\mathbf{p}^{3} \tag{3.155}$$

The last two equations plus equation (3.150) can be rearranged into

$$\begin{bmatrix}\mathbf{H}^{1} & \mathbf{0} & \mathbf{H}^{12} & -\mathbf{G}^{12} & \mathbf{H}^{13} & -\mathbf{G}^{13}\\ \mathbf{0} & \mathbf{0} & \mathbf{H}^{21} & \mathbf{G}^{21} & \mathbf{0} & \mathbf{0}\\ \mathbf{0} & \mathbf{H}^{3} & \mathbf{0} & \mathbf{0} & \mathbf{H}^{31} & \mathbf{G}^{31}\end{bmatrix}\begin{Bmatrix}\mathbf{u}^{1}\\ \mathbf{u}^{3}\\ \mathbf{u}^{12}\\ \mathbf{p}^{12}\\ \mathbf{u}^{13}\\ \mathbf{p}^{13}\end{Bmatrix} = \begin{bmatrix}\mathbf{G}^{1} & \mathbf{0}\\ \mathbf{0} & \mathbf{0}\\ \mathbf{0} & \mathbf{G}^{3}\end{bmatrix}\begin{Bmatrix}\mathbf{p}^{1}\\ \mathbf{p}^{3}\end{Bmatrix} \tag{3.156}$$

The above system of equations may be solved once the boundary conditions on Γ^1 and Γ^3 are prescribed. The total number of unknowns is equal to the number of nodal degrees of freedom over the external boundaries plus twice the number of nodal degrees of freedom over the internal boundaries.

The subdivision of the region into several zones may be used also in homogeneous media as a way of avoiding numerical problems or improving computational efficiency. For instance problems that include cracks or notches, as the one shown in figure 3.13, present numerical difficulties when the boundary is discretized due to the proximity of some of the nodes. The difficulties disappear if the region is divided into two zones in a way that the nodes that are very close belong to different regions. Another situation where the subdivision of a homogeneous region may be useful corresponds to problems with a large number of unknowns, as the one shown in figure 3.14. In those cases the subdivision transforms the fully populated matrices into banded matrices, which are more convenient from a computational point of view. In those cases, the increase in the number of unknowns because of the internal boundaries must be small, otherwise the subdivision will be worthless.

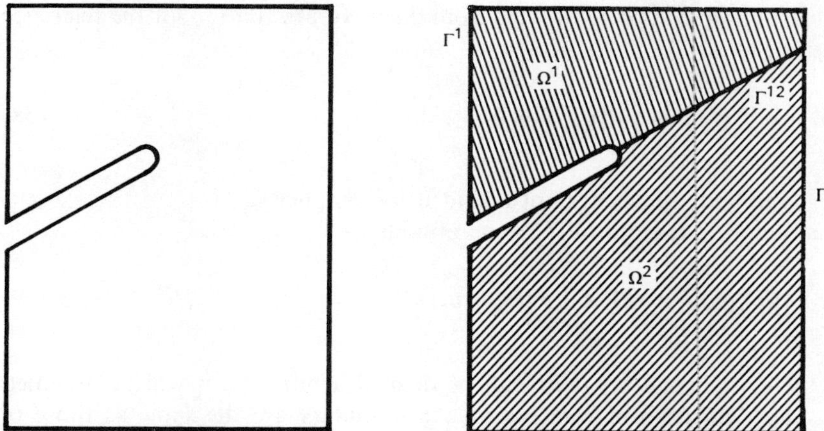

Figure 3.13 Subdivision to avoid numerical difficulties in a problem with a notch

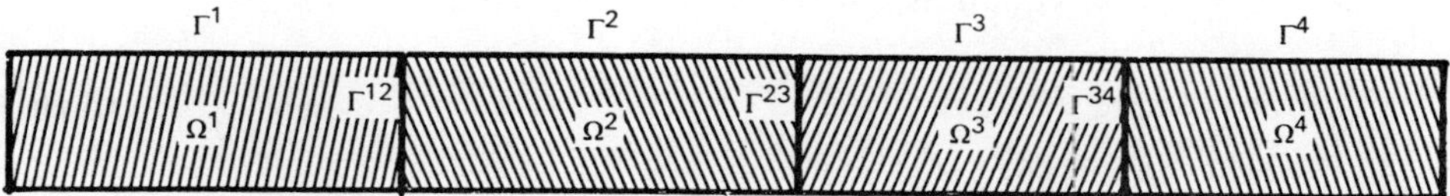

Figure 3.14 Subdivision to increase computational efficiency

3.8 Indirect Formulations

The fundamental boundary integral equation used throughout this chapter is known as the direct integral representation and gives displacements at internal and boundary points in terms of boundary tractions and boundary displacements. There are some other integral representations where the displacements are written in function of variables which are not explicitly the boundary displacements or tractions. Those representations are the basis of the so-called indirect boundary element formulations.

Let us consider a state σ_{ij}, ε_{ij}, u_j, p_j over the domain Ω of figure 3.15(a). We can now define another state over the complementary domain Ω' (figure 3.15(b)) with variables σ'_{ij}, ε'_{ij}, u'_j, p'_j. The tractions over the external region of Ω' will be referred to the normal n of the internal domain as shown in the figure.

It is easy to establish a reciprocity relationship for the complementary domain between the fundamental solution applied at a point in Ω and the complementary state indicated by primes. This gives,

$$\int_\Gamma u^*_{lk} p'_k \, d\Gamma - \int_\Gamma p^*_{lk} u'_k \, d\Gamma = 0 \tag{3.157}$$

This relationship can be subtracted from the integral equation for the reference state, as given by (3.71), i.e.

$$u_l^i - \int_\Gamma u_{lk}^* p_k \, d\Gamma + \int_\Gamma p_{lk}^* u_k \, d\Gamma = 0 \tag{3.158}$$

Notice that the body forces are not included for simplicity.

Subtracting (3.157) from (3.158) one obtains

$$u_l^i = \int_\Gamma u_{lk}^* (p_k - p_k') \, d\Gamma + \int_\Gamma p_{lk}^* (u_k' - u_k) \, d\Gamma \tag{3.159}$$

Since the complementary state can be defined arbitrarily, it will be assumed that it is such that its displacements on the boundary are the same as those of the original solution, i.e.

$$u_k' = u_k \qquad \text{on } \Gamma \tag{3.160}$$

Calling $\sigma_k = p_k - p_k'$ equation (3.159) can be rewritten as

$$u_l^i = \int_\Gamma u_{lk}^* \sigma_k \, d\Gamma \tag{3.161}$$

This equation can be interpreted as the displacement at a point 'i' inside Ω and can be obtained by the summation of displacements due to loads $\sigma_k \, d\Gamma$ applied at every $d\Gamma$ when Ω is considered to be part of the complete region. The integral representation in equation (3.161) is known as the single layer potential with density σ_k. As can be seen from equation (3.160) the displacements are considered to be continuous on the boundary while the tractions are discontinuous.

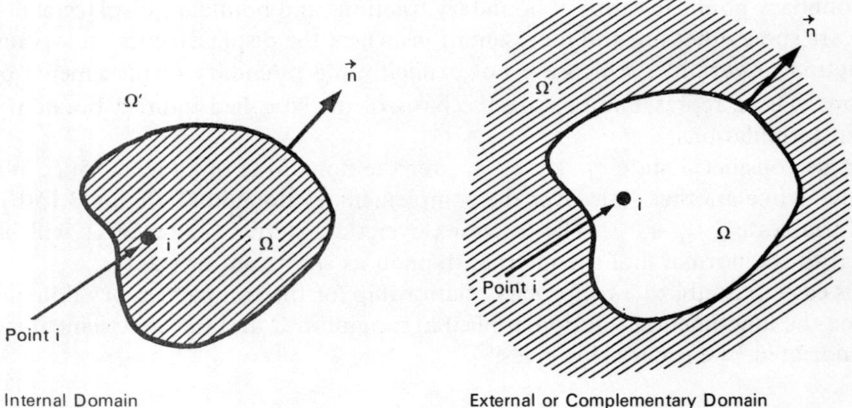

Figure 3.15 Internal and external domains

Tractions at any point can then be computed by carrying out the derivatives of (3.161) and in this case give

$$p_l = \int_\Gamma p^*_{lk} \sigma_k \, d\Gamma \tag{3.162}$$

Another possibility is to assume that the complementary state is such that

$$p'_j = p_j \qquad \text{on } \Gamma \tag{3.163}$$

Calling $\omega_k = u'_k - u_k$ equation (3.159) can be written as,

$$u^i_l = \int_\Gamma p^*_{lk} \omega_k \, d\Gamma \tag{3.164}$$

The above integral representation is known as double layer potential with intensity ω_k. As can be seen by the definition of ω, the double layer approach produces displacements which are discontinuous on γ while the tractions are continuous. Equation (3.164) can also be physically interpreted as a superposition of the displacements at i when dislocations $\omega_k \, d\Gamma$ are applied at every $d\Gamma$, with Ω considered to be part of the complete region. Tractions are calculated by carrying out derivatives in (3.164) which has the disadvantage over the previous indirect formulation of producing a higher order singularity.

It has been shown that internal displacements can be represented by a single layer potential (equation (3.161)), a double layer potential (equation (3.164)) or a combination of both, which is the basis of the direct formulation.

In general, the internal stresses and displacements computed by means of the discretized form of the integral representations are more accurate than those obtained using other numerical methods and similar discretizations. This is a consequence of the fact that the internal values are obtained by integration of fundamental solutions that are exact and only the boundary densities of the potentials are approximated. According to St. Venant's principle the local errors of this approximation may be expected to damp quickly.

The method based on the solution of the above integral equations by boundary discretization are known as Indirect Boundary Element Methods. Sometimes this name is used only for the single layer potential representation while the method based on the double layer potential is called the Displacement Discontinuity Method.

Example 3.3 Foundations

One of the most important aspects of the design of foundations is the computation of their stiffness or impedance. This is represented by the matrix **K** which relates applied loads to displacements,

$$\mathbf{F} = \mathbf{KU} \tag*{((a)}$$

Although here we will only refer to the static stiffness of foundations, it is important to point out that their dynamic response may be important in many other applications. This problem can also be solved using boundary elements but its discussion is beyond the scope of this text.

Figure 3.16 shows a transversal section of a foundation embedded in the soil. In many cases the behaviour of the soil can be assumed to be isotropic and linear and represented by a half-space. As the stiffness of the foundation is much larger this is assumed to behave as a rigid body and moments or forces applied at the top to simulate the loadings. When applying the Kelvin fundamental solution to the three dimensional soil domain, one needs to discretize not only the soil-foundation interface but also the free surface of the soil. This usually introduces an approximation because the discretization of the free surface is only carried out up to a certain distance from the foundations or because one develops an approximate infinite element. Figure 3.17 shows such a discretization for a square embedded foundation with level of embedment $E/B = 4/3$ and an amount of free field defined by $A/B = 2.5$. The analysis was carried out using constant elements and the axis of rotation for the rocking motion was considered to be on the soil-foundation interface. Over the free surface elements the tractions were considered to be zero.

The variation of stiffness with number of elements along half of the side of the bottom of the embedded foundation is shown in figure 3.18 (the discretization of the lateral walls is varied consistently). The figure shows that the boundary element mesh does not need to be very refined to obtain accurate results, mainly because in this case we are interested in the integrated tractions – i.e. resultant forces and moments – rather than the stresses along the foundation. Results for $N = 6$ compared favourably with those obtained with finer meshes.

The effect of the amount of soil free surface that is discretized can be seen in figure 3.19 for the same foundation. The study has been carried out adding successively lines of constant elements and the previous discretization of the soil free surface. Results converge rapidly.

The discretizaton of the soil free surface may be avoided by using Mindlin's fundamental solution [9] instead of Kelvin. This solution corresponds to the point load in an elastic half-space and reduces the number of elements required to run this type of problem. The computer time per integration over a boundary element

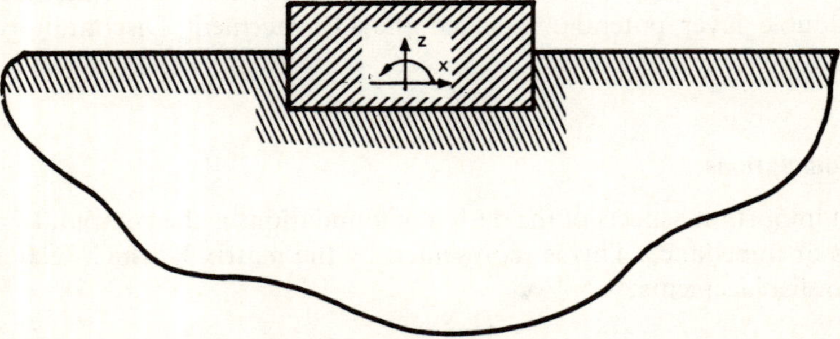

Figure 3.16 Rigid foundation embedded in the soil

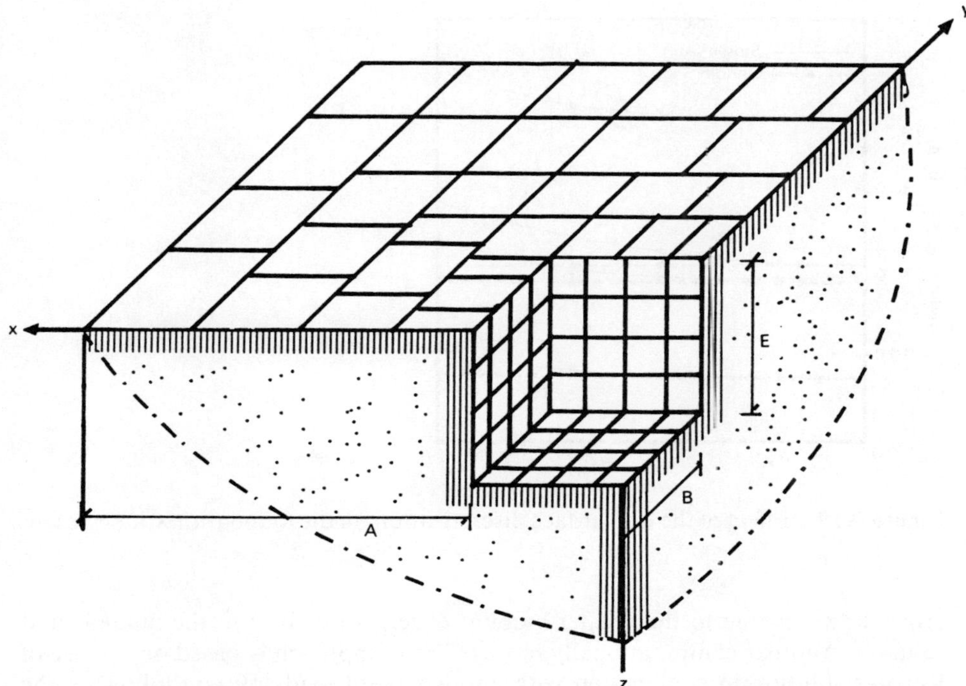

Figure 3.17 Discretization for one quarter of a square embedded foundation

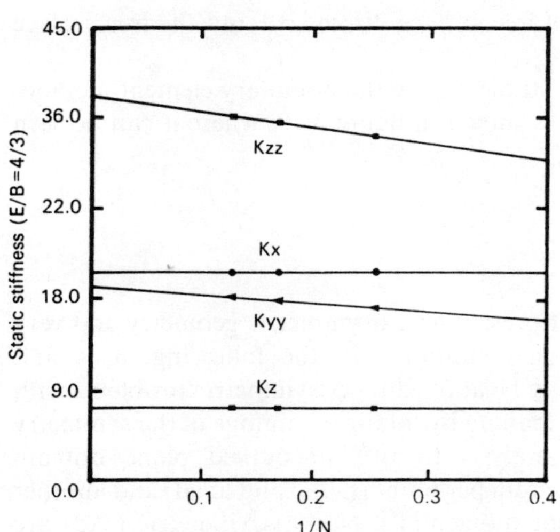

Figure 3.18 Effect of the interface discretization on the foundation stiffness. (Results are normalized as follows: Horizontal stiffness: $K_{x\,norm} = K_x(2 - \nu)/GB$; Vertical stiffness: $K_{z\,norm} = K_z(1 - \nu)/GB$; Rocking stiffness: $K_{yy\,norm} = K_{yy}(1 - \nu)/GB^3$; and Torsional stiffness: $K_{zz\,norm} = K_{zz}/GB^3$.)

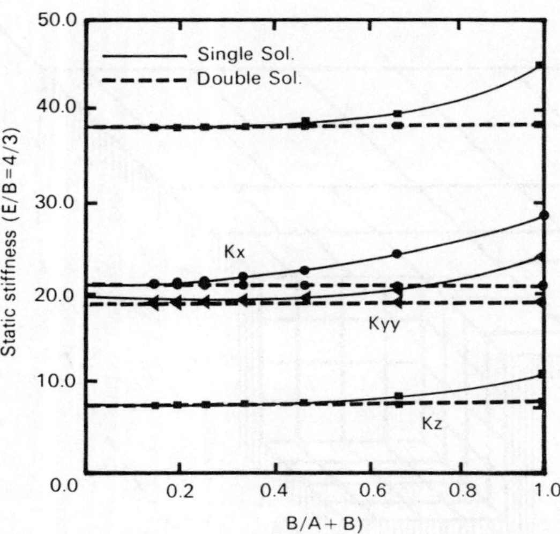

Figure 3.19 Effect of the free surface discretization on the foundation stiffnesses

grows however due to the greater amount of terms involved in the fundamental solution. Another computationally more efficient approach is based on the use of Kelvin's solution in conjunction with another point load solution following the same direction but located at the image point of the first with respect to the soil surface. The use of this solution reduces to zero the surface tractions with the exception of p_{13}, p_{23} and p_{31}, p_{32} which are expected to have a small effect in soils. Figure 3.19 shows how when the double solution is used the result becomes almost independent of the amount of free surface discretized and the free surface does not have to be considered.

Many other soil problems can be studied using the boundary element method. Some applications in zoned media are shown in figure 3.20, where it can be seen that the soil can be inhomogeneous.

3.9 Axisymmetric Problems

There are many elastic problems that present an axisymmetric geometry and very frequently also axisymmetric loading conditions. In the following, it is first explained how boundary elements may be applied to axisymmetric problems with respect to both the geometry and the loading by taking advantage of the symmetry that reduces a three-dimensional analysis to two uncoupled plane domain problems; one with two degrees of freedom per point (radial and axial) and another with one degree of freedom per point (tangential). Non-axisymmetric loads are studied later. The first boundary elements formulation for axisymmetric elastic problems was published in 1975 [11], [12]. It was based on the Somigliana's identity obtained from the application of the reciprocity theorem between the actual axisymmetric problem and the fundamental solutions corresponding to a radial

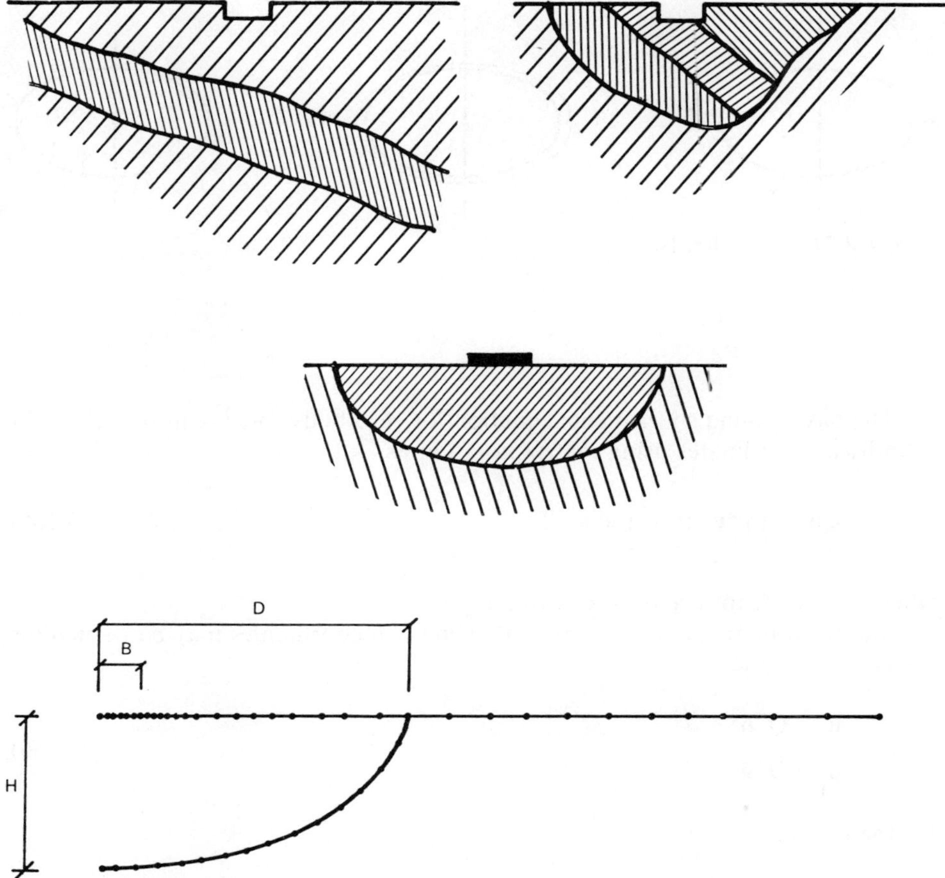

Figure 3.20 Zoned soil models and boundary discretization

ring load and an axial ring load, for one part of the problem, and a tangential ring load for the other (torsion) part of the problem, as indicated in figure 3.21.

The basic equation has now the same expression as in previous cases (equation (3.87)) with $\mathbf{u}^*$ and $\mathbf{p}^*$ being the displacements and tractions due to the ring loads. Those fundamental solutions were obtained by different procedures by Kermanidis [11], Mayr [12], Cruse *et al.* [13], and Dominguez and Abascal [9]. They are written in terms of Legendre functions or complete elliptic integrals, which makes their integration along the boundary elements rather involved. Explicit expressions of the ring loads fundamentals solutions may be found, for instance, in [11], [13]. The procedure is basically the same as that presented in section 2.15 for axisymmetric potential problems. An alternative and more general approach based on the three dimensional formulation is presented in what follows.

The three dimensional fundamental solution will be used and numerical integration of $\mathbf{p}^*$ and $\mathbf{u}^*$ performed on the axisymmetric elements.

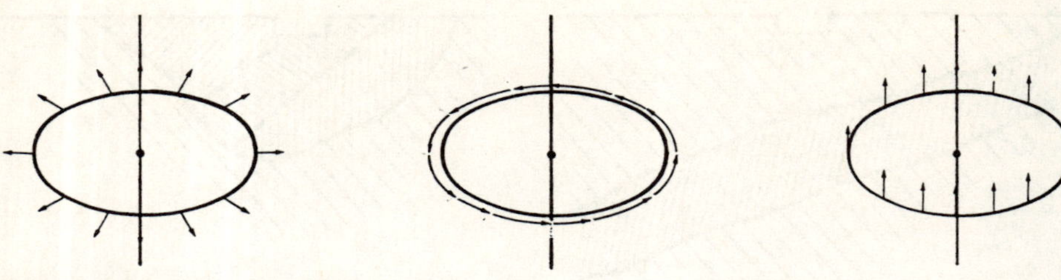

Figure 3.21 Ring loads

The basic boundary elements equation for zero body forces can be written in cylindrical coordinates using matrix notation as,

$$\mathbf{c}_c^i\mathbf{u}_c^i + \int_\Gamma \mathbf{p}_c^*\mathbf{u}_c \, d\Gamma = \int_\Gamma \mathbf{u}_c^*\mathbf{p}_c \, d\Gamma \tag{3.165}$$

where the subscript c stands for cylindrical.

The relation between cartesian and cylindrical coordinates may be written for vectors $\mathbf{u}$ and $\mathbf{p}$ as

$$\begin{aligned} \mathbf{u}_c &= \mathbf{Q}^T\mathbf{u} \\ \mathbf{p}_c &= \mathbf{Q}^T\mathbf{p} \end{aligned} \tag{3.166}$$

for the matrix $\mathbf{c}^i$

$$\mathbf{c}_c^i = \mathbf{Q}^{i,T}\mathbf{c}^i\mathbf{Q}^i \tag{3.167}$$

and for matrices $\mathbf{p}^*$ and $\mathbf{u}^*$

$$\begin{aligned} \mathbf{u}_c^* &= \mathbf{Q}^{i,T}\mathbf{u}^*\mathbf{Q} \\ \mathbf{p}_c^* &= \mathbf{Q}^{i,T}\mathbf{p}^*\mathbf{Q} \end{aligned} \tag{3.168}$$

where (figure 3.22)

$$\mathbf{Q} = \begin{bmatrix} \cos\theta & -\sin\theta & 0 \\ \sin\theta & \cos\theta & 0 \\ 0 & 0 & 1 \end{bmatrix} \tag{3.169}$$

Notice that $\mathbf{u}^*$ and $\mathbf{p}^*$ relate the collocation point i and the integration point.

Since i is the collocation point, it may be assumed that $\theta^i = 0$ (figure 3.22), which makes the transformation matrix $\mathbf{Q}^i = \mathbf{I}$. The kernels of the integrals in

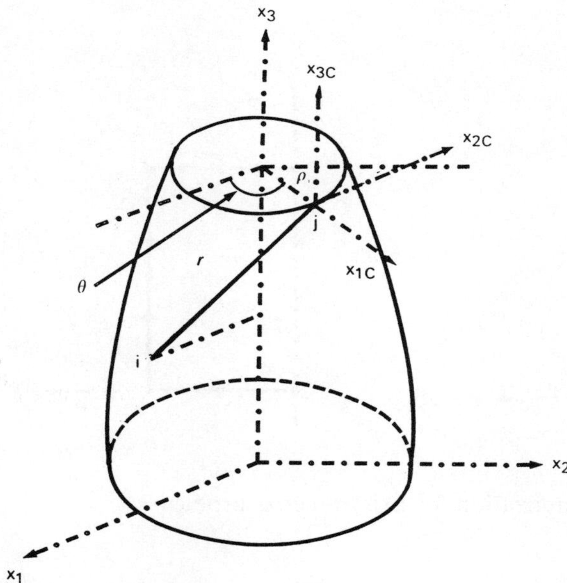

Figure 3.22 Cylindrical coordinates

equation (3.165) have the form

$$\mathbf{u}_c^* = \mathbf{u}^*\mathbf{Q} = \begin{bmatrix} u_{11}^* \cos\theta + u_{12}^* \sin\theta & -u_{11}^* \sin\theta + u_{12}^* \cos\theta & u_{13}^* \\ u_{21}^* \cos\theta + u_{22}^* \sin\theta & -u_{21}^* \sin\theta + u_{22}^* \cos\theta & u_{23}^* \\ u_{31}^* \cos\theta + u_{32}^* \sin\theta & -u_{31}^* \sin\theta + u_{32}^* \cos\theta & u_{33}^* \end{bmatrix} \tag{3.170}$$

One half of a meridional section of the body is discretized into elements (figure 3.23) and equation (3.165) may be written for the boundary node i as

$$\mathbf{c}^i\mathbf{u}^i + \sum_{j=1}^{N} \left\{ \int_{\Gamma_j} \rho \left[\int_0^{2\pi} \mathbf{p}_c^* \, d\theta \right] \boldsymbol{\phi} \, d\Gamma \right\} \mathbf{u}^j = \sum_{j=1}^{N} \left\{ \int_{\Gamma_j} \rho \left[\int_0^{2\pi} u_c^* \, d\theta \right] \boldsymbol{\phi} \, d\Gamma \right\} \mathbf{p}^j \tag{3.171}$$

where ρ is the radius shown in figure 3.22, Γ_j are the boundary elements, $\boldsymbol{\phi}$ the usual shape functions for two dimensional problems and the subindex c has been dropped out for simplicity.

The integration along θ may be easily done by numerical procedures. The submatrices $\mathbf{H}^{ij}$ and $\mathbf{G}^{ij}$ that relate two nodes i and j have the pattern

$$\begin{bmatrix} * & 0 & * \\ 0 & * & 0 \\ * & 0 & * \end{bmatrix} \begin{matrix} \leftarrow \rho \\ \leftarrow \theta \\ \leftarrow z \end{matrix} \tag{3.172}$$

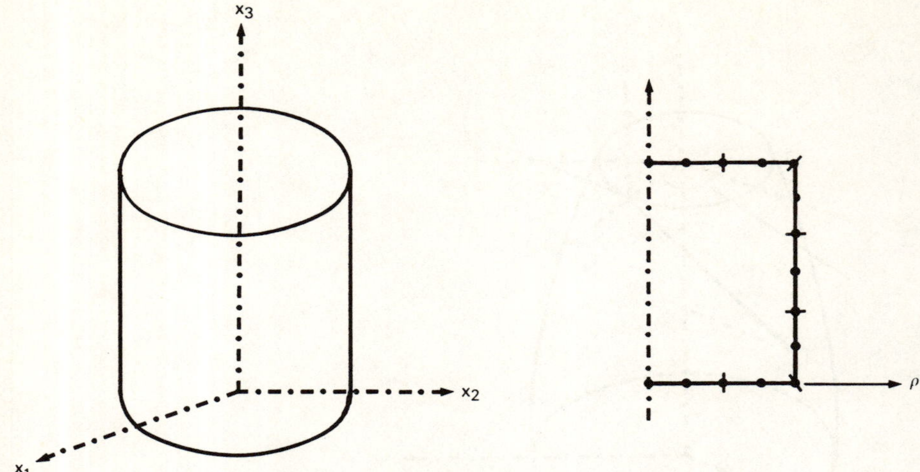

Figure 3.23 Quadratic BE discretization for axisymmetric problem

where the zeros denote elements that are null due to the skewsymmetry of the corresponding terms in equation (3.170). It is clear that the torsion and the radial-axial problems are uncoupled and both may be studied on a plane domain.

A special numerical integration scheme for axisymmetric problems was proposed by Gomez-Lera *et al.* [14] for axisymmetric problems using the three-dimensional point load fundamental solution. When Kelvin's solution is integrated around the axis, a Gauss quadrature may be applied to every semi-ring; however, its accuracy is easily improved by increasing the density of integration points near the collocation points by means of a parabolic transformation of the circumferential coordinate

$$\theta = \frac{\pi}{4}(\xi + 1)^2; \qquad -1 \leqslant \xi \leqslant 1 \tag{3.173}$$

It should be noticed that since a rigid body motion along the radial coordinate of an axisymmetric body is meaningless, the corresponding terms of $\mathbf{c}^i$ have to be computed from the rest of the terms of the **G** and **H** matrices giving a uniform shrinking of the ξ coordinate or using the analytical expression given for plane problems in the next chapter.

When the boundary conditions are not axisymmetric, the problem may still be analysed by means of a plane model. The problem is divided into a number of uncoupled plane problems by representing the prescribed loading or displacement by a Fourier series along the tangential coordinate [15]. Each term of the series produces displacements and stresses in the same Fourier mode and if the prescribed values do not vary very rapidly around the axis, a few modes will be enough for an accurate solution. The Fourier expansion is of the form

$$u_\rho = \sum_{n=0}^{\infty} (u_{n\rho}^s \cos n\theta + u_{n\rho}^a \sin n\theta)$$

$$u_\theta = \sum_{n=0}^{\infty} (-u_{n\theta}^s \sin n\theta + u_{n\theta}^a \cos n\theta) \tag{3.174}$$

$$u_z = \sum_{n=0}^{\infty} (u_{nz}^s \cos n\theta + u_{nz}^a \sin n\theta)$$

where 's' indicates the symmetric terms and 'a' the antisymmetric ones.

For each Fourier mode amplitude a discretized boundary equation like equation (3.171) may be written with the only difference being that $\mathbf{p}_c^*$ and $\mathbf{u}_c^*$ are now weighted by a sine or a cosine function and the integrals around the axis are of the form

$$\int_0^{2\pi} \mathbf{u}_c^* \sin n\theta \, d\theta; \qquad \int_0^{2\pi} \mathbf{u}_c^* \cos n\theta \, d\theta \tag{3.175}$$

It is worth noting that since $\sin n\theta$ has zero value at $\theta = 0$, the i point cannot be located at $\theta^i = 0$ to compute the amplitude of those terms of the Fourier series. One only has to move i to a point, for instance $\theta^i = -\pi/2n$,where the amplitude is not zero. This change is easily taken into account by a shift of the origin of θ in equation (3.170).

3.10 Anisotropic Elasticity

The constitutive relationships described in section 3.2 are valid for isotropic cases for which the behaviour of the material can be described in terms of only two constants. In more general cases the material can be orthotropic or generaly anisotropic. For a three dimensional anisotropic case the number of different elastic constants is 21 and they can be expressed in matrix form as,

$$\boldsymbol{\sigma} = \mathbf{D}\boldsymbol{\varepsilon} \tag{3.176}$$

where $\mathbf{D}$ is a 6×6 matrix. The $\boldsymbol{\sigma}$ and $\boldsymbol{\varepsilon}$ are the usual six stress and strain components. The above elastic relationship can also be written in index form which gives a notation more consistent with the previous sections, i.e.

$$\sigma_{ij} = d_{ijkl}\varepsilon_{kl} \tag{3.177}$$

d_{ijkl} are called the rigidity coefficients. The inverse of (3.176) produces the elastic compliance, c_{ijkl}, i.e.

$$\varepsilon_{kij} = c_{ijkl}\sigma_{kl} \tag{3.178}$$

Although there are in general 21 constants for an elastic material, their number is reduced when the material structure has one or more planes of symmetry.

Materials with the orthogonal planes of symmetry are said to be orthotropic and the number of independent coefficients reduces to 9.

For two dimensional problems, the plane $x_1 - x_2$ is a plane of symmetry for the material structure. In this case the number of independent coefficients reduces to 13. Furthermore if either the x_1 or x_2 are axes of symmetry, the material is orthotropic and the constants reduce to 9 as before.

Fundamental solutions exist for the case of two and three dimensional anisotropic cases but they are difficult to use because of the complexity of their mathematical formulation or the need to find part of the solution numerically, which may be inefficient.

Because of this a different technique will be described in this section which can be applied to any anisotropic material. The approach consists of using a reference isotropic fundamental solution and iterating to find the correct result. The procedure will be explained in what follows.

Consider again the starting weighted residual statement, in terms of the equilibrium equation (3.66), i.e.

$$\int_{\Omega} (\sigma_{kj,j} + b_k) u_k^* \, d\Omega = \int_{\Gamma_2} (p_k - \bar{p}_k) u_k^* \, d\Gamma + \int_{\Gamma_1} (\bar{u}_k - u_k) p_k^* \, d\Gamma \tag{3.179}$$

The next step is to consider that the material is anisotropic but use the isotropic fundamental solution corresponding to a reference elastic model, whose properties may be found by averaging the anisotropic constants. Hence the original coefficients can be expressed as,

$$d_{kjmn} = d^0_{kjmn} + \hat{d}_{kjmn} \tag{3.180}$$

where d^0 indicates the reference state and $\hat{d}$ the residual or difference between the actual and the isotropic elastic constants.

Integrating by parts (3.179) one finds that

$$\begin{aligned} -\int_{\Gamma} \sigma_{kj} \varepsilon_{kj}^* \, d\Omega + \int_{\Omega} b_k u_k^* \, d\Omega \\ = -\int_{\Gamma_2} \bar{p}_k u_k^* \, d\Gamma - \int_{\Gamma_1} p_k u_k^* \, d\Gamma + \int_{\Gamma_1} (\bar{u}_k - u_k) p_k^* \, d\Gamma \end{aligned} \tag{3.181}$$

Next, one substitutes the following expression

$$\sigma_{kj} = d_{kjmn} \varepsilon_{mn} = (d^0_{kjmn} + \hat{d}_{kjmn}) \varepsilon_{mn} = \sigma^0_{kj} + \hat{\sigma}_{kj} \tag{3.182}$$

while the fundamental solution obeys the following constitutive equation

$$\sigma_{kj}^* = \overset{0}{d}_{kjmn} \varepsilon_{mn}^*$$

Hence formula (3.181) becomes

$$-\int_{\Omega} \sigma^0_{kj}\varepsilon^*_{kj}\, d\Omega - \int_{\Omega} \hat{\sigma}_{jk}\varepsilon^*_{jk}\, d\Omega + \int_{\Omega} b_k u^*_k\, d\Omega$$

$$= -\int_{\Gamma_2} \bar{p}_k u^*_k\, d\Gamma - \int_{\Gamma_1} p_k u^*_k\, d\Gamma + \int_{\Gamma_1} (\bar{u}_k - u_k) p^*_k\, d\Gamma \tag{3.183}$$

After noticing that

$$\int_{\Omega} \sigma^0_{kj}\varepsilon^*_{kj}\, d\Omega = \int_{\Omega} \sigma^*_{kj}\varepsilon_{kj}\, d\Omega \tag{3.184}$$

one can integrate by parts the first terms of (3.183) to produce

$$\int_{\Omega} \sigma^*_{kj,j} u_k\, d\Omega - \int_{\Omega} \hat{\sigma}_{kj}\varepsilon^*_{kj}\, d\Omega + \int_{\Omega} b_k u^*_k\, d\Omega$$

$$= -\int_{\Gamma_2} \bar{p}_k u^*_k\, d\Gamma - \int_{\Gamma_1} p_k u^*_k\, d\Gamma + \int_{\Gamma_2} u_k p^*_k\, d\Gamma + \int_{\Gamma_1} \bar{u}_k p^*_k\, d\Gamma \tag{3.185}$$

Taking into consideration that the fundamental solution can be applied inside the domain or on the boundary, equation (3.185) gives the following integral statement,

$$c^i_{lk} u^i_k + \int_{\Gamma} p^*_{lk} u_k\, d\Gamma = \int_{\Gamma} u^*_{lk} p_k\, d\Gamma + \int_{\Omega} u^*_{lk} b_k\, d\Omega - \int_{\Omega} \hat{\sigma}_{kj}\varepsilon^*_{lkj}\, d\Omega \tag{3.186}$$

where $\Gamma = \Gamma_1 + \Gamma_2$.

Notice that this formula is similar to (3.79) with only the addition of a new domain term. As before 'l' represents the direction in which the fundamental solution component is acting.

Equation (3.186) contains two types of domain terms. One is due to the body force components b_k and can sometimes be taken to the boundary as described in section 3.6. The other term can be integrated on the domain using cells or transformed into a 'body force' type term and then taken to the boundary. Let us consider the term on its own, i.e.

$$\int_{\Omega} \hat{\sigma}_{kj}\varepsilon^*_{lkj}\, d\Omega \tag{3.187}$$

Next one integrate by parts (3.187) which gives

$$-\int_{\Omega} \hat{\sigma}_{kj,j} u^*_{lk}\, d\Omega + \int_{\Gamma} \hat{p}_k u^*_{lk}\, d\Gamma \tag{3.188}$$

The first integral can now be interpreted in function of a fictitious body force such that

$$\hat{\sigma}_{kj,j} = \hat{b}_k \tag{3.189}$$

Hence equation (3.188) and (3.189) can now be substituted into (3.186) to give

$$c^i_{lk}u^i_k + \int_\Gamma p^*_{lk}u_k \, d\Gamma = \int_\Gamma u^*_{lk}(p_k + \hat{p}_k) \, d\Gamma + \int_\Omega u^*_{lk}(b_k + \hat{b}_k) \, d\Omega \tag{3.190}$$

$\hat{p}_k$ and $\hat{b}_k$ are acting now as out of balance forces which represents the anisotropic effect. The problem can be solved iteratively by finding the first solution with $\hat{p}_k$ and $\hat{b}_k \equiv 0$, and then computing their values and resolving the system as many times as required. The influence matrices **H** and **G** are always the same. In matrix form equation (3.190) can be written as,

$$\mathbf{HU} = \mathbf{GP} + \mathbf{G\hat{P}} + \mathbf{DB} + \mathbf{D\hat{B}} \tag{3.191}$$

The equations at the beginning of the iteration process are simply,

$$\mathbf{HU} = \mathbf{GP} + \mathbf{DB} \tag{3.192}$$

and from then on one solves the following equation

$$\mathbf{HU} = \mathbf{GP} + \mathbf{DB} + \mathbf{B}' \tag{3.193}$$

where $\mathbf{B}' = \mathbf{G\hat{P}} + \mathbf{D\hat{B}}$, where $\hat{\mathbf{P}}$, $\hat{\mathbf{B}}$ terms result from the previous step.

While the b_k terms may be reduced to the boundary as shown earlier the main problem remains how to convert $\hat{b}_k$ terms into boundary integrals. If this is not possible one will need to divide the domain into cells as explained in section 3.5.

A general way of reducing body force terms to the boundary is by using particular solutions. This has been generalized in references [16] to [21] by using a technique called the dual reciprocity method (DRM). Although this is an interesting development with many applications in boundary elements it falls beyond the scope of this book and for the further information the reader is referred to the work of Brebbia and his collaborators Nardini, Wrobel and Tang given in the above articles.

References

[1] Alarcon, E., Brebbia, C. A. and Dominguez, J. The Boundary Element Method in Elasticity, *Intern. Jnl. of Mechanical Sciences*, **20**, 625–639, 1978.

[2] Rizzo, F. J. An Integral Equation Approach to Boundary Value Problems of Classical Elastostatics, *Quarterly of Applied Mathematics*, **25**, 83–95, 1967.

[3] Love, A. E. H. *A Treatise on the Mathematical Theory of Elasticity*, Dover, New York, 1944.

[4] Cruse, T. A. and Rizzo, F. J. A Direct Fomulation and Numerical Solution of the General Transient Elastodynamic Problem I, *Journal of Math. Analysis and Applications*, **22**, 244, 1968.

[5] Cruse, T. A. A Direct Formulation and Numerical Solution of the General Transient Elastodynamic Problem II, *Journal of Math. Analysis and Applications*, **22**, 341, 1968.

[6] Brebbia, C. A. *The Boundary Element Method for Engineers*, Pentech Press London, 1978. Computational Mechanics Publications, Boston, 1984.
[7] Brebbia, C. A., Telles, J. C. F. and Wrobel, L. C. *Boundary Element Techniques*, Springer-Verlag, Berlin and NY, 1984.
[8] Jaswon, M. A. and Symm, G. T. *Integral Equation Methods in Potential Theory and Elastostatics*, Academic Press, London, 1977.
[9] Dominguez, J. and Abascal, R. On Fundamental Solutions for the Boundary Integral Equations Method in Static and Dynamic Elasticity, *Engineering Analysis*, **1** 128–134, 1984.
[10] Hartmann, F. '*Elastostatics*', *Progress in Boundary Element Methods*, C. A. Brebbia, Ed., Pentech Press, London, 1981.
[11] Kermanidis, T. A. Numerical Solution for Axially Symmetrical Elasticity Problems, *Journal of Solids and Structures*, **11**, 493–500, 1975.
[12] Mayr, M. Ein Integralgleichungsverfahren Losung Rotationssymetrischer Elastizitatsprobleme, *Dissertation*, Technical University Munchen, 1975.
[13] Cruse, T. A., Snow, D. A. and Wilson, R. B. Numerical Solutions in Axisymmetric Elasticity, *Computers and Structures*, **8**, 445–451, 1977.
[14] Gomez-Lera, M. S., Dominguez, J. and Alarcon, E. On the Use of a 3-D Fundamental Solution for Axisymmetric Steady-State Dynamic Problems, *Boundary Elements VII*, Proceedings of the 7th International Conference, Italy, C. A. Brebbia and G. Maier Eds, Springer-Verlag, Berin and NY, 1985.
[15] Wilson, E. Structural Analysis of Axisymmetric Solids, *AIAA Journal*, **3**, 2269–2274, 1965.
[16] Brebbia, C. A. and Nardini, D. Dynamic Analysis in Solid Mechanics by an Alternative Boundary Element Formulation, *Int. Jnl. Soil Dynamics and Earthquake Engineering*, **2**, 1983.
[17] Nardini, D. and Brebbia, C. A. Boundary Integral Formulation of Mass Matrices for Dynamic Analysis, Chapter 7 in *Topics in Boundary Element Research*, *Vol. 2* (Ed. C. A. Brebbia), Springer-Verlag, Berlin and NY, 1985.
[18] Wrobel, L. C., Telles, J. C. F. and Brebbia, C. A. A Dual Reciprocity Boundary Element Formulation for Axisymmetric Diffusion Problems, in *BETECH/86* (Eds. J. J. Connor and C. A. Brebbia), Computational Mechanics Publications, Southampton and Boston, 1986.
[19] Wrobel, L., Brebbia, C. A. and Nardini, D. Analysis of Transient Thermal Problems in the BEASY System, in *BETECH/86* (Eds. J. J. Connor and C. A. Brebbia), Computational Mechanics Publications, Southampton and Boston, 1986.
[20] Brebbia, C. A. and Wrobel, L. Nonlinear Transient Thermal Analysis using the Dual Reciprocity Method, in *Boundary Element Techniques*: *Applications in Stress Analysis and Heat Transfer* (Eds. C. A. Brebbia and W. S. Venturini), Computational Mechanics Publications, Southampton and Boston, 1987.
[21] Tang, W. and Brebbia, C. A. Critical Comparison of Two Transformation Methods for taking BEM Domain Integrals to the Boundary, to appear in *Engineering Analysis*.

Exercises

3.1. Verify that the equation $\nabla^2(\nabla^2 G_l) + \frac{1}{\mu}\Delta^i e_l = 0$ is obtained by substitution of displacements in terms of Galerkin's vector into Navier's equation for the fundamental solution (equation (3.34)).

3.2. Show that $\lim_{\varepsilon \to 0} \int_{\Gamma_\varepsilon} p_k u_{lk}^* \, d\Gamma = 0$, for any part or complete spherical surface of radius ε around the collocation point.

3.3. Compute the terms of the 3×3 matrix $\mathbf{c}^i$ for a point on a quarter of a sphere edge.

3.4. Obtain the value of the kernel D_{kij} of the internal point stress representation by derivation of the fundamental solution u_{lk}^*.

3.5. Is the matrix $\mathbf{u}^*$ (equation (3.86)) relating two points of the boundless domain symmetric? Answer the same question for matrix $\mathbf{p}^*$ (equation (3.85)). Explain the reasons.

3.6. Assume a rectangular constant element with a centred node for three dimensional problems. Which terms of the matrices $\mathbf{H}^{ii}$ and $\mathbf{G}^{ii}$ relating the node with itself are zero?

3.7. Give an explicit formula, in terms of internal coordinates, for the numerical integration of $\mathbf{u}^*$ and $\mathbf{p}^*$ over a constant rectangular element when the collocation point is not inside the element. The element is defined by the cartesian coordinates of its four corners. 3×3 Gauss quadrature points should be used.

3.8. In section 3.6 it is said that a rigid body motion may be dropped from the fundamental solution when solving two-dimensional elasticity problems. Give the reason why this is possible for bounded regions.

3.9. When will the above question be true for unbounded regions?

3.10. Obtain the 3×3 fundamental solution matrix $\mathbf{u}_c^* = \mathbf{Q}^{iT}\mathbf{u}^*\mathbf{Q}^*$ in cylindrical coordinates when the collocation point is at $\theta' = -\pi/2$.

3.11. When a circular foundation resting on the surface of an elastic half-space is under the effects of a horizontal force, the motion and traction components at any point of the half-space surface may be written as:

$$u_\rho = u'_\rho \cos\theta \qquad t_\rho = t'_\rho \cos\theta$$

$$u_\theta = -u'_\theta \sin\theta \qquad t_\theta = -t'_\theta \sin\theta$$

$$u_z = u'_z \cos\theta \qquad t_z = t'_z \cos\theta$$

Write the integral equation relating u'_ρ, u'_θ, u'_z and t'_ρ, t'_θ, t'_z for boundary points using equation (3.170) and the formula derived in exercise 3.10, for the θ component.

Chapter 4

Two Dimensional Elastostatics

4.1 Introduction

This chapter deals with the applications of boundary elements to solve two dimensional elastostatics problems. The basic equations of elasticity are reviewed, first pointing out that boundary elements for these cases are based on the plane strain approach but can be extended to plane stress if the elastic coefficients are replaced by the corresponding equivalent values, as will be seen below.

The boundary element formulation of elastostatics is substantially different from the one for potential problems as its unknowns are in vector rather than scalar form. This means that at a particular point there are two (or three in three dimensions) components of displacements or tractions, rather than one of each as discussed in Chapter 3. The associated fundamental solution also represents a point load acting in a given direction. All this implies that the boundary element formulation for elastostatics is more complex than the one described in Chapter 2 for potential problems although it involves the same basic steps.

It seems appropriate to start by developing the necessary theory and then the computer code corresponding to the simplest case, i.e. the one with constant elements. After this the development of a quadrilateral element and associated code is described, without presenting the linear case in detail. This approach has been preferred as the linear case has already been discussed in Chapter 2 for potential problems and is comparatively simple to extend the formulation to elastostatics. The quadratic elasticity code is also more interesting from the point of view of applications, as many engineering problems are difficult to solve accurately using constant elements and the linear codes do not converge rapidly for cases involving bending, for instance. Hence while constant or linear elements can be satisfactorily applied for many potential problems they – particularly the constant ones – are generally not sufficiently accurate for stress analysis applications.

The chapter presents a series of examples describing the excellent results which can be obtained by using boundary elements in elastostatics. This chapter in conjunction with the previous one attempts to be an introduction to the use of boundary elements for stress analysis. For further developments and more complex applications the reader is advised to consult the relevant publications listed in Appendix C.

4.2 Plate Stretching – Plane Strain Problems

Two dimensional problems are divided into two types, plate stretching (sometimes also called plane stress) and plane strain problems, depending on how the solid is restrained in the direction perpendicular to the plane under study.

To understand the difference between these two states consider the prismatic homogeneous solid shown in figure 4.1. The end surfaces are defined by the planes at $x_3 = \pm h/2$, and the cylindrical surface by $x_1 = x_1(\Gamma)$, $x_2 = x_2(\Gamma)$, where Γ is the arc length along the boundary curve.

Plate stretching The basic assumptions for plate stretching are

(i) that the body is thin, i.e. h is small by comparison with the representative dimensions along x_1 or x_2,
(ii) there are no tractions acting at the end surfaces, i.e. at $x_3 = \pm h/2$, $p_j = 0$,
(iii) the body forces are acting on $x_1 - x_2$ planes and independent of $\mathbf{x}_3$, i.e. $b_3 = 0$ and b_1, b_2 are functions of $x_1 x_2$ only,
(iv) the forces acting on the cylindrical body are planar and independent of x_3, i.e. $p_3 = 0$ and $p_1 p_2$ are functions of x_1 and x_2.

Under these assumptions it is assumed that σ_{33}, σ_{31} and σ_{32} are all small in comparison with $\sigma_{11}\sigma_{22}\sigma_{12}$ and that the variation of the latter with respect to x_3 is negligible. Hence one assumes:

$$\sigma_{33} = \sigma_{31} = \sigma_{32} = 0 \tag{4.1}$$

and σ_{11}, σ_{22} and σ_{12} are functions of x_1, x_2 only.

$$\sigma_{11}(x_1, x_2); \qquad \sigma_{22}(x_1, x_2); \qquad \sigma_{12}(x_1, x_2)$$

It should be noticed however that although these assumptions are reasonable in engineering practice they are only approximate as they violate the compatibility equations.

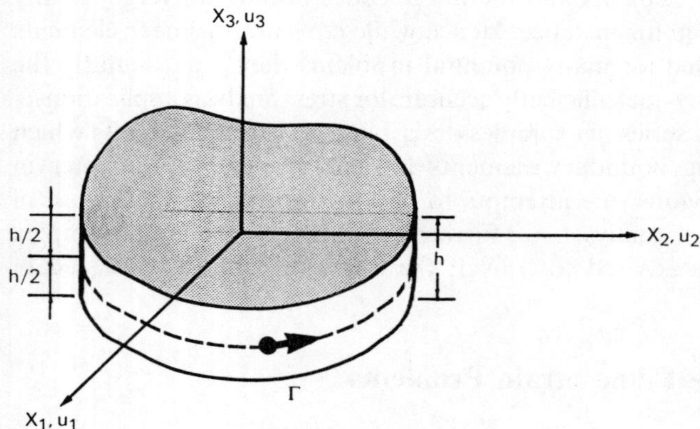

Figure 4.1 Prismatic solid

Equation (4.1) is written in function of displacements and one can have an alternative statement, i.e. the displacement components are functions of $x_1 x_2$ only, i.e.

$$u_1 = u_1(x_1, x_2); \qquad u_2 = u_2(x_1, x_2) \tag{4.2}$$

Notice that $u_3 \neq 0$ and the ε_{33} strains can be determined in function of the σ_{11}, σ_{22}, σ_{12} stresses.

Plane Strain This case usually represents the behaviour of long structures such as tunnels and for this the displacements in the normal direction can be assumed to be zero (i.e. on the end faces).

The the plane strain assumptions are

(ii) The end faces displacements u_3 are zero as they are restrained to move normally because the thickness is large in comparison with the representative dimensions in $x_1 x_2$ directions.

(ii) The body and surface forces acting on the cylindrical surface have no x_3 component and are independent of x_3.

In this case in addition to $u_3 = 0$, one can assume that the in-plane displacements u_1 and u_2 are independent of x_3. Hence

$$u_1 = u_1(x_1, x_2); \qquad u_2 = u_2(x_1, x_2); \qquad u_3 = 0 \tag{4.3}$$

This means that some of the resulting strains will also be zero, i.e.

$$\varepsilon_{33} = \varepsilon_{31} = \varepsilon_{32} = 0 \tag{4.4}$$

and the others independent of x_3, i.e. $\varepsilon_{11}(x_1, x_2)$, $\varepsilon_{22}(x_1, x_2)$, $\varepsilon_{12}(x_1, x_2)$.

For this case $\sigma_{33} \neq 0$ and can be determined from the value of the other components.

Constitutive Relations

One can now expand the three dimensional stress-strain relationships for an isotropic body in terms of the non-zero components for plane strain or plane stress, i.e.

$$\begin{aligned}
\varepsilon_{11} &= \frac{1}{E}(\sigma_{11} - \nu\sigma_{22} - \nu\sigma_{33}) \\
\varepsilon_{22} &= \frac{1}{E}(-\nu\sigma_{11} + \sigma_{22} - \nu\sigma_{33}) \\
\varepsilon_{33} &= \frac{1}{E}(-\nu\sigma_{11} - \nu\sigma_{22} + \sigma_{33}) \\
\varepsilon_{12} &= \frac{1}{2G}\sigma_{12}
\end{aligned} \tag{4.5}$$

The relationships for plate stretching can be obtained by setting $\sigma_{33} \equiv 0$ in (4.5) and considering only the planar components $\varepsilon_{11}, \varepsilon_{22}$ and ε_{12}, i.e.

$$\begin{aligned} \varepsilon_{11} &= \frac{1}{E}(\sigma_{11} - \nu\sigma_{22}) \\ \varepsilon_{22} &= \frac{1}{E}(-\nu\sigma_{11} + \sigma_{22}) \\ \varepsilon_{12} &= \frac{1}{2G}\sigma_{12} \end{aligned} \tag{4.6}$$

The value of ε_{33} can be obtained a posteriori from the third relationship in (4.5), i.e.

$$\varepsilon_{33} = \frac{1}{E}(-\nu\sigma_{11} - \nu\sigma_{22}) \tag{4.7}$$

Equations (4.6) can be inverted to produce,

$$\begin{aligned} \sigma_{11} &= \frac{E}{1-\nu^2}(\varepsilon_{11} + \nu\varepsilon_{22}) \\ \sigma_{22} &= \frac{E}{1-\nu^2}(\nu\varepsilon_{11} + \varepsilon_{22}) \\ \sigma_{12} &= 2G\varepsilon_{12} \end{aligned} \tag{4.8}$$

The plane strain equations can be found by first eliminating ε_{33} from (4.5), i.e.

$$\varepsilon_{33} = \frac{1}{E}(-\nu\sigma_{11} - \nu\sigma_{22} + \sigma_{33}) \equiv 0 \tag{4.9}$$

which gives

$$\sigma_{33} = \nu(\sigma_1 + \sigma_{22}) \tag{4.10}$$

Substituting for σ_{33} in equation (4.5) one obtains the following expressions,

$$\begin{aligned} \varepsilon_{11} &= \frac{1}{E}[(1-\nu^2)\sigma_{11} - \nu(1+\nu)\sigma_{22}] \\ \varepsilon_{22} &= \frac{1}{E}[-\nu(1+\nu)\sigma_{11} + (1-\nu^2)\sigma_{22}] \\ \varepsilon_{12} &= \frac{1}{2G}\sigma_{12} \end{aligned} \tag{4.11}$$

The inverse of these expressions is

$$\sigma_{11} = \frac{E}{(1+\nu)(1-2\nu)}[(1-\nu)\varepsilon_{11} + \nu\varepsilon_{22}]$$

$$\sigma_{22} = \frac{E}{(1+\nu)(1-2\nu)}[\nu\varepsilon_{11} + (1-\nu)\varepsilon_{22}] \tag{4.12}$$

$$\sigma_{12} = 2G\varepsilon_{12}$$

The value of ε_{33} if required can be obtained from (4.10).

Notice that it is sometimes more convenient to express the first two relationships above using Lamé's constant $\lambda = \nu E/[(1+\nu)(1-2\nu)]$, i.e.

$$\sigma_{11} = \frac{\lambda}{\nu}[(1-\nu)\varepsilon_{11} + \nu\varepsilon_{22}]$$

$$\sigma_{22} = \frac{\lambda}{\nu}[(1-\nu)\varepsilon_{22} + \nu\varepsilon_{11}] \tag{4.13}$$

One can pass from equations (4.8) for plate stretching to equations (4.12) for plane strain simply by replacing E and ν in the first equation by two equivalent values, E' and ν', such that

$$E' = \frac{E}{1-\nu^2}; \qquad \nu' = \frac{\nu}{1-\nu} \tag{4.14}$$

where the value of G remains the same.

This interesting relationship means that one can implement a plate stretching program and by transforming the elastic constant data in accordance to (4.14) solve also a plane strain problem. This is done in finite element analysis. Conversely in boundary elements one works with plane strain problems as the fundamental solution is known for this type of problem and then plate stretching problems can be solved using the inverse relationship to (4.14), i.e.

$$E' = (1-\nu^2)E; \qquad \nu' = \frac{\nu}{1+\nu} \tag{4.15}$$

4.3 Boundary Element Formulation

The basic relationships for boundary elements in elastostatics have been developed in Chapter 3. In what follows some of the corresponding equations will be reviewed and specialized for two dimensional cases with a view to applying them in two computer codes, one using constant element and the other quadratic elements, the latter more appropriate for elasticity problems. The development of an elastostatics linear code if required, is left to the reader as it follows similar lines as the linear code for potential problems.

One can start with the integral expression (3.79), i.e.

$$c^i_{lk}u^i_k + \int_\Gamma u_k p^*_{lk}\, d\Gamma = \int_\Gamma p_k u^*_{lk}\, d\Gamma + \int_\Omega b_k u^*_{lk}\, d\Omega \tag{4.16}$$

where the fundamental solution has been assumed to satisfy the following equation,

$$\sigma^*_{jk,j} + \Delta^i = 0 \tag{4.17}$$

Equation (4.16) applies for points on the boundary or internal points (with $c^i_{lk} = \delta_{lk}$). Smooth surfaces give $c^i_{lk} = \frac{1}{2}\delta_{lk}$ and corners produce a different type of c^i_{lk} array as will be discussed shortly.

The fundamental solution for an isotropic material in plane strain has been given in Chapter 3 but will be repeated here for completeness.

$$\begin{aligned} u^*_{lk} &= \frac{1}{8\pi G(1-\nu)}\left[(3-4\nu)\ln\left(\frac{1}{r}\right)\delta_{lk} + \frac{\partial r}{\partial x_l}\frac{\partial r}{\partial x_k}\right] \\ p^*_{lk} &= -\frac{1}{4\pi(1-\nu)r}\left[\frac{\partial r}{\partial n}\left\{(1-2\nu)\delta_{kl} + 2\frac{\partial r}{\partial x_k}\frac{\partial r}{\partial x_l}\right\}\right. \\ &\qquad \left. -(1-2\nu)\left(\frac{\partial r}{\partial x_l}n_k - \frac{\partial r}{\partial x_k}n_l\right)\right] \end{aligned} \tag{4.18}$$

p^*_{lk} and u^*_{lk} represent the tractions and displacements in the k direction due to a unit load in the l direction acting at 'i'.

One can now write equation (4.16) in matrix form by defining the following arrays. The fundamental solution components can be written as two 2×2 matrices with elements u^*_{lk} and p^*_{lk}, i.e.

$$\mathbf{u}^* = \begin{bmatrix} u^*_{11} & u^*_{12} \\ u^*_{21} & u^*_{22} \end{bmatrix}; \quad \mathbf{p}^* = \begin{bmatrix} p^*_{11} & p^*_{12} \\ p^*_{21} & p^*_{22} \end{bmatrix} \tag{4.19}$$

The displacement, tractions and body forces vectors are

$$\mathbf{u} = \begin{Bmatrix} u_1 \\ u_2 \end{Bmatrix}; \quad \mathbf{p} = \begin{Bmatrix} p_1 \\ p_2 \end{Bmatrix}; \quad \mathbf{b} = \begin{Bmatrix} b_1 \\ b_2 \end{Bmatrix} \tag{4.20}$$

Hence the basic equation becomes,

$$\mathbf{c}^i\mathbf{u}^i + \int_\Gamma \mathbf{p}^*\mathbf{u}\, d\Gamma = \int_\Gamma \mathbf{u}^*\mathbf{p}\, d\Gamma + \int_\Omega \mathbf{u}^*\mathbf{b}\, d\Omega \tag{4.21}$$

where the $\mathbf{u}^i$ defines the displacements at the – internal or boundary – point i

where the load is applied. $\mathbf{c}^i$ is a 2×2 array of constant which values depend on the type of point under consideration. If 'i' is an internal point

$$\mathbf{c}^i = \begin{bmatrix} 1 & 0 \\ 0 & 1 \end{bmatrix} \tag{4.22}$$

If 'i' is a boundary point on a smooth surface then,

$$\mathbf{c}^i = \begin{bmatrix} \frac{1}{2} & 0 \\ 0 & \frac{1}{2} \end{bmatrix} \tag{4.23}$$

If 'i' is a corner, we will have

$$\mathbf{c}^i = \begin{bmatrix} c_{11} & c_{12} \\ c_{21} & c_{22} \end{bmatrix} \tag{4.24}$$

where the c_{lk} values will depend on the type of corner under consideration as we will see shortly.

4.4 Constant Element Formulation

Consider now that the surface of the boundary under study is discretized using constant elements (figure 4.2). This implies that the values of $\mathbf{u}$ and $\mathbf{p}$ are assumed to be constant on each element and equal to the value at the mid-node of the element. One can also discretize the interiors of the domain in a number of cells which are required for the integration of the body force term in (4.21). These cells are used only for numerical integration of the body force terms and in certain cases – as seen in Chapter 3 – they can be taken to the boundary. Consider here that the body Ω is discretized into N boundary elements and M internal cells, hence formula (4.21) can be written as,

$$\mathbf{c}^i\mathbf{u}^i + \sum_{j=1}^{N} \left\{ \int_{\Gamma_j} \mathbf{p}^* \, d\Gamma \right\} \mathbf{u}^j = \sum_{j=1}^{N} \left\{ \int_{\Gamma_j} \mathbf{u}^* \, d\Gamma \right\} \mathbf{p}^j + \sum_{s=1}^{M} \left\{ \int_{\Omega_s} \mathbf{u}^*\mathbf{b} \, d\Omega \right\} \tag{4.25}$$

The above equation corresponds to the particular node 'i' where the unit forces are assumed to be acting.

Notice that terms such as $\int_{\Gamma_j} \mathbf{u} \, d\Gamma$ and $\int_{\Gamma_j} \mathbf{p} \, d\Gamma$ relate node 'i' to the element or node 'j'. They produce a type of influence coefficient. After integration the integrals produce two 2×2 submatrices called $\hat{\mathbf{H}}^{ij}$ and $\mathbf{G}^{ij}$. Numerical integration of the body forces term can be carried out as follows,

$$\sum_{s=1}^{M} \left\{ \int_{\Omega_s} \mathbf{u}^*\mathbf{b} \, d\Omega \right\} = \sum_{s=1}^{M} \left\{ \sum_{p=1}^{t} (\mathbf{u}^*\mathbf{b})_p w_p \right\} \Omega_s = \sum_{s=1}^{M} \mathbf{B}^{is} \tag{4.26}$$

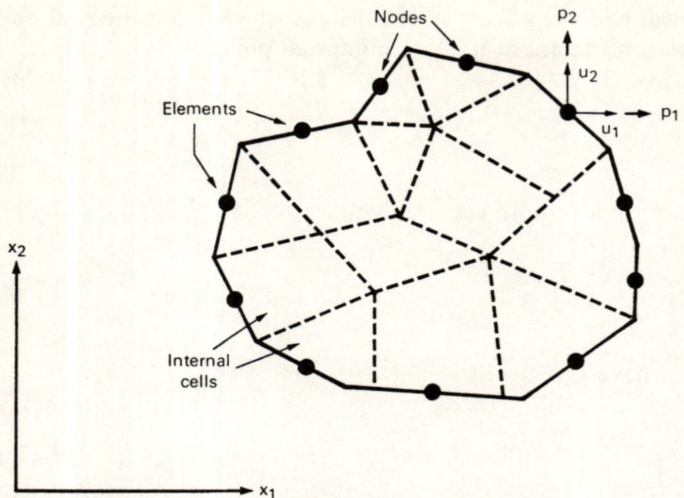

Figure 4.2 Two-dimensional body divided into boundary elements and internal cells

Notice that this produces two components of $\mathbf{B}^{is}$, i.e. B_1^{is} and B_2^{is} after the numerical integration has taken place. w_p are the weighting coefficients and Ω_s the area of the cell under consideration. The function $(\mathbf{u}^*\mathbf{b})$ has to be calculated at the p integration points, where p varies from 1 to r.

Equation (4.25) can now be written

$$\mathbf{c}^i\mathbf{u}^i + \sum_{j=1}^{N} \hat{\mathbf{H}}^{ij}\mathbf{u}^j = \sum_{j=1}^{N} \mathbf{G}^{ij}\mathbf{p}^j + \sum_{s=1}^{M} \mathbf{B}^{is} \tag{4.27}$$

This equation relates the values of **u** at node 'i' with the values of **u**'s and **p**'s at all the nodes on the boundary, including 'i'. Notice that in this case – smooth boundary – $\mathbf{c}^i$ is a 2×2 matrix with only $\frac{1}{2}$ on the diagonal.

Equation (4.27) can be written in a more compact manner if we define

$$\begin{aligned} \mathbf{H}^{ij} &= \hat{\mathbf{H}}^{ij} \qquad \text{when } i \neq j \\ \mathbf{H}^{ij} &= \hat{\mathbf{H}}^{ij} + \mathbf{c}^i \qquad \text{when } i = j \end{aligned} \tag{4.28}$$

Formula (4.27) has then the following form

$$\sum_{j=1}^{N} \mathbf{H}^{ij}\mathbf{u}^j = \sum_{j=1}^{N} \mathbf{G}^{ij}\mathbf{p}^j + \sum_{s=1}^{M} \mathbf{B}^{is} \tag{4.29}$$

If one applies (4.29) to all boundary points the result can also be written in matrix form, i.e.

$$\mathbf{HU} = \mathbf{GP} + \mathbf{B} \tag{4.30}$$

where **H** and **G** are $2N \times 2N$ matrices (N number of boundary nodes).

Equation (4.30) has to be rearranged when applying the boundary conditions. The process consists of moving to the left hand side all columns multiplied by an unknown and accumulating on the right hand side vector **F** all the values obtained by multiplying the known boundary conditions by the terms in the corresponding columns. This produces the following system of equations:

$$\mathbf{AX} = \mathbf{F} + \mathbf{B} \tag{4.31}$$

The vector **X** represents all unknowns – displacements or tractions – in the problem. Once (4.31) is solved all boundary values are found.

Internal Points Results

Once the values of displacements and tractions are known on the boundary it is possible to calculate the displacements and stresses at any interior point. The displacements are given by formula (4.25) with $\mathbf{c}^i = \mathbf{I}$ (**I** is unit diagonal matrix), i.e.

$$\mathbf{u}^i = \int_\Gamma \mathbf{u}^*\mathbf{p}\, d\Gamma - \int_\Gamma \mathbf{p}^*\mathbf{u}\, d\Gamma + \int_\Omega \mathbf{u}^*\mathbf{b}\, d\Omega \tag{4.32}$$

This expression can be discretized

$$\mathbf{u}^i = \sum_{j=1}^{N} \left\{ \int_{\Gamma_j} \mathbf{u}^*\, d\Gamma \right\} \mathbf{p}^j - \sum_{j=1}^{N} \left\{ \int_{\Gamma_j} \mathbf{p}^*\, d\Gamma \right\} \mathbf{u}^j + \sum_{s=1}^{M} \left\{ \int_{\Omega_s} \mathbf{u}^*\mathbf{b}\, d\Omega \right\} \tag{4.33}$$

and then integrated numerically, analytically or by a combination of both techniques.

The internal stresses can be found using the formula (3.109), i.e.

$$\sigma_{kl} = \int_\Gamma \mathbf{D}_{kl}\mathbf{p}\, d\Gamma - \int_\Gamma \mathbf{S}_{kl}\mathbf{u}\, d\Gamma + \int_\Omega \mathbf{D}_{kl}\mathbf{b}\, d\Omega \tag{4.34}$$

where

$$\mathbf{D}_{kl} = [D_{1kl}, D_{2kl}]$$

$$\mathbf{S}_{kl} = [S_{1kl}, S_{2kl}]$$

Formula (4.34) can be written in discretized form as follows,

$$\sigma_{kl} = \sum_{j=1}^{N} \left\{ \int_{\Gamma_j} \mathbf{D}_{kl}\, d\Gamma \right\} \mathbf{p}^j - \sum_{j=1}^{N} \left\{ \int_{\Gamma_j} \mathbf{S}_{kl}\, d\Gamma \right\} \mathbf{u}^j + \sum_{s=1}^{M} \left\{ \int_{\Omega_s} \mathbf{D}_{kl}\mathbf{b}\, d\Omega \right\} \tag{4.35}$$

Integration

As we will see for the case of the quadratic elements, all integrals in the above expressions can be done numerically. For the case of constant elements however it is simpler and more exact to carry out some integrations analytically, particularly those over the element with the singularity, i.e. for the case $i=j$. Everywhere else the values of the integrals in $\mathbf{H}^{ij}$ and $\mathbf{G}^{ij}$ have been computed using a four-point Gauss quadrature formula. Notice that values of the submatrices $\mathbf{H}^{ij}$ (for $i=j$) are easy to calculate using rigid body considerations as shown in 3.5. The terms in $\mathbf{G}^{ii}$ are the only coefficients to be computed analytically. The integrals are carried out for all the elements in $\mathbf{G}^{ii}$, i.e.

$$\mathbf{G}^{ii}=\begin{bmatrix} G_{11} & G_{12} \\ G_{21} & G_{22} \end{bmatrix}^{ii} \tag{4.36}$$

Substituting the fundamental solution into the corresponding integrals one can define three terms as,

$$G_{11}=\frac{1}{8\pi\mu(1-\nu)}\left[(3-4\nu)\int_{\Gamma_i}\ln\left(\frac{1}{r}\right)d\Gamma+\int_{\Gamma_i}\left(\frac{\partial r}{\partial x_1}\right)^2 d\Gamma\right] \tag{4.37}$$

$$G_{12}=G_{21}=\frac{1}{8\pi\mu(1-\nu)}\left[\int_{\Gamma_i}\frac{\partial r}{\partial x_1}\frac{\partial r}{\partial x_2}\,d\Gamma\right] \tag{4.38}$$

$$G_{22}=\frac{1}{8\pi\mu(1-\nu)}\left[(3-4\nu)\int_{\Gamma_i}\ln\left(\frac{1}{r}\right)d\Gamma+\int_{\Gamma_i}\left(\frac{\partial r}{\partial x_2}\right)^2 d\Gamma\right] \tag{4.39}$$

Notice that Γ_i refers to element 'i' over which the singularity is acting. The derivatives of r in formulae (4.37) to (4.39) are given, for the general case, by (figure 4.3(a)).

$$\frac{\partial r}{\partial x_1}=\frac{r_1}{r}=\frac{x_1^k-x_1^i}{|r|}$$
$$\frac{\partial r}{\partial x_2}=\frac{r_2}{r}=\frac{x_2^k-x_2^i}{|r|} \tag{4.40}$$

where $|r|$ is the magnitude of the distance vector $\vec{r}$.

With the above definitions, one can now concentrate on the case for which the integration is carried out over the element 'i' which contains the singularity (figure 4.3(b)). Let us consider that the element starts at extreme point (1) and finishes at extreme point (2) shown in figure 4.3(a). The distances from these points to the singularity located at i in the centre of the element is R, while small r is the variable distance from i to any point over the element, i.e. r is equivalent to Γ.

Using the $\theta-r$ system defined in figure 4.3(b) one finds the following relationship for (4.40)

$$\frac{\partial r}{\partial x_1}=\frac{r_1}{r}=\cos\theta$$
$$\frac{\partial r}{\partial x_2}=\frac{r_2}{r}=\sin\theta \tag{4.41}$$

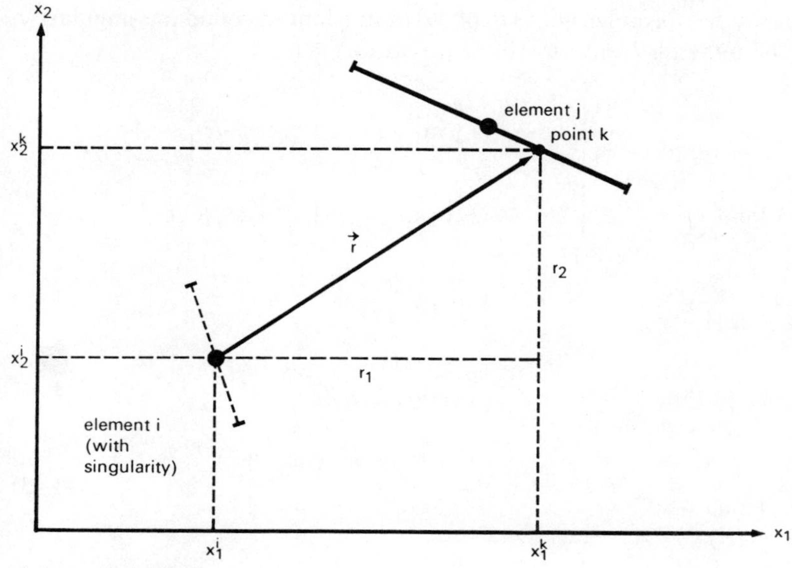

(a) Geometrical definition

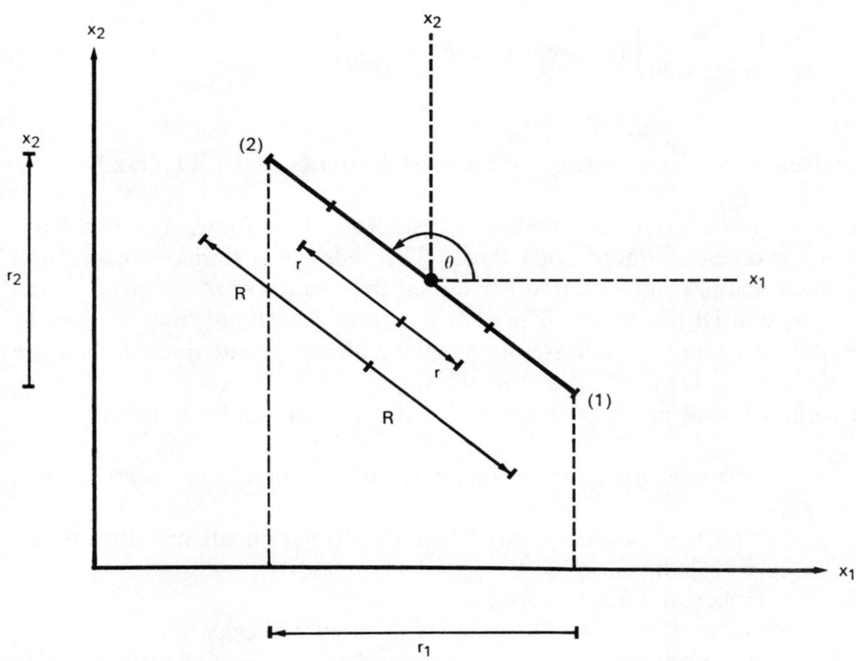

(b) Definition for integration over collocation element

Figure 4.3

One can now write formulae (4.37) to (4.39) taking limits around the singularity. (Notice that the integrals from i to (1) or (2) are the same.)

$$G_{11} = \lim_{\varepsilon \to 0} \left\{ \frac{1}{8\pi\mu(1-\nu)} \left[2(3-4\nu) \int_{\varepsilon}^{R} \ln \frac{1}{r} \, dr + 2 \int_{\varepsilon}^{R} \cos^2 \theta \, dr \right] \right\}$$

$$= \lim_{\varepsilon \to 0} \left\{ \frac{1}{8\pi\mu(1-\nu)} \left[2(3-4\nu) \left\{ \left[r \ln \frac{1}{r} \right]_{\varepsilon}^{R} + [r]_{\varepsilon}^{R} \right\} + 2R \cos^2 \theta \right] \right\}$$

$$= \frac{2R}{8\pi\mu(1-\nu)} \left[(3-4\nu)(1-\ln R) + \frac{(r_1)^2}{4R^2} \right] \quad (4.42)$$

$$G_{12} = G_{21} = \lim_{\varepsilon \to 0} \left\{ \frac{1}{8\pi\mu(1-\nu)} \left[2 \int_{\varepsilon}^{R} \sin \theta \cos \theta \, d\Gamma \right] \right\}$$

$$= \frac{1}{8\pi\mu(1-\nu)} [2R \sin\theta \cos \theta] = \frac{2R}{8\pi\mu(1-\nu)} \left[\frac{r_1 r_2}{4R^2} \right] \quad (4.43)$$

$$G_{22} = \lim_{\varepsilon \to 0} \left\{ \frac{1}{8\pi\mu(1-\nu)} \left[2(3-4\nu) \int_{\varepsilon}^{R} \ln \frac{1}{r} \, dr + 2 \int_{\varepsilon}^{R} \sin^2 \theta \, dr \right] \right\}$$

$$= \lim_{\varepsilon \to 0} \left\{ \frac{1}{8\pi\mu(1-\nu)} \left[2(3-4\nu)[r \ln \frac{1}{r}]_{\varepsilon}^{R} + [r]_{\varepsilon}^{R} + 2R \sin^2 \theta \right] \right\}$$

$$= \frac{2R}{8\pi\mu(1-\nu)} \left[(3-4\nu)(1-\ln R) + \frac{(r_2)^2}{4R^2} \right] \quad (4.44)$$

4.5 Elastostatics Code using Constant Elements (ELCONBE)

This section describes a computer code for the solution of two dimensional isotropic elastostatics problems without body forces. The code has a similar organization as those described in Chapter 2 and because of this the routines are given names similar to those used previously. The code is written for plane strain but can be used for plate stretching as well, simply by giving an equivalent value of ν (notice that the shear modulus μ or G remains the same).

The variables used in the program and their meaning are listed below,

N: Number of boundary elements (equal to number of nodes for this case).

L: Number of internal points where the displacements and stresses are to be calculated.

M: Number of different surfaces (maximum 5).

NC(): Number of the last node of each different surface.

GE: Shear modulus.

XNU: Poisson's ratio for plane strain; use a fictitious Poisson's ratio XNU $= \nu/(1+\nu)$ for plane stress, ν being the Poisson's ratio.

X,Y: One dimensional arrays with x_1 and x_2 coordinates of the extreme points of the boundary elements.
XM,YM: One dimensional arrays with the coordinates of the nodes.
G: Matrix defined in equation (4.30). After application of the boundary conditions the matrix **A** is stored in the same location (equation (4.31)).
H: Matrix defined by equation (4.30).
KODE: One dimensional array indicating the type of boundary condition at the element nodes. KODE = 0 means that a displacement is prescribed and KODE = 1 that a traction is prescribed.
FI: Vector where the prescribed values of boundary conditions are stored. Each element is associated with a value of KODE.
DFI: Right hand side vector in the global system. After solution it contains the values of the unknowns.
CX,CY: One dimensional vectors with coordinates of internal points (maximum number 20).
DSOL: Values of displacements at internal points (2 displacements per point).
SSOL: Values of stresses at internal points (3 stresses per point).

Body forces have not been introduced for simplicity.

1. Main Program

The FORTRAN listing for the main program is as follows:

```
C-----------------------------------------------------------------------
C
      PROGRAM ELCONBE
C
C  PROGRAM 28
C
C  THIS PROGRAM SOLVES TWO-DIMENSIONAL (EL)ASTIC PROBLEMS
C  USING (CON)STANT (B)OUNDARY (E)LEMENTS
C
      CHARACTER*10 FILEIN,FILEOUT
C
      COMMON/MATG/ G(100,100)
      COMMON/MATH/ H(100,100)
      COMMON N,L,NC(5),M,GE,XNU,INP,IPR
      DIMENSION X(51),Y(51),XM(50),YM(50),FI(100),DFI(100)
      DIMENSION KODE(100),CX(20),CY(20),SSOL(60),DSOL(40)
C
C  SET MAXIMUN DIMENSION OF THE SYSTEM OF EQUATIONS (NX)
C  (THIS NUMBER MUST BE EQUAL OR SMALLER THAN THE DIMENSION OF G AND H)
C
      NX=100
C
C  ASSIGN NUMBERS FOR INPUT AND OUTPUT FILES
C
      INP=5
      IPR=6
C
C  READ NAMES AND OPEN FILES FOR INPUT AND UOTPUT
C
      WRITE(*,' (A) ') ' NAME OF THE INPUT FILE (MAX. 10 CHART.)'
      READ(*,' (A) ') FILEIN
      OPEN(INP,FILE=FILEIN,STATUS='OLD')
```

```
      WRITE(*,' (A) ') ' NAME OF THE OUTPUT FILE (MAX.10 CHART.)'
      READ(*,' (A) ') FILEOUT
      OPEN(IPR,FILE=FILEOUT,STATUS='NEW')
C
C  READ DATA
C
      CALL INPUTEC(CX,CY,X,Y,KODE,FI)
C
C  COMPUTE H AND G MATRICES AND FORM SYSTEM (A X = F)
C
      CALL GHMATEC(X,Y,XM,YM,G,H,FI,DFI,KODE,NX)
C
C  SOLVE SYSTEM OF EQUATIONS
C
      NN=2*N
      CALL SLNPD(G,DFI,D,NN,NX)
C
C  COMPUTE STRESS AND DISPLACEMENT AT INTERNAL POINTS
C
      CALL INTEREC(FI,DFI,KODE,CX,CY,X,Y,SSOL,DSOL)
C
C  PRINT RESULTS AT BOUNDARY NODES AND INTERNAL POINTS
C
      CALL OUTPTEC(XM,YM,FI,DFI,CX,CY,SSOL,DSOL)
      CLOSE (INP)
      CLOSE (IPR)
      STOP
      END
```

2. Routine INPUTEC

This subroutine reads the input required by the code. A file whose name is requested from the user by the main program will contain the following lines of input.

(i) *Title Line* Contains the title of the problem.
(ii) *Basic parameter line* Contains the number of elements, number of surfaces, last nodes of each different surface, the shear modulus and Poisson ratio or fictitious Poisson ratio for plane stress problems $v/(1+v)$.
(iii) *Extreme Points of Boundary Elements Lines* The coordinates of the extreme point of an element are read in the counterclockwise direction for external surfaces and clockwise for internal ones. The coordinates are read in free FORMAT.
(iv) *Boundary Conditions Lines* As many lines as boundary nodes giving the values of the known variable in x_1 and x_2 directions. The variables are displacements if KODE = 0 or tractions if KODE = 1.
(v) *Internal Points Coordinate Lines* Internal nodes are defined, each with the $x_1 x_2$ coordinates of the point. The coordinates are read in free FORMAT.

The subroutine prints first the name of the job and the basic parameters. Then the coordinates of the extreme points of the elements and the boundary conditions given by node number, codes and prescribed values are printed. The internal point coordinates will be printed in the subroutine OUTPTEC. Its listing is as follows:

```
C-----------------------------------------------------------------------
      SUBROUTINE INPUTEC(CX,CY,X,Y,KODE,FI)
C
C  PROGRAM 29
C
```

```
      CHARACTER*80 TITLE
      DIMENSION CX(1),CY(1),X(1),Y(1),KODE(1),FI(1)
      COMMON N,L,NC(5),M,GE,XNU,INP,IPR
C
C   N= NUMBER OF BOUNDARY NODES (= NUMBER OF ELEMENTS)
C   L= NUMBER OF INTERNAL POINTS WHERE DISPLACEMENT AND STRESS
C      ARE CALCULATED
C   M= NUMBER OF DIFFERENT BOUNDARIES
C   NC(I)= LAST NODE OF BOUNDARY I
C   GE= SHEAR MODULUS
C   XNU= POISSON MODULUS
C
      WRITE(IPR,100)
  100 FORMAT(/' ',79('*'))
C
C   READ JOB TITLE
C
      READ(INP,'(A)') TITLE
      WRITE(IPR,'(A)') TITLE
C
C   READ NUMBER OF NODES, INTERNAL POINTS AND DIFFERENT BOUNDARIES;
C   READ LAST NODES OF THESE BOUNDARIES AND MATERIAL PROPERTIES
C
      READ(INP,*)N,L,M,(NC(K),K=1,5),GE,XNU
      WRITE(IPR,300)N,L,GE,XNU
  300 FORMAT(//' DATA'//2X,'NUMBER OF BOUNDARY ELEMENTS =',I3/2X,
     1'NUMBER OF INTERNAL POINTS =',I3/2X,
     2'SHEAR MODULUS =',E14.7/2X,'POISSON RATIO =',E14.7)
      IF(M)40,40,30
   30 WRITE(IPR,999)M,(NC(K),K=1,M)
  999 FORMAT(2X,'NUMBER OF DIFFERENT BOUNDARIES=',I3/2X,
     1'LAST NODES OF THESE BOUNDARIES =',5(2X,I3))
C
C   READ COORDINATES OF EXTREME POINTS OF THE BOUNDARY
C   ELEMENTS IN ARRAYS X AND Y
C
   40 WRITE(IPR,500)
  500 FORMAT(//2X,'COORDINATES OF THE EXTREME POINTS OF'
     1' THE BOUNDARY ELEMENTS'//4X,'POINT',10X,'X',18X,'Y')
      READ(INP,*) (X(I),Y(I),I=1,N)
      DO 10 I=1,N
   10 WRITE(IPR,700)I,X(I),Y(I)
  700 FORMAT(5X,I3,2(5X,E14.7))
C
C   READ BOUNDARY CONDITIONS IN FI(I) VECTOR, IF KODE(I)=0 THE FI(I)
C   VALUE IS A KNOWN DISPLACEMENT; IF KODE(I)=1 THE FI(I) VALUE IS A
C   KNOWN TRACTION.
C
      WRITE(IPR,800)
  800 FORMAT(//2X,'BOUNDARY CONDITIONS'//15X,'PRESCRIBED VALUE',15X,
     1'PRESCRIBED VALUE'/5X,'NODE',9X,'X DIRECTION',8X,'CODE',8X,
     2'Y DIRECTION',8X,'CODE')
      DO 20 I=1,N
      READ(INP,*) KODE(2*I-1),FI(2*I-1),KODE(2*I),FI(2*I)
   20 WRITE(IPR,950)I,FI(2*I-1),KODE(2*I-1),FI(2*I),KODE(2*I)
  950 FORMAT(5X,I3,8X,E14.7,8X,I1,8X,E14.7,8X,I1)
C
C   READ COORDINATES OF THE INTERNAL POINTS
C
      IF(L.EQ.0) GO TO 50
      READ(INP,*) (CX(I),CY(I),I=1,L)
   50 RETURN
      END
```

3. Routine GHMATEC

This routine computes the system matrices **H** and **G** by calling the routines EXTINEC and LOCINEC, i.e.

(i) *EXTINEC* Computes the $\mathbf{H}^{ij}$ and $\mathbf{G}^{ij}$ submatrices using numerical integration for the cases $i \neq j$.

(ii) *LOCINEC* Only calculates the submatrices $\mathbf{G}^{ii}$ applying formulae (4.42) to (4.44).

Notice that the diagonal submatrices of **H** for a smooth boundary as in constant elements, are given by

$$\mathbf{H}^{ii} = \begin{bmatrix} \frac{1}{2} & 0 \\ 0 & \frac{1}{2} \end{bmatrix}$$

After computing **H** and **G** matrices the routine arranges the equations in accordance with the boundary conditions and forms the **A** matrix of equations (4.31). When rearranging the columns those of **G** are multiplied by the shear modulus (called GE in the code) so that all elements of A are of the same order. This helps to increase the numerical accuracy of the solution.

The right hand side vector F of equation (4.31) also found in this routine is stored in array DFI.

GHMATEC listing is as follows:

```
C-----------------------------------------------------------------------------
      SUBROUTINE GHMATEC(X,Y,XM,YM,G,H,FI,DFI,KODE,NX)
C
C  PROGRAM 30
C
C  THIS SUBROUTINE COMPUTES THE G AND H MATRICES AND
C  FORMS THE SYSTEM OF EQUATIONS A X = F
C
      DIMENSION G(NX,NX),H(NX,NX)
      DIMENSION X(1),Y(1),XM(1),YM(1),FI(1)
      DIMENSION KODE(1),DFI(1)
      COMMON N,L,NC(5),M,GE,XNU,INP,IPR
C
C  COMPUTE THE NODAL COORDINATES AND STORE IN ARRAYS XM AND YM
C
      X(N+1)=X(1)
      Y(N+1)=Y(1)
      DO 10 I=1,N
      XM(I)=(X(I)+X(I+1))/2
   10 YM(I)=(Y(I)+Y(I+1))/2
      IF(M-1)15,15,12
   12 XM(NC(1))=(X(NC(1))+X(1))/2
      YM(NC(1))=(Y(NC(1))+Y(1))/2
      DO 13 K=2,M
      XM(NC(K))=(X(NC(K))+X(NC(K-1)+1))/2
   13 YM(NC(K))=(Y(NC(K))+Y(NC(K-1)+1))/2
C
C  COMPUTE THE COEFICIENTS OF G AND H MATRICES
C
   15 DO 30 I=1,N
      DO 30 J=1,N
      IF(M-1)16,16,17
   17 IF(J-NC(1))19,18,19
   18 KK=1
      GO TO 23
   19 DO 22 K=2,M
      IF(J-NC(K))22,21,22
   21 KK=NC(K-1)+1
      GO TO 23
   22 CONTINUE
   16 KK=J+1
   23 IF(I-J)20,25,20
```

```
   20 CALL EXTINEC(XM(I),YM(I),X(J),Y(J),X(KK),Y(KK),H((2*I-1),(2*J-1)),
     1H((2*I-1),(2*J)),H((2*I),(2*J-1)),H((2*I),(2*J)),G((2*I-1),
     2(2*J-1)),G((2*I-1),(2*J)),G((2*I),(2*J)))
      G((2*I),(2*J-1))=G((2*I-1),(2*J))
      GO TO 26
   25 CALL LOCINEC(X(J),Y(J),X(KK),Y(KK),G((2*I-1),(2*J-1)),
     1G((2*I-1),(2*J)),G((2*I),(2*J)))
      H((2*I-1),(2*J-1))=0.5
      H((2*I),(2*J))=0.5
      H((2*I-1),(2*J))=0.
      H((2*I),(2*J-1))=0.
      G((2*I),(2*J-1))=G((2*I-1),(2*J))
   26 CONTINUE
   30 CONTINUE
C
C  REORDER THE COLUMNS OF THE SYSTEM OF EQUATIONS IN ACCORDANCE
C  WITH THE BOUNDARY CONDITIONS AND FORM SYSTEM MATRIX A WHICH
C  IS STORED IN G
C
      NN=2*N
      DO 50 J=1,NN
      IF(KODE(J))43,43,40
   40 DO 42 I=1,NN
      CH=G(I,J)
      G(I,J)=-H(I,J)
   42 H(I,J)=-CH
      GO TO 50
   43 DO 45 I=1,NN
   45 G(I,J)=G(I,J)*GE
   50 CONTINUE
C
C  FORM THE RIGHT HAND SIDE VECTOR F WHICH IS STORED IN DFI
C
      DO 60 I=1,NN
      DFI(I)=0.
      DO 60 J=1,NN
      DFI(I)=DFI(I)+H(I,J)*FI(J)
   60 CONTINUE
      RETURN
      END
```

4. Routine EXTINEC

This routine computes the $\mathbf{G}^{ij}$ and $\mathbf{H}^{ij}$ submatrices (for $i \neq j$) using numerical integration. Notice that $\mathbf{G}^{ij}$ is symmetric but $\mathbf{H}^{ij}$ is not. The values of the coefficients are computed using four points Gaussian quadrature.

```
C-----------------------------------------------------------------------
      SUBROUTINE EXTINEC(XP,YP,X1,Y1,X2,Y2,H11,H12,H21,H22,G11,G12,G22)
C
C  PROGRAM 31
C
C  THIS SOUBROUTINE COMPUTES THE G AND H MATRICES
C  COEFFICIENTS THAT RELATE A COLLOCATION POINT WITH A DIFFERENT
C  ELEMENT USING GAUSS QUADRATURE
C
C  DIST= DISTANCE FROM THE COLOCATION POINT TO THE
C        LINE TANGENT TO THE ELEMENT
C  RA= DISTANCE FROM THE COLOCATION POINT TO THE
C       GAUSS INTEGRATION POINT AT THE BOUNDARY ELEMENT
C
      DIMENSION XCO(4),YCO(4),GI(4),OME(4)
      COMMON N,L,NC(5),M,GE,XNU,INP,IPR
      DATA GI/0.86113631,-0.86113631,0.33998104,-0.33998104/
      DATA OME/0.34785485,0.34785485,0.65214515,0.65214515/
C
      AX=(X2-X1)/2
      BX=(X2+X1)/2
```

```
      AY=(Y2-Y1)/2
      BY=(Y2+Y1)/2
      ETA1=(Y2-Y1)/(2*SQRT(AX**2+AY**2))
      ETA2=(X1-X2)/(2*SQRT(AX**2+AY**2))
C
C  COMPUTE THE DISTANCE FROM THE POINT TO THE LINE OF THE ELEMENT
C
      IF(AX)10,20,10
   10 TA=AY/AX
      DIST=ABS((TA*XP-YP+Y1-TA*X1)/SQRT(TA**2+1))
      GO TO 30
   20 DIST=ABS(XP-X1)
C
C  DETERMINE THE DIRECTION OF THE OUTWARD NORMAL
C
   30 SIG=(X1-XP)*(Y2-YP)-(X2-XP)*(Y1-YP)
      IF(SIG)31,32,32
   31 DIST=-DIST
   32 H11=0.
      H12=0.
      H21=0.
      H22=0.
      G11=0.
      G12=0.
      G22=0.
C
C  COMPUTE G AND H COEFFICIENTS
C
      DE=4*3.141592*(1-XNU)
      DO 40 I=1,4
      XCO(I)=AX*GI(I)+BX
      YCO(I)=AY*GI(I)+BY
      RA=SQRT((XP-XCO(I))**2+(YP-YCO(I))**2)
      RD1=(XCO(I)-XP)/RA
      RD2=(YCO(I)-YP)/RA
      G11=G11+((3-4*XNU)*ALOG(1./RA)+RD1**2)*OME(I)*SQRT(AX**2+AY**2)/(2
     1*DE*GE)
      G12=G12+RD1*RD2*OME(I)*SQRT(AX**2+AY**2)/(2*DE*GE)
      G22=G22+((3-4*XNU)*ALOG(1./RA)+RD2**2)*OME(I)*SQRT(AX**2+AY**2)/(2
     1*DE*GE)
      H11=H11-DIST*((1-2*XNU)+2*RD1**2)/(RA**2*DE)*OME(I)*SQRT(AX**2+AY*
     1*2)
      H12=H12-(DIST*2*RD1*RD2/RA+(1-2*XNU)*(ETA1*RD2-ETA2*RD1))*OME(I)*S
     1QRT(AX**2+AY**2)/(RA*DE)
      H21=H21-(DIST*2*RD1*RD2/RA+(1-2*XNU)*(ETA2*RD1-ETA1*RD2))*OME(I)*S
     1QRT(AX**2+AY**2)/(RA*DE)
   40 H22=H22-DIST*((1-2*XNU)+2*RD2**2)*OME(I)*SQRT(AX**2+AY**2)/(RA**2*
     1DE)
      RETURN
      END
```

5. Routine LOCINEC

It calculates the G^{ii} submatrix in accordance with formulae (4.42) to (4.44), which give the analytical results.

```
C-----------------------------------------------------------------------
      SUBROUTINE LOCINEC(X1,Y1,X2,Y2,G11,G12,G22)
C
C  PROGRAM 32
C
C  THIS SUBROUTINE COMPUTES THE VALUES OF THE MATRIX G COEFFICIENTS
C  THAT RELATE AN ELEMENT WITH ITSELF
C
      COMMON N,L,NC(5),M,GE,XNU,INP,IPR
C
      AX=(X2-X1)/2
      AY=(Y2-Y1)/2
      SR=SQRT(AX**2+AY**2)
```

```
      DE=4*3.141592*GE*(1-XNU)
      G11=SR*((3-4*XNU)*(1-ALOG(SR))+(X2-X1)**2/(4*SR**2))/DE
      G22=SR*((3-4*XNU)*(1-ALOG(SR))+(Y2-Y1)**2/(4*SR**2))/DE
      G12=(X2-X1)*(Y2-Y1)/(4*SR*DE)
      RETURN
      END
```

6. Routine SLNPD

This is the same solver as used in previous codes.

7. Routine INTEREC

This subroutine computes the displacements and stresses at internal points. First it reorders the vectors DFI and FI. Notice that the first contains the values of **X** in equation (4.31) as given by the solver and the second the boundary conditions. After reorganization the boundary displacements are all stored in FI and the tractions in DFI.

Formulae (4.33) and (4.35) are applied to compute internal stresses and displacements. This requires the integration of terms with S and D coefficients which are carried out numerically in a routine called SIGMAEC.

The FORTRAN Listing is as follows:

```
C-----------------------------------------------------------------------
      SUBROUTINE INTEREC(FI,DFI,KODE,CX,CY,X,Y,SSOL,DSOL)
C
C  PROGRAM 33
C
C  THIS SUBROUTINE COMPUTES THE VALUES OF THE STRESS AND DISPLACEMENT
C  COMPONENTS AT INTERNAL POINTS
C
      DIMENSION CX(1),CY(1),SSOL(1),DSOL(1)
      DIMENSION FI(1),DFI(1),KODE(1),X(1),Y(1)
      COMMON N,L,NC(5),M,GE,XNU,INP,IPR
C
C  REARRANGE FI AND DFI ARRAYS TO STORE ALL THE VALUES
C  OF THE DISPLACEMENT IN FI AND ALL THE VALUES OF THE TRACTIONS IN DFI
C
      NN=2*N
      DO 20 I=1,NN
      IF(KODE(I)) 15,15,10
   10 CH=FI(I)
      FI(I)=DFI(I)
      DFI(I)=CH
      GO TO 20
   15 DFI(I)=DFI(I)*GE
   20 CONTINUE
C
C  COMPUTE THE VALUES OF STRESSES AND DISPLACEMENTS
C  AT INTERNAL POINTS.
C
      IF(L.EQ.0) GO TO 50
      DO 40 K=1,L
      DSOL(2*K-1)=0.
      DSOL(2*K)=0.
      SSOL(3*K-2)=0.
      SSOL(3*K-1)=0.
      SSOL(3*K)=0.
      DO 30 J=1,N
      IF(M-1)28,28,22
   22 IF(J-NC(1))24,23,24
```

```
23 KK=1
   GO TO 29
24 DO 26 LK=2,M
   IF(J-NC(LK))26,25,26
25 KK=NC(LK-1)+1
   GO TO 29
26 CONTINUE
28 KK=J+1
29 CALL EXTINEC(CX(K),CY(K),X(J),Y(J),X(KK),Y(KK),H11,H12,H21,H22,
  1 G11,G12,G22)
   DSOL(2*K-1)=DSOL(2*K-1)+DFI(2*J-1)*G11+DFI(2*J)*G12-FI(2*J-1)*H11-
  1FI(2*J)*H12
   DSOL(2*K)=DSOL(2*K)+DFI(2*J-1)*G12+DFI(2*J)*G22-FI(2*J-1)*H21-FI(2
  1*J)*H22
   CALL SIGMAEC(CX(K),CY(K),X(J),Y(J),X(KK),Y(KK),D111,D211,D112,
  1 D212,D122,D222,S111,S211,S112,S212,S122,S222)
   SSOL(3*K-2)=SSOL(3*K-2)+DFI(2*J-1)*D111+DFI(2*J)*D211-FI(2*J-1)*S1
  111-FI(2*J)*S211
   SSOL(3*K-1)=SSOL(3*K-1)+DFI(2*J-1)*D112+DFI(2*J)*D212-FI(2*J-1)*S1
  112-FI(2*J)*S212
30 SSOL(3*K)=SSOL(3*K)+DFI(2*J-1)*D122+DFI(2*J)*D222-FI(2*J-1)*S122-F
  1I(2*J)*S222
40 CONTINUE
50 RETURN
   END
```

8. Routine SIGMAEC

This subroutine computes the integrals with S and D terms described in formulae (4.35) and needed to calculate the internal stresses. The integrations are carried out using a four points Gaussian integration scheme.

```
C-----------------------------------------------------------------------
      SUBROUTINE SIGMAEC(XP,YP,X1,Y1,X2,Y2,D111,D211,D112,D212,D122,
     1 D222,S111,S211,S112,S212,S122,S222)
C
C   PROGRAM 34
C
C   THIS SUBROUTINE COMPUTES THE VALUES OF THE S AND D MATRICES
C   USING GAUSS QUADRATURE IN ORDER TO COMPUTE THE STRESSES
C   AT ANY INTERNAL POINT
C
C   RA= DISTANCE FROM THE POINT TO THE GAUSS INTEGRATION POINTS
C      ON THE BOUNDARY ELEMENTS
C   DIST=    DISTANCE FROM THE POINT TO THE LINE TANGENT
C            TO THE ELEMENT
C   RD1,RD2=    DERIVATIVES OF RA
C   ETA1,ETA2= COMPONENTS OF THE UNIT NORMAL TO THE ELEMENT
C
      DIMENSION XCO(4),YCO(4),GI(4),OME(4)
      COMMON N,L,NC(5),M,GE,XNU,INP,IPR
      DATA GI/0.86113631,-0.86113631,0.33998104,-0.33998104/
      DATA OME/0.34785485,0.34785485,0.65214515,0.65214515/
C
      AX=(X2-X1)/2
      BX=(X2+X1)/2
      AY=(Y2-Y1)/2
      BY=(Y2+Y1)/2
      ETA1=(Y2-Y1)/(2*SQRT(AX**2+AY**2))
      ETA2=(X1-X2)/(2*SQRT(AX**2+AY**2))
C
C   COMPUTE THE DISTANCE FROM THE POINT TO THE LINE OF THE ELEMENT
C
      IF(AX)10,20,10
   10 TA=AY/AX
      DIST=ABS((TA*XP-YP+Y1-TA*X1)/SQRT(TA**2+1))
      GO TO 30
   20 DIST=ABS(XP-X1)
```

```
C
C   DETERMINE THE DIRECTION OF THE OUTWARD NORMAL
C
   30 SIG=(X1-XP)*(Y2-YP)-(X2-XP)*(Y1-YP)
      IF(SIG)31,32,32
   31 DIST=-DIST
   32 D111=0.
      D211=0.
      D112=0.
      D212=0.
      D122=0.
      D222=0.
      S111=0.
      S211=0.
      S112=0.
      S212=0.
      S122=0.
      S222=0.
C
C   COMPUTE D AND S COEFFICIENTS
C
      FA=1-4*XNU
      AL=1-2*XNU
      DE=4*3.141592*(1-XNU)
      DO 40 I=1,4
      XCO(I)=AX*GI(I)+BX
      YCO(I)=AY*GI(I)+BY
      RA=SQRT((XP-XCO(I))**2+(YP-YCO(I))**2)
      RD1=(XCO(I)-XP)/RA
      RD2=(YCO(I)-YP)/RA
      D111=D111+(AL*RD1+2*RD1**3)*OME(I)*SQRT(AX**2+AY**2)/(DE*RA)
      D211=D211+(2*RD1**2*RD2-AL*RD2)*OME(I)*SQRT(AX**2+AY**2)/(DE*RA)
      D112=D112+(AL*RD2+2*RD1**2*RD2)/(DE*RA)*OME(I)*SQRT(AX**2+AY**2)
      D212=D212+(AL*RD1+2*RD1*RD2**2)/(DE*RA)*OME(I)*SQRT(AX**2+AY**2)
      D122=D122+(2*RD1*RD2**2-AL*RD1)/(DE*RA)*OME(I)*SQRT(AX**2+AY**2)
      D222=D222+(AL*RD2+2*RD2**3)/(DE*RA)*OME(I)*SQRT(AX**2+AY**2)
      S111=S111+(2*DIST/RA*(AL*RD1+XNU*2*RD1-4*RD1**3)+4*XNU*ETA1*RD1**2
     1+AL*(2*ETA1*RD1**2+2*ETA1)-FA*ETA1)*2*GE/(DE*RA**2)*OME(I)*SQRT(AX
     2**2+AY**2)
      S211=S211+(2*DIST/RA*(AL*RD2-4*RD1**2*RD2)+4*XNU*ETA1*RD1*RD2+AL*2
     1*ETA2*RD1**2-FA*ETA2)*2*GE/(DE*RA**2)*OME(I)*SQRT(AX**2+AY**2)
      S112=S112+(2*DIST/RA*(XNU*RD2-4*RD1**2*RD2)+2*XNU*(ETA1*RD2*RD1+ET
     1A2*RD1**2)+AL*(2*ETA1*RD1*RD2+ETA2))*2*GE/(DE*RA**2)*OME(I)*SQRT(A
     2X**2+AY**2)
      S212=S212+(2*DIST/RA*(XNU*RD1-4*RD1*RD2**2)+2*XNU*(ETA1*RD2**2+ETA
     12*RD1*RD2)+AL*(2*ETA2*RD1*RD2+ETA1))*2*GE/(DE*RA**2)*OME(I)*SQRT(A
     2X**2+AY**2)
      S122=S122+(2*DIST/RA*(AL*RD1-4*RD1*RD2**2)+4*XNU*ETA2*RD1*RD2+AL*2
     1*ETA1*RD2**2-FA*ETA1)*2*GE/(DE*RA**2)*OME(I)*SQRT(AX**2+AY**2)
   40 S222=S222+(2*DIST/RA*(AL*RD2+2*XNU*RD2-4*RD2**3)+4*XNU*ETA2*RD2**2
     1+AL*(2*ETA2*RD2**2+2*ETA2)-FA*ETA2)*2*GE/(DE*RA**2)*OME(I)*SQRT(AX
     2**2+AY**2)
      RETURN
      END
```

9. Routine OUTPTEC

This subroutine prints the results in the following order.

(ii) Boundary nodes values with coordinates $x_1 x_2$, values of $u_1 u_2$ displacements and $p_1 p_2$ tractions.

(ii) Internal node results with coordinates $x_1 x_2$, values of $u_1 u_2$ displacements and $\sigma_{11} \sigma_{12} \sigma_{22}$ stresses.

Notice that stresses on the boundary are not produced. If required they will need to be computed using formulae (3.5) with the derivatives of displacements tangential to the surface computed by considering the nodes of displacements and adjacent elements and applying a finite different type approach.

The listing of OUTPT is as follows:

```
C-----------------------------------------------------------------------
      SUBROUTINE OUTPTEC(XM,YM,FI,DFI,CX,CY,SSOL,DSOL)
C
C   PROGRAM 35
C
C   THIS SUBROUTINE PRINTS THE VALUES OF THE DISPLACEMENTS
C   AND TRACTIONS AT BOUNDARY NODES. IT ALSO PRINTS THE VALUES
C   OF DISPLACEMENTS AND STRESSES AT INTERNAL POINTS
C
      DIMENSION XM(1),YM(1),FI(1),DFI(1)
      DIMENSION CX(1),CY(1),SSOL(1),DSOL(1)
      COMMON N,L,NC(5),M,GE,XNU,INP,IPR
C
      WRITE(IPR,100)
  100 FORMAT(' ',79('*')//1X,'RESULTS'//2X,'BOUNDARY NODES'//6X
     1,'X',12X,'Y',9X,'DISPL. X',5X,'DISPL. Y',4X,
     2'TRACTION X',3X,'TRACTION Y'/)
      DO 10 I=1,N
   10 WRITE(IPR,200) XM(I),YM(I),FI(2*I-1),FI(2*I),DFI(2*I-1),DFI(2*I)
  200 FORMAT(6(1X,E12.5))
C
      IF(L.EQ.0.) GO TO 30
      WRITE(IPR,300)
  300 FORMAT(//2X,'INTERNAL POINTS DISPLACEMENTS'//8X,'X',15X,'Y',10X,
     1'DISPLACEMENT X',5X,'DISPLACEMENT Y')
      DO 20 K=1,L
   20 WRITE(IPR,400)CX(K),CY(K),DSOL(2*K-1),DSOL(2*K)
      WRITE(IPR,350)
  350 FORMAT(//2X,'INTERNAL POINTS STRESSES'//8X,'X',15X,'Y',12X,
     1'SIGMA X',10X,'TAU XY',9X,'SIGMA Y')
      DO 25 K=1,L
   25 WRITE(IPR,450) CX(K),CY(K),SSOL(3*K-2),SSOL(3*K-1),SSOL(3*K)
  400 FORMAT(2(2X,E14.7),2(5X,E14.7))
  450 FORMAT(5(2X,E14.7))
   30 WRITE(IPR,500)
  500 FORMAT(' ',79('*'))
      RETURN
      END
```

Example 4.1

Figure 4.4 describes a circular cavity under internal pressure in an infinite medium. Numerical results will be compared against the known analytical solution, first using constant elements and then with the quadratic elements of section 4.7 and 4.8.

The boundary is divided here into 24 constant elements and 10 internal nodes are defined at which displacements and stresses will be found.

Notice that in order to stop all rigid body movements one can suppress the displacements in the x_2 directions at node 18.

The input needed to run the ELCONBE code is as follows.

CIRCULAR CAVITY (DATA)

```
  CIRCULAR CAVITY UNDER INTERNAL PRESSURE (24 CONSTANT ELEMENTS)
24 5  0 0 0 0 0 0 94500. 0.1
-.3916 -2.9743
-1.1481 -2.7716
-1.8263 -2.3801
-2.3801 -1.8263
-2.7716 -1.1480
-2.9743 -.3916
-2.9743  .3916
```

```
-2.7716  1.1481
-2.3801  1.8263
-1.8263  2.3801
-1.1480  2.7716
-.3916   2.9743
.3916   2.9743
1.1481 2.7716
1.8263 2.3801
2.3801 1.8263
2.7716 1.1480
2.9743 .3916
2.9743 -.3916
2.7716 -1.1481
2.3801 -1.8263
1.8263 -2.3801
1.1480 -2.7716
.3916   -2.9743
1 -25.88 1 -96.59
1 -50.    1 -86.6
1 -70.71 1 -70.71
1 -86.6  1 -50.
1 -96.59 1 -25.88
1 -100.   1 0.
1 -96.59 1 25.88
1 -86.6  1 50.
1 -70.71 1 70.71
1 -50.    1 86.6
1 -25.88 1 96.59
0  0.     1 100.
1 25.88 1 96.59
1 50.   1 86.6
1 70.71 1 70.71
1 86.6  1 50.
1 96.59 1 25.88
1 100.  0 0.
1 96.59 1 -25.88
1 86.6  1 -50.
1 70.71 1 -70.71
1 50.   1 -86.6
1 25.88 1 -96.59
0  0.   1 -100.
4. 0. 2.82843 2.82843 -4. 0. 6. 0. 10. 0.
```

This produces the output given below. Notice that the results are very close to the theoretical results as shown by the values of radial stresses at internal points. Stresses and displacements decay with increasing distance from the cavity as is to be expected. Radial and hoop stresses at internal points have the same absolute value but with different signs as expected.

Table 4.1 Radial Stresses at Internal Points. 24 Constant Elements Discretization

Distance to the centre of the cavity	Constant boundary element discretization	Elasticity theory
4	-57.234	-56.250
6	-25.295	-25.000
10	-9.106	-9.000
20	-2.276	-2.250
50	-0.364	-0.360
200	-0.227×10^{-1}	-0.225×10^{-1}
1000	-0.991×10^{-3}	-0.9×10^{-3}

The computer output is as follows,

CIRCULAR CAVITY (OUTPUT)

```
******************************************************************************
 CIRCULAR CAVITY UNDER INTERNAL PRESSURE (24 CONSTANT ELEMENTS)


DATA

 NUMBER OF BOUNDARY ELEMENTS = 24
 NUMBER OF INTERNAL POINTS =  5
 SHEAR MODULUS = 0.9450000E+05
 POISSON RATIO = 0.1000000E+00


 COORDINATES OF THE EXTREME POINTS OF' THE BOUNDARY ELEMENTS

   POINT          X                  Y
     1      -0.3916000E+00     -0.2974300E+01
     2      -0.1148100E+01     -0.2771600E+01
     3      -0.1826300E+01     -0.2380100E+01
     4      -0.2380100E+01     -0.1826300E+01
     5      -0.2771600E+01     -0.1148000E+01
     6      -0.2974300E+01     -0.3916000E+00
     7      -0.2974300E+01      0.3916000E+00
     8      -0.2771600E+01      0.1148100E+01
     9      -0.2380100E+01      0.1826300E+01
    10      -0.1826300E+01      0.2380100E+01
    11      -0.1148000E+01      0.2771600E+01
    12      -0.3916000E+00      0.2974300E+01
    13       0.3916000E+00      0.2974300E+01
    14       0.1148100E+01      0.2771600E+01
    15       0.1826300E+01      0.2380100E+01
    16       0.2380100E+01      0.1826300E+01
    17       0.2771600E+01      0.1148000E+01
    18       0.2974300E+01      0.3916000E+00
    19       0.2974300E+01     -0.3916000E+00
    20       0.2771600E+01     -0.1148100E+01
    21       0.2380100E+01     -0.1826300E+01
    22       0.1826300E+01     -0.2380100E+01
    23       0.1148000E+01     -0.2771600E+01
    24       0.3916000E+00     -0.2974300E+01


 BOUNDARY CONDITIONS

               PRESCRIBED VALUE                        PRESCRIBED VALUE
    NODE         X DIRECTION          CODE               Y DIRECTION          CODE
     1         -0.2588000E+02          1              -0.9659000E+02           1
     2         -0.5000000E+02          1              -0.8660000E+02           1
     3         -0.7071000E+02          1              -0.7071000E+02           1
     4         -0.8660000E+02          1              -0.5000000E+02           1
     5         -0.9659000E+02          1              -0.2588000E+02           1
     6         -0.1000000E+03          1               0.0000000E+00           1
     7         -0.9659000E+02          1               0.2588000E+02           1
     8         -0.8660000E+02          1               0.5000000E+02           1
     9         -0.7071000E+02          1               0.7071000E+02           1
    10         -0.5000000E+02          1               0.8660000E+02           1
    11         -0.2588000E+02          1               0.9659000E+02           1
    12          0.0000000E+00          0               0.1000000E+03           1
    13          0.2588000E+02          1               0.9659000E+02           1
    14          0.5000000E+02          1               0.8660000E+02           1
    15          0.7071000E+02          1               0.7071000E+02           1
    16          0.8660000E+02          1               0.5000000E+02           1
    17          0.9659000E+02          1               0.2588000E+02           1
    18          0.1000000E+03          1               0.0000000E+00           0
    19          0.9659000E+02          1              -0.2588000E+02           1
    20          0.8660000E+02          1              -0.5000000E+02           1
    21          0.7071000E+02          1              -0.7071000E+02           1
    22          0.5000000E+02          1              -0.8660000E+02           1
    23          0.2588000E+02          1              -0.9659000E+02           1
    24          0.0000000E+00          0              -0.1000000E+03           1
******************************************************************************
```

RESULTS

BOUNDARY NODES

X	Y	DISPL. X	DISPL. Y	TRACTION X	TRACTION Y
-0.76985E+00	-0.28730E+01	-0.42449E-03	-0.15842E-02	-0.25880E+02	-0.96590E+02
-0.14872E+01	-0.25759E+01	-0.82002E-03	-0.14204E-02	-0.50000E+02	-0.86600E+02
-0.21032E+01	-0.21032E+01	-0.11597E-02	-0.11597E-02	-0.70710E+02	-0.70710E+02
-0.25759E+01	-0.14872E+01	-0.14204E-02	-0.82001E-03	-0.86600E+02	-0.50000E+02
-0.28730E+01	-0.76980E+00	-0.15842E-02	-0.42449E-03	-0.96590E+02	-0.25880E+02
-0.29743E+01	0.00000E+00	-0.16401E-02	-0.73851E-08	-0.10000E+03	0.00000E+00
-0.28730E+01	0.76985E+00	-0.15842E-02	0.42448E-03	-0.96590E+02	0.25880E+02
-0.25759E+01	0.14872E+01	-0.14204E-02	0.82002E-03	-0.86600E+02	0.50000E+02
-0.21032E+01	0.21032E+01	-0.11597E-02	0.11597E-02	-0.70710E+02	0.70710E+02
-0.14872E+01	0.25759E+01	-0.82001E-03	0.14204E-02	-0.50000E+02	0.86600E+02
-0.76980E+00	0.28730E+01	-0.42449E-03	0.15842E-02	-0.25880E+02	0.96590E+02
0.00000E+00	0.29743E+01	0.00000E+00	0.16401E-02	0.14446E-02	0.10000E+03
0.76985E+00	0.28730E+01	0.42449E-03	0.15842E-02	0.25880E+02	0.96590E+02
0.14872E+01	0.25759E+01	0.82002E-03	0.14204E-02	0.50000E+02	0.86600E+02
0.21032E+01	0.21032E+01	0.11597E-02	0.11597E-02	0.70710E+02	0.70710E+02
0.25759E+01	0.14872E+01	0.14204E-02	0.82001E-03	0.86600E+02	0.50000E+02
0.28730E+01	0.76980E+00	0.15842E-02	0.42449E-03	0.96590E+02	0.25880E+02
0.29743E+01	0.00000E+00	0.16401E-02	0.00000E+00	0.10000E+03	-0.20253E-02
0.28730E+01	-0.76985E+00	0.15842E-02	-0.42448E-03	0.96590E+02	-0.25880E+02
0.25759E+01	-0.14872E+01	0.14204E-02	-0.82002E-03	0.86600E+02	-0.50000E+02
0.21032E+01	-0.21032E+01	0.11597E-02	-0.11597E-02	0.70710E+02	-0.70710E+02
0.14872E+01	-0.25759E+01	0.82001E-03	-0.14204E-02	0.50000E+02	-0.86600E+02
0.76980E+00	-0.28730E+01	0.42449E-03	-0.15842E-02	0.25880E+02	-0.96590E+02
0.00000E+00	-0.29743E+01	0.00000E+00	-0.16401E-02	-0.12447E-02	-0.10000E+03

INTERNAL POINTS DISPLACEMENTS

X	Y	DISPLACEMENT X	DISPLACEMENT Y
0.4000000E+01	0.0000000E+00	0.1204821E-02	0.7319613E-08
0.2828430E+01	0.2828430E+01	0.8519205E-03	0.8519267E-03
-0.4000000E+01	0.0000000E+00	-0.1204821E-02	-0.4833055E-08
0.6000000E+01	0.0000000E+00	0.8029994E-03	0.6300070E-08
0.1000000E+02	0.0000000E+00	0.4817977E-03	0.6126356E-08

INTERNAL POINTS STRESSES

X	Y	SIGMA X	TAU XY	SIGMA Y
0.4000000E+01	0.0000000E+00	-0.5723444E+02	-0.5717874E-03	0.5711818E+
0.2828430E+01	0.2828430E+01	-0.5605986E-01	-0.5717675E+02	-0.5633205E-
-0.4000000E+01	0.0000000E+00	-0.5723446E+02	-0.6064773E-03	0.5711818E+
0.6000000E+01	0.0000000E+00	-0.2529478E+02	-0.7735193E-04	0.2529438E+
0.1000000E+02	0.0000000E+00	-0.9106032E+01	-0.1208484E-04	0.9105983E+

****

4.6 Linear Elements

In this section we will consider the development of linear elements similar to those discussed in Chapter 2 section 5, for potential problems. The difference is that now we need to interpolate two values for u's and two for p's (figure 4.5).

$$\mathbf{u} = \begin{Bmatrix} u_1 \\ u_2 \end{Bmatrix} = \begin{bmatrix} \phi_1 & 0 & \phi_2 & 0 \\ 0 & \phi_1 & 0 & \phi_2 \end{bmatrix} \begin{Bmatrix} u_1^1 \\ u_2^1 \\ u_1^2 \\ u_2^2 \end{Bmatrix}^j = \mathbf{\Phi u}^j \tag{4.45}$$

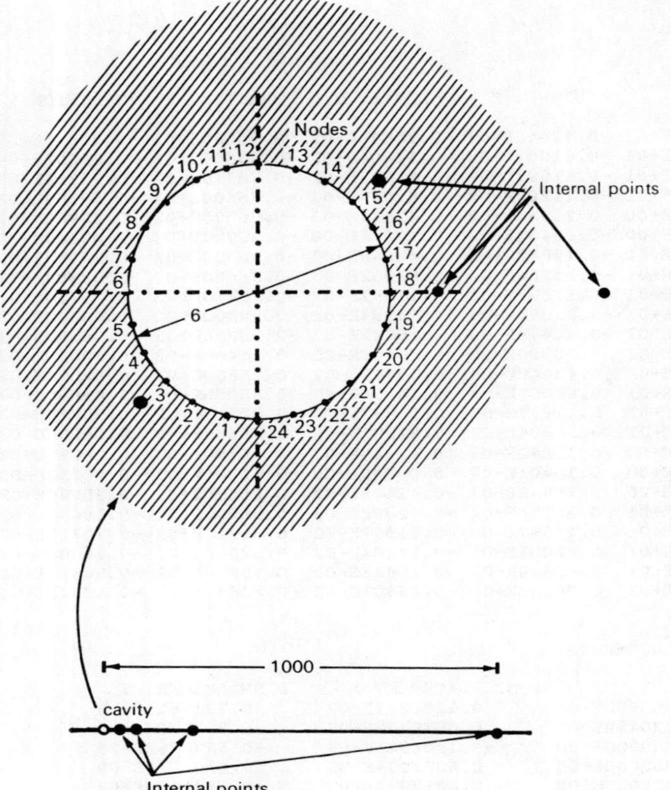

Figure 4.4 Circular cavity under internal pressure boundary element mesh and internal point description

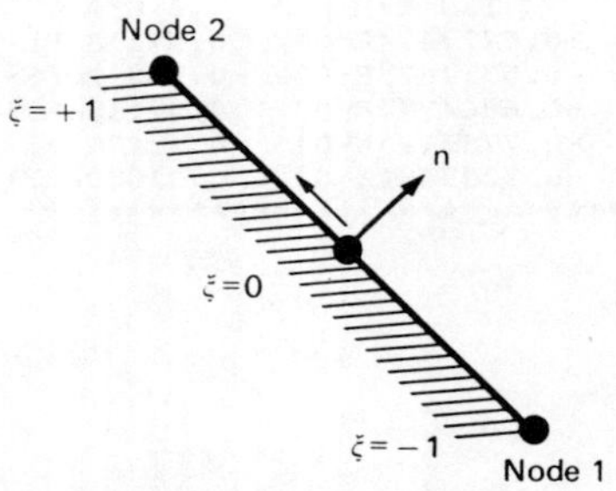

Figure 4.5 Linear element

$$\mathbf{p} = \begin{Bmatrix} p_1 \\ p_2 \end{Bmatrix} = \begin{bmatrix} \phi_1 & 0 & \phi_2 & 0 \\ 0 & \phi_1 & 0 & \phi_2 \end{bmatrix} \begin{Bmatrix} p_1^1 \\ p_2^1 \\ p_1^2 \\ p_2^2 \end{Bmatrix}^j = \mathbf{\Phi} \mathbf{p}^j \tag{4.46}$$

where $\mathbf{u}^j$ and $\mathbf{p}^j$ refer to the nodal components of elements j. The components of these vectors are u_l^k and p_l^k where k represents the node under consideration within the element and l defines the component of displacements or tractions in the l direction.

The functions ϕ_i are linear interpolation functions, such that,

$$\begin{aligned} \phi_1 &= -\tfrac{1}{2}(\xi - 1) \\ \phi_2 &= \tfrac{1}{2}(\xi + 1) \end{aligned} \tag{4.47}$$

If we consider N linear boundary elements the governing equation after neglecting boundary forces for simplicity, becomes,

$$\mathbf{c}^i\mathbf{u}^i + \sum_{j=1}^{N} \left\{ \int_{\Gamma_j} \mathbf{p}^*\mathbf{\Phi}\, d\Gamma \right\} \mathbf{u}^j = \sum_{j=1}^{N} \left\{ \int_{\Gamma_j} \mathbf{u}^*\mathbf{\Phi}\, d\Gamma \right\} \mathbf{p}^j \tag{4.48}$$

Some of these integrals can be evaluated using numerical integration, others – those with the singularity – can still be computed analytically. Once they are all found, all the elements contribution can be assembled in much the same way as for the linear potential problem (section 2.5). A major difference with the linear potential problem is that now we have two unknowns per node instead of one.

Corner Points

Another important difference with the linear potential problem is the type of $\mathbf{c}^i$ coefficients (equation (4.48)) required in the elasticity solution at corners. While these coefficients in the linear potential case were associated with the value of the solid angle at the corner, those for the elasticity problems are more complicated to find.

For smooth boundaries the $\mathbf{c}^i$ is simply a diagonal matrix with $\frac{1}{2}$ on the diagonal. When the point i is at a corner however (figure 4.6) the limit of the fundamental solution tractions, i.e.

$$[I] = \lim_{\varepsilon \to 0} \left\{ \int_{\Gamma_\varepsilon} \mathbf{p}^*\, d\Gamma \right\} \tag{4.49}$$

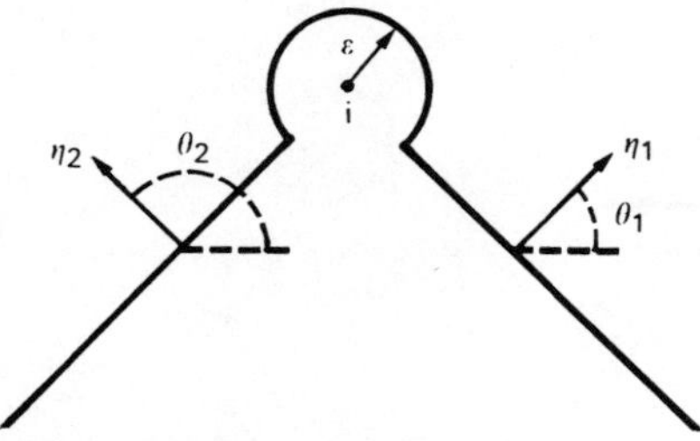

Figure 4.6 Corner point

gives a different result. For instance, for the two dimensional problems discussed here the above limit instead of a $-\frac{1}{2}$ diagonal matrix gives the following result,

$$[I] = \frac{-1}{8\pi(1-\nu)} \begin{bmatrix} 4(1-\nu)(\pi+\theta_2-\theta_1)+\sin 2\theta_1-\sin 2\theta_2, & \cos 2\theta_2-\cos 2\theta_1 \\ \cos 2\theta_2-\cos 2\theta_1, & 4(1-\nu)(\pi+\theta_2-\theta_1)+\sin 2\theta_2-\sin 2\theta_1 \end{bmatrix}$$

so that,

$$c_{lk} = \delta_{lk} + I_{lk} \tag{4.50}$$

It is much more complex to obtain a general expression for I_{lk} in three dimensions since the slope discontinuity may be of different types. In principle however, one may always do the integration over the corresponding portion of the spherical surface.

The simplest way of computing the diagonal submatrices of the **H** matrix, which includes the **c** submatrices is using rigid body consideration as shown in Chapter 3 for bounded domains or regions tending to infinity.

Boundary Conditions

As discussed in section 3.5 different strategies can be followed to define boundary conditions at corners or points of discontinuity. The simplest way to solve this problem is by introducing the concept of discontinuous elements (figure 4.7) which implies moving the points inside the element when the number of unknowns at

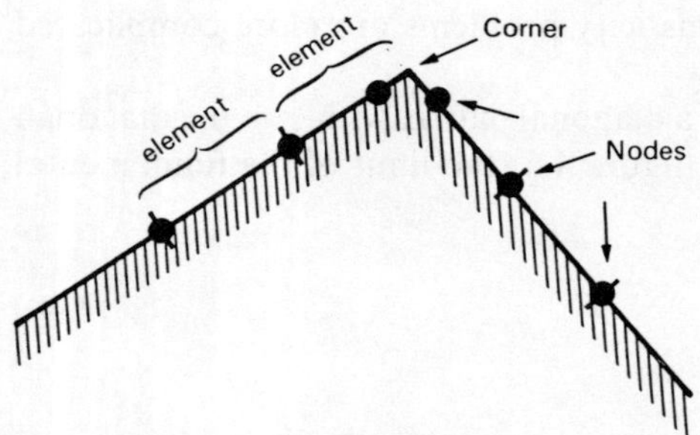

(a) Use of discontinuous elements at corners

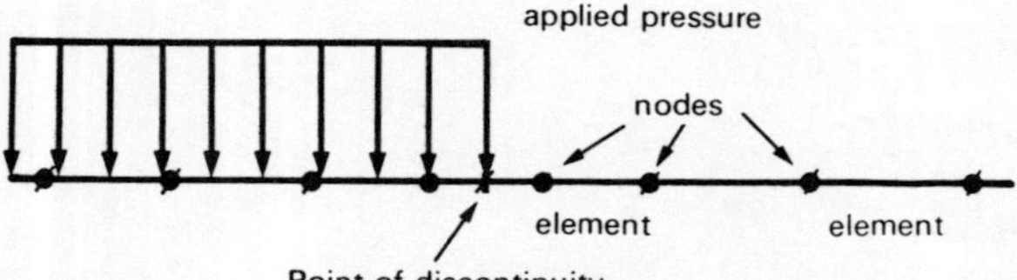

(b) Use of discontinuous elements to simulate a sudden change in boundary conditions

Figure 4.7 Linear discontinuous elements

the corner – or point of discontinuity – is more than one in each direction. This allows the user to apply all types of boundary conditions in adjacent elements in a rather simple fashion. Furthermore the approach can be used even in the presence of high stress concentration or singularities at the corner giving in this case also acceptable results when the discretization around the singularity is sufficiently fine.

4.7 Quadratic Elements

Constant and linear boundary elements are well suited to solve many plane elasticity problems including those with infinite as well as finite domains and some problems which present regions of stress concentration. Their main limitation is that they can not represent properly curved geometries. Problems involving flexure also require the use of higher order elements as their deformations are difficult to model using linear elements (constant elements for this type of problem give very poor results).

The simplest and more versatile type of curved boundary element is the quadratic for which the displacements and tractions can be represented as (figure 4.8)

$$u = \begin{Bmatrix} u_1 \\ u_2 \end{Bmatrix} = \begin{bmatrix} \phi_1 & 0 & \phi_2 & 0 & \phi_3 & 0 \\ 0 & \phi_1 & 0 & \phi_2 & 0 & \phi_3 \end{bmatrix} \begin{Bmatrix} u_1^1 \\ u_2^1 \\ u_1^2 \\ u_2^2 \\ u_1^3 \\ u_2^3 \end{Bmatrix}^j = \mathbf{\Phi} \mathbf{u}^j \tag{4.51}$$

$$p = \begin{Bmatrix} p_1 \\ p_2 \end{Bmatrix} = \begin{bmatrix} \phi_1 & 0 & \phi_2 & 0 & \phi_3 & 0 \\ 0 & \phi_1 & 0 & \phi_2 & 0 & \phi_3 \end{bmatrix} \begin{Bmatrix} p_1^1 \\ p_2^1 \\ p_1^2 \\ p_2^2 \\ p_1^3 \\ p_2^3 \end{Bmatrix}^j = \mathbf{\Phi} \mathbf{p}^j \tag{4.52}$$

The ϕ_i are quadratic interpolation functions such that,

$$\begin{aligned} \phi_1 &= \tfrac{1}{2}\xi(\xi - 1) \\ \phi_2 &= (1 - \xi^2) \\ \phi_3 &= \tfrac{1}{2}\xi(\xi + 1) \end{aligned} \tag{4.53}$$

where ξ is the dimensionless coordinate along the element (figure 4.7).

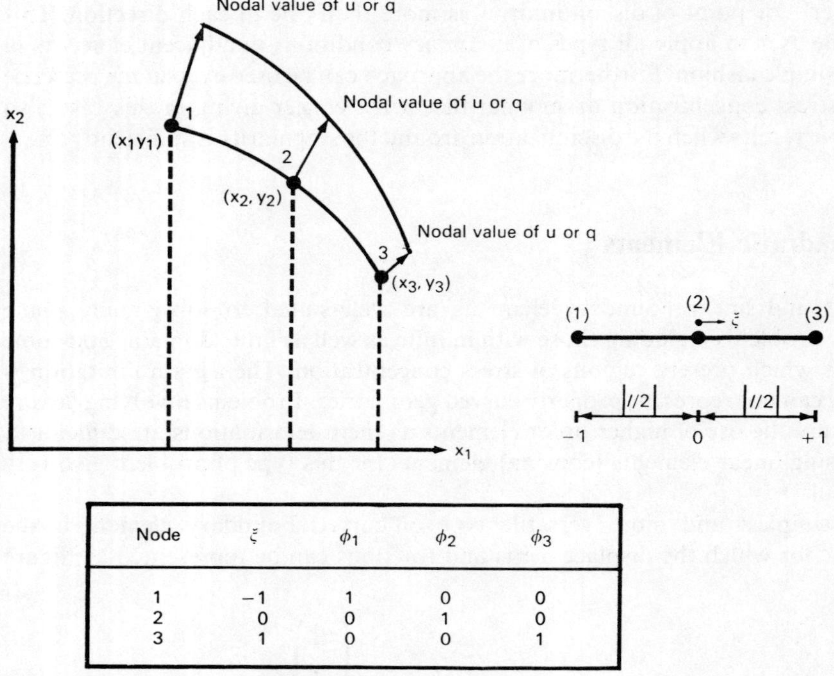

Node	ξ	ϕ_1	ϕ_2	ϕ_3
1	−1	1	0	0
2	0	0	1	0
3	1	0	0	1

Figure 4.8 Quadratic element

The geometry of the element can also be considered as quadratic and is represented by the nodes coordinates and the same interpolation functions ϕ_i used for displacements and traction components, i.e.

$$\mathbf{x} = \begin{Bmatrix} x_1 \\ x_2 \end{Bmatrix} = \begin{bmatrix} \phi_1 & 0 & \phi_2 & 0 & \phi_3 & 0 \\ 0 & \phi_1 & 0 & \phi_2 & 0 & \phi_3 \end{bmatrix} \begin{Bmatrix} x_1^1 \\ x_2^1 \\ x_1^2 \\ x_2^2 \\ x_1^3 \\ x_2^3 \end{Bmatrix}^j = \mathbf{\Phi}\mathbf{x}^j \tag{4.54}$$

The discretized boundary integral equations can be written as follows,

$$\mathbf{c}^i\mathbf{u}^i + \sum_{j=1}^{NE} \left\{ \int_{\Gamma_j} \mathbf{p}^*\mathbf{\Phi}\, d\Gamma \right\} \mathbf{u}^j = \sum_{j=1}^{NE} \left\{ \int_{\Gamma_j} \mathbf{u}^*\mathbf{\Phi}\, d\Gamma \right\} \mathbf{p}^j \tag{4.55}$$

where NE is the number of elements.

The integrals along $d\Gamma$ need now to be transformed to the homogeneous coordinate system ξ as was done in section 2.7 for potential problems. Hence the element integrals in (4.55) become,

$$\int_{\Gamma_j} \mathbf{p}^*\mathbf{\Phi}\, d\Gamma = \int_{-1}^{+1} \mathbf{p}^*\mathbf{\Phi}|G|\, d\xi$$
$$\int_{\Gamma_j} \mathbf{u}^*\mathbf{\Phi}\, d\Gamma = \int_{-1}^{+1} \mathbf{u}^*\mathbf{\Phi}|G|\, d\xi \tag{4.56}$$

where the Jacobian $|G|$ is,

$$|G| = \frac{d\Gamma}{d\xi} = \left\{\left(\frac{dx_1}{d\xi}\right)^2 + \left(\frac{dx_2}{d\xi}\right)^2\right\}^{1/2} \tag{4.57}$$

where $dx_1/d\xi$ and $dx_2/d\xi$ are easily obtained by differentiating (4.54) in terms of ξ.

4.8 Elastostatics Code using Quadratic Elements (ELQUABE)

In what follows a FORTRAN code for isotropic elasticity is described. The program has the same organization as the constant element code previously studied.

All integer variables in this code have the same meaning as for the constant element program (ELCONBE). The same applies to all real arrays with the exception of XM and YM which are not required now. Also FI and DFI have a slightly different form. The dimension of FI is $(2N)$, N being the number of nodes and that of DFI is $(3N)$ or what is the same $(6NE)$, where NE is the number of elements. The prescribed boundary conditions are read in DFI (six per element). FI is used as the right hand side vector for the solution of the system of equations. Vectors FI and DFI are reordered after solution in such a way that all values of displacements are stored in FI and tractions in DFI.

The program allows for the traction values at both sides of the nodes connecting two elements to be different. This is a similar treatment as was done for linear and quadratic potential elements (codes POLINBE and POQUABE, sections 2.6 and 2.8). For each particular x_1 or x_2 direction, (i) when both tractions are prescribed as different at both sides of the node the displacement is the only unknown; (ii) when the displacement and one traction is prescribed one on each side the other traction is the unknown and (iii) if only the displacement is prescribed, one value of the traction is the only unknown and this will be the same at both sides of the node.

In problems with only one region the case of having more unknowns than boundary conditions at a corner point seldom occurs and does not present difficulties. If the displacement is known along the two boundary elements that converge into a corner, the derivatives along these two directions are known as well, thus strains, stresses and tractions are known. One only needs to prescribe two variables along each direction and leave the third (displacement, traction before the node or after the node) as a check that it coincides with its known value. Only in cases of discontinuities of the stress tensor at certain corners one needs to implement the problem of corners in a different way. The same happens for non-homogeneous media. An accurate and simple procedure is the use of discontinuous elements, i.e. elements that have one of the node shifted inside the element in such a way that a different node will be created for each element converging into the corner.

1. Main Program

This program follows the same structure as the constant elasticity code in section 4.5.

Its present dimensions allow for 50 elements and 100 nodes.

The listing is as follows:

```
C---------------------------------------------------------------------------

      PROGRAM ELQUABE
C
C   PROGRAM 36
C
C
C   THIS PROGRAM SOLVES TWO DIMENSIONAL (EL)ASTIC PROBLEMS
C   USING (QUA)DRATIC (B)OUNDARY (E)LEMENTS
C
      CHARACTER*10 FILEIN,FILEOUT
C
      COMMON/MATG/ G(100,150)
      COMMON/MATH/ H(100,100)
      COMMON N,L,GE,XNU,INP,IPR
      DIMENSION X(51),Y(51)
      DIMENSION DFI(150),FI(100),KODE(150)
      DIMENSION CX(20),CY(20),SSOL(60),DSOL(40)
C
C   SET MAXIMUN DIMENSION OF THE SYSTEM OF EQUATIONS (NX)
C   NX=  2*MAXIMUN NUMBER OF NODES= 4*MAXIMUN NUMBER OF ELEMENTS
C   NX1= 3*MAXIMUN NUMBER OF NODES= 6*MAXIMUN NUMBER OF ELEMENTS
C
      NX=100
      NX1=150
C
C   ASSIGN NUMBERS FOR INPUT AND OUTPUT FILES
C
      INP=5
      IPR=6
C
C   READ NAMES AND OPEN  FILES FOR INPUT AND OUTPUT
C
      WRITE(*,' (A) ') ' NAME OF INPUT FILE (MAX. 10 CHART.)'
      READ(*,'  (A) ')FILEIN
      OPEN(INP,FILE=FILEIN,STATUS='OLD')
      WRITE(*,' (A) ') ' NAME OF OUTPUT FILE (MAX. 10 CHART.)'
      READ(*,' (A) ')FILEOUT
      OPEN(IPR,FILE=FILEOUT,STATUS='NEW')
C
C   READ DATA
C
      CALL INPUTEQ(CX,CY,X,Y,KODE,DFI)
C
C   COMPUTE G AND H MATRICES AND FORM SYSTEM (A X = F)
C
      CALL GHMATEQ(X,Y,G,H,FI,DFI,KODE,NX,NX1)
C
```

```
C   SOLVE SYSTEM OF EQUATIONS
C
      NN=2*N
      CALL SLNPD(H,FI,D,NN,NX)
C
C   COMPUTE STRESS AND DISPLACEMENT VALUES AT INTERNAL POINTS.
C
      CALL INTEREQ(FI,DFI,KODE,CX,CY,X,Y,SSOL,DSOL)
C
C   PRINT RESULTS AT BOUNDARY NODES AND INTERNAL POINTS
C
      CALL OUTPTEQ(X,Y,FI,DFI,CX,CY,SSOL,DSOL)
C
C    CLOSE INPUT AND OUTPUT FILES
C
      CLOSE (INP)
      CLOSE (IPR)
      STOP
      END
```

2. Routine INPUTEQ

This subroutine reads all the input required by the program and requests a file from the user containing the following lines:

(i) *Title Line* Contains the title of the problem.
(ii) *Basic Parameter Lines* Contains the number of elements, number of internal points, the shear modulus and the Poisson ratio (or fictitious Poisson's ratio for plane stress problems for which $\nu' = \nu/(1+\nu)$).
(iii) *Boundary Nodes Coordinates Lines* The coordinates are read counter-clockwise for external boundaries and clockwise for internal ones. Free FORMAT is used.
(iv) *Boundary Conditions Lines* As many lines as boundary elements. Six values of KODE and the known variables are read for each element, corresponding to the three nodes and two directions per node. In this way, a value of a traction may be prescribed for a node as part of one element and a different value as part of the other element. The displacement however must be unique for any node. The variables are displacement if KODE = 0 or traction if KODE = 1.
(v) *Internal Points Coordinate Lines* The $x_1 x_2$ coordinates for each point are read in free format. There will be one or more lines, if necessary.

This subroutine first prints the name of the run and the basic parameters. Then the coordinates of the nodes and the boundary conditions given by element, with codes and prescribed values, are printed. The internal point coordinates will be printed in the subroutine OUTPTEQ.

The FORTRAN listing of INPUTEQ is as follows:

```
C-----------------------------------------------------------------------
       SUBROUTINE INPUTEQ(CX,CY,X,Y,KODE,DFI)
C
C    PROGRAM 37
C
C    NE= NUMBER OF BOUNDARY ELEMENTS
C    L = NUMBER OF INTERNAL POINTS
C    GE= SHEAR MODULUS
```

```
C  XNU= POISSON RATIO
C
      CHARACTER*80 TITLE
      COMMON N,L,GE,XNU,INP,IPR
      DIMENSION KODE(1),X(1),Y(1),CX(1),CY(1),DFI(1)
      WRITE(IPR,100)
  100 FORMAT(' ',79('*'))
C
C  READ JOB TITLE
C
      READ(INP,'(A)') TITLE
      WRITE(IPR,'(A)') TITLE
C
C  READ NUMBER OF BOUNDARY ELEMENTS AND INTERNAL POINTS;
C  READ MATERIAL PROPERTIES
C
      READ(INP,*)NE,L,GE,XNU
      WRITE(IPR,210)NE,L,GE,XNU
  210 FORMAT(//2X,'DATA'/2X,'NUMBER OF BOUNDARY ELEMENTS=',
     1I3/2X,'NUMBER OF INTERNAL POINTS=',I3/2X,'SHEAR MODULUS=',
     2E14.7/2X,'POISSON RATIO=',E14.7)
      N=2*NE
C
C  READ BOUNDARY NODES COORDINATES
C
      N=2*NE
      WRITE(IPR,500)
      READ(INP,*) (X(I),Y(I),I=1,N)
      DO 10 I=1,N
   10 WRITE(IPR,240) I,X(I),Y(I)
  500 FORMAT(//2X,'BOUNDARY NODES COORDINATES'///4X,
     1'NODE',10X,'X',18X,'Y'/)
  240 FORMAT(5X,I3,2(5X,E14.7))
C
C  READ BOUNDARY CONDITIONS IN DFI(I) VECTOR, IF KODE(I)=0
C  THE DFI(I) VALUE IS A KNOWN DISPLACEMENT; IF KODE(I)=1 THE
C  DFI(I) VALUE IS A KNOWN TRACTION.
C  SIX BOUNDARY CONDITIONS ARE READ PER ELEMENT.
C  NODES BETWEEN TWO ELEMENTS MAY HAVE TWO DIFFERENT VALUES
C  OF THE TRACTION BUT ONLY ONE VALUE OF THE DISPLACEMENT.
C
      WRITE(IPR,800)
  800 FORMAT(//2X,'BOUNDARY CONDITIONS'//35X,'PRESCRIBED VALUES' /
     17X,'-----FIRST NODE------',3X,'-----SECOND NODE-----',3X,
     2'-----THIRD NODE------'/
     31X,'ELE',3X,'X DIR.',2X,'C',3X,'Y DIR.',2X,'C',3X,
     4'X DIR.',2X,'C',3X,'Y DIR.',2X,'C',3X,'X DIR.',
     52X,'C',3X,'Y DIR.',2X,'C'/)
      DO 20 I=1,NE
      READ(INP,*) (KODE(6*I-6+J),DFI(6*I-6+J),J=1,6)
   20 WRITE(IPR,950)I,(DFI(6*I-6+J),KODE(6*I-6+J),J=1,6)
  950 FORMAT(1X,I3,6(F9.3,2X,I1))
C
C  READ INTERNAL POINTS COORDINATES
C
      IF(L.EQ.0) GO TO 30
      READ(INP,*) (CX(I),CY(I),I=1,L)
   30 RETURN
      END
```

3. Routine GHMATEQ

This subroutine computes the G and H system matrices by calling routines EXTINEQ and LOCINEQ.

EXTINEQ: Computes the GW and HW (2×6) submatrices which relates a collocation point with an element as defined by its three nodes.

LOCINEQ: Computes the GW (2×6) submatrix for the case when the collocation point is one of the nodes within the element under consideration (i.e. the singularity is in the same element). Notice that the corresponding HW (2×6) is computed using EXTINEQ except for that part that relates a node with itself which is computed using rigid body considerations, which results in adding row coefficients.

The resulting GW and HW submatrices are assembled in the **G** and **H** system matrices. Matrix G is now rectangular since each extreme node of an element may have different tractions, i.e. one 'before' and another 'after' the node.

Once the matrices H and G are assembled, the system of equations needs to be reordered in accordance with the boundary conditions to form

$$\mathbf{AX} = \mathbf{F}$$

where **X** is a $(2N)$ vector of unknowns, N being the number of nodes; **A** is a $(2N \times 2N)$ matrix whose columns are a combination of columns of **H** or **G** depending on the boundary conditions or of two consecutive columns of **G** when the unknown is the *unique* value of the tractions at both sides of the extreme node of an element; **F** is a known vector computed by multiplying the prescribed boundary conditions by the corresponding row terms of **G** or **H**.

At the end of the subroutine GHMATEQ and after rearranging H contains the matrix **A**, and **FI** the **F** vector.

The FORTRAN listing of GHMATEQ is as follows.

```
C-----------------------------------------------------------------------
      SUBROUTINE GHMATEQ(X,Y,G,H,FI,DFI,KODE,NX,NX1)
C
C     PROGRAM 38
C
C     THIS SUBROUTINE COMPUTES THE G AND H MATRICES AND FORMS
C     THE SYSTEM A X = F READY TO BE SOLVED
C     H IS A SQUARE MATRIX (4*NE,4*NE); G IS RECTANGULAR (4*NE,6*NE)
C
      DIMENSION X(1),Y(1),G(NX,NX1),H(NX,NX),HW(2,6),GW(2,6)
      DIMENSION FI(1),DFI(1),KODE(1)
      COMMON N,L,GE,XNU,INP,IPR
      NN=2*N
      NE=N/2
      DO 20 I=1,NN
      DO 11 J=1,NN
   11 H(I,J)=0.
      DO 12 J=1,3*N
   12 G(I,J)=0.
   20 CONTINUE
      X(N+1)=X(1)
      Y(N+1)=Y(1)
C
C     COMPUTE THE GW AND HW MATRICES FOR EACH COLLOCATION
C     POINT AND EACH BOUNDARY ELEMENT
C
      DO 40 LL=1,N
      DO 40 I=1,N-1,2
      IF((LL-I)*(LL-I-1)*(LL-I-2)*(LL-I+N-2)) 22,21,22
   21 NODO=LL-I+1
      IF((LL.EQ.1).AND.(I.EQ.N-1)) NODO=NODO+N
      CALL EXTINEQ(X(LL),Y(LL),X(I),Y(I),X(I+1),Y(I+1),X(I+2),Y(I+2),
     *HW,GW)
```

```
      CALL LOCINEQ(X(I),Y(I),X(I+1),Y(I+1),X(I+2),Y(I+2),GW,NODO)
      GO TO 34
   22 CALL EXTINEQ(X(LL),Y(LL),X(I),Y(I),X(I+1),Y(I+1),X(I+2),Y(I+2),
     *HW,GW)
C
C  PLUG THE GW AND HW MATRICES INTO THE GENERAL G AND H MATRICES.
C
   34 DO 39 K=1,2
      DO 38 J=1,6
      G(2*LL-2+K,3*I-3+J)=G(2*LL-2+K,3*I-3+J)+GW(K,J)
      IF(I-N+1) 37,35,37
   35 IF(J-5) 37,36,36
   36 H(2*LL-2+K,J-4)=H(2*LL-2+K,J-4)+HW(K,J)
      GO TO 38
   37 H(2*LL-2+K,2*I-2+J)=H(2*LL-2+K,2*I-2+J)+HW(K,J)
   38 CONTINUE
   39 CONTINUE
   40 CONTINUE
C
C  COMPUTE THE DIAGONAL COEFFICIENTS OF THE H MATRIX
C
      DO 70 I=1,N
      H(2*I-1,2*I-1)=0.
      H(2*I,2*I-1)=0.
      H(2*I-1,2*I)=0.
      H(2*I,2*I)=0.
      DO 60 J=1,N
      IF(I.EQ.J) GO TO 60
      H(2*I-1,2*I-1)=H(2*I-1,2*I-1)-H(2*I-1,2*J-1)
      H(2*I,2*I-1)=H(2*I,2*I-1)-H(2*I,2*J-1)
      H(2*I-1,2*I)=H(2*I-1,2*I)-H(2*I-1,2*J)
      H(2*I,2*I)=H(2*I,2*I)-H(2*I,2*J)
   60 CONTINUE
C
C  ADD ONE TO THE DIAGONAL COEFFICIENTS FOR
C  EXTERNAL PROBLEMS.
C
      IF(H(2*I-1,2*I-1)) 65,70,70
   65 H(2*I-1,2*I-1)=1.+H(2*I-1,2*I-1)
      H(2*I,2*I)=1.+H(2*I,2*I)
   70 CONTINUE
C
C  REORDER THE COLUMNS OF THE SYSTEM OF EQUATIONS IN ACCORDANCE
C  WITH THE BOUNDARY CONDITIONS AND FORM SYSTEM MATRIX A WHICH
C  IS STORED IN H
C
      DO 180 I=1,NE
      DO 170 J=1,6
      IF(KODE(6*I-6+J)) 110,110,170
  110 IF((I-NE).NE.0 .OR. J.LT.5) GO TO 125
      IF(KODE(J-4)) 115,115,113
  113 DO 114 K=1,NN
      CH=H(K,J-4)
      H(K,J-4)=-G(K,6*I-6+J)*GE
  114 G(K,6*I-6+J)=-CH
      GO TO 170
  115 DO 116 K=1,NN
      H(K,J-4)=H(K,J-4)-G(K,6*I-6+J)*GE
  116 G(K,6*I-6+J)=0.
      GO TO 170
  125 IF(I.EQ.1 .OR. J.GT.2 .OR. KODE(6*I-8+J).EQ.1) GO TO 130
      DO 129 K=1,NN
      H(K,4*I-4+J)=H(K,4*I-4+J)-G(K,6*I-6+J)*GE
  129 G(K,6*I-6+J)=0.
      GO TO 170
  130 DO 132 K=1,NN
      CH=H(K,4*I-4+J)
      H(K,4*I-4+J)=-G(K,6*I-6+J)*GE
  132 G(K,6*I-6+J)=-CH
  170 CONTINUE
  180 CONTINUE
C
C  FORM THE RIGHT HAND SIDE VECTOR F WHICH IS STORED IN FI
```

```
C
      DO 190 I=1,NN
      FI(I)= 0.
      DO 185 J=1,6*NE
  185 FI(I)=FI(I)+G(I,J)*DFI(J)
  190 CONTINUE
      RETURN
      END
```

4. Routine EXTINEQ

This subroutine computes using numerical integration, the (2×6) submatrices GW and HW that corresponds to an element when the collocation point is at a node other than any of those 3 in the element. The coordinates of the collocation points are XP and YP. The integrals are of the type (4.56), i.e.

$$\mathbf{HW} = \int_{\Gamma_j} \mathbf{p}^* \boldsymbol{\phi} \, d\xi = \int_{-1}^{+1} \mathbf{p}^* \boldsymbol{\phi} |G| \, d\xi \tag{4.58}$$

$$\mathbf{GW} = \int_{\Gamma_j} \mathbf{u}^* \boldsymbol{\phi} \, d\Gamma = \int_{-1}^{+1} \mathbf{u}^* \boldsymbol{\phi} |G| \, d\xi \tag{4.59}$$

They can be expanded as follows,

$$\mathbf{HW} = \int_{-1}^{+1} \begin{bmatrix} p_{11}^* & p_{12}^* \\ p_{21}^* & p_{22}^* \end{bmatrix} \begin{bmatrix} \phi_1 & 0 & \phi_2 & 0 & \phi_3 & 0 \\ 0 & \phi_1 & 0 & \phi_2 & 0 & \phi_3 \end{bmatrix} |G| \, d\xi \tag{4.60}$$

$$\mathbf{GW} = \int_{-1}^{+1} \begin{bmatrix} u_{11}^* & u_{12}^* \\ u_{21}^* & u_{22}^* \end{bmatrix} \begin{bmatrix} \phi_1 & 0 & \phi_2 & 0 & \phi_3 & 0 \\ 0 & \phi_1 & 0 & \phi_2 & 0 & \phi_3 \end{bmatrix} |G| \, d\xi \tag{4.61}$$

The Jacobians are calculated by taking derivatives of the expressions for the x_1 and x_2 coordinates, in the same way that was done in section 2.8 for potential problems using quadratic elements.

The FORTRAN listing is as follows:

```
C-----------------------------------------------------------------------
      SUBROUTINE EXTINEQ(XP,YP,X1,Y1,X2,Y2,X3,Y3,HW,GW)
C
C   PROGRAM 39
C
C   THIS SUBRUOTINE COMPUTES THE HW AND GW MATRICES
C   THAT RELATE A NODE (XP,YP) WITH A BOUNDARY
C   ELEMENT USING GAUSS QUADRATURE.
C
C   RA             = RADIUS
C   RD1,RD2,RDN    = RADIUS DERIVATIVES
C   ETA1,ETA2      = COMPONENTS OF THE UNIT NORMAL TO THE ELEMENT
C   XCO,YCO        = INTEGRATION POINT ALONG THE ELEMENT
C   XJA            = JACOBIAN
C
      COMMON N,L,GE,XNU,INP,IPR
      DIMENSION GW(2,6),HW(2,6)
      DIMENSION GI(10),OME(10)
      DATA GI/0.9739065285,-0.9739065285,0.8650633666,-0.8650633666
```

```
     @,0.6794095683,-0.6794095682,0.4333953941,-0.4333953941,
     @0.1488743389,-0.1488743389/
      DATA OME/0.0666713443,0.0666713443,0.1494513491,0.1494513491
     @,0.2190863625,0.2190863625,0.2692667193,0.2692667193,
     @0.2955242247,0.2955242247/
      DO 30 I=1,2
      DO 20 J=1,6
      HW(I,J)=0.
   20 GW(I,J)=0.
   30 CONTINUE
      A=X3-2*X2+X1
      B=(X3-X1)/2
      C=Y3-2*Y2+Y1
      D=(Y3-Y1)/2
      DE=4*3.141592*(1-XNU)
      DO 40 I=1,10
C
C  COMPUTE THE VALUES OF THE SHAPE FUNCTIONS AT THE
C  INTEGRATION POINTS
C
      F1=GI(I)*(GI(I)-1)*0.5
      F2=1.-GI(I)**2
      F3=GI(I)*(GI(I)+1)*0.5
C
C  COMPUTE GEOMETRICAL PROPERTIES AT THE INTEGRATION POINTS
C
      XCO=X1*F1+X2*F2+X3*F3
      YCO=Y1*F1+Y2*F2+Y3*F3
      XJA=SQRT((GI(I)*A+B)**2+(GI(I)*C+D)**2)
      ETA1=(GI(I)*C+D)/XJA
      ETA2=-(GI(I)*A+B)/XJA
      RA=SQRT((XP-XCO)**2+(YP-YCO)**2)
      RD1=(XCO-XP)/RA
      RD2=(YCO-YP)/RA
      RDN=RD1*ETA1+RD2*ETA2
C
C  COMPUTE GW AND HW MATRICES
C
      GW(1,1)=GW(1,1)+((3-4*XNU)*ALOG(1./RA)+RD1**2)*OME(I)*XJA*F1/
     1(2*DE*GE)
      GW(1,2)=GW(1,2)+RD1*RD2*OME(I)*XJA*F1/(2*DE*GE)
      GW(2,1)=GW(1,2)
      GW(2,2)=GW(2,2)+((3-4*XNU)*ALOG(1./RA)+RD2**2)*OME(I)*XJA*F1/
     1(2*DE*GE)
      HW(1,1)=HW(1,1)-RDN*((1-2*XNU)+2*RD1**2)/(RA*DE)*OME(I)*XJA*F1
      HW(1,2)=HW(1,2)-(RDN*2*RD1*RD2+(1-2*XNU)*(ETA1*RD2-ETA2*RD1))*
     1OME(I)*XJA*F1/(RA*DE)
      HW(2,1)=HW(2,1)-(RDN*2*RD1*RD2+(1-2*XNU)*(ETA2*RD1-ETA1*RD2))*
     1OME(I)*XJA*F1/(RA*DE)
      HW(2,2)=HW(2,2)-RDN*((1-2*XNU)+2*RD2**2)*OME(I)*XJA*F1/(RA*DE)
      GW(1,3)=GW(1,3)+((3-4*XNU)*ALOG(1./RA)+RD1**2)*OME(I)*XJA*F2/
     1(2*DE*GE)
      GW(1,4)=GW(1,4)+RD1*RD2*OME(I)*XJA*F2/(2*DE*GE)
      GW(2,3)=GW(1,4)
      GW(2,4)=GW(2,4)+((3-4*XNU)*ALOG(1./RA)+RD2**2)*OME(I)*XJA*F2/
     1(2*DE*GE)
      HW(1,3)=HW(1,3)-RDN*((1-2*XNU)+2*RD1**2)/(RA*DE)*OME(I)*XJA*F2
      HW(1,4)=HW(1,4)-(RDN*2*RD1*RD2+(1-2*XNU)*(ETA1*RD2-ETA2*RD1))*
     1OME(I)*XJA*F2/(RA*DE)
      HW(2,3)=HW(2,3)-(RDN*2*RD1*RD2+(1-2*XNU)*(ETA2*RD1-ETA1*RD2))*
     1OME(I)*XJA*F2/(RA*DE)
      HW(2,4)=HW(2,4)-RDN*((1-2*XNU)+2*RD2**2)*OME(I)*XJA*F2/(RA*DE)
      GW(1,5)=GW(1,5)+((3-4*XNU)*ALOG(1./RA)+RD1**2)*OME(I)*XJA*F3/
     1(2*DE*GE)
      GW(1,6)=GW(1,6)+RD1*RD2*OME(I)*XJA*F3/(2*DE*GE)
      GW(2,5)=GW(1,6)
      GW(2,6)=GW(2,6)+((3-4*XNU)*ALOG(1./RA)+RD2**2)*OME(I)*XJA*F3/
     1(2*DE*GE)
      HW(1,5)=HW(1,5)-RDN*((1-2*XNU)+2*RD1**2)/(RA*DE)*OME(I)*XJA*F3
      HW(1,6)=HW(1,6)-(RDN*2*RD1*RD2+(1-2*XNU)*(ETA1*RD2-ETA2*RD1))*
     1OME(I)*XJA*F3/(RA*DE)
      HW(2,5)=HW(2,5)-(RDN*2*RD1*RD2+(1-2*XNU)*(ETA2*RD1-ETA1*RD2))*
```

```
     1OME(I)*XJA*F3/(RA*DE)
  40 HW(2,6)=HW(2,6)-RDN*((1-2*XNU)+2*RD2**2)*OME(I)*XJA*F3/(RA*DE)
     RETURN
     END
```

5. Routine LOCINEQ

This subroutine computes using standard Gauss quadrature and a special quadrature formula, the (2×6) submatrix GW when the collocation point is one of the nodes within the element under consideration.

The integration process is completely analogous to that shown for potential problems in section 2.8. The integrals are split into two parts: one without singularity which is integrated by means of the standard formula; and the other, the loarithmic part, that is integrated using a special quadrature formula. More details of the integration process are given in section 2.8. The only difference is now that kernels to be integrated have some more terms that in the case of potential problems. Also GW is now (2×6) instead of (3).

The listing for this subroutine is as follows:

```
C-----------------------------------------------------------------------
      SUBROUTINE LOCINEQ(XG1,YG1,XG2,YG2,XG3,YG3,GW,NODO)
C
C  PROGRAM 40
C
C  THIS SUBROUTINE COMPUTES THE GW MATRIX WHEN THE COLLOCATION
C  POINT IS ONE OF THE NODES OF THE INTEGRATION ELEMENT.
C  THE COEFFICIENTS ARE COMPUTED BY NUMERICAL INTEGRATION:
C  THE NON SINGULAR PART IS COMPUTED USING STANDARD GAUSS QUADRATURE,
C  THE LOGARITHMIC PART IS COMPUTED USING A SPECIAL QUADRATURE FORMULA.
C
      COMMON N,L,GE,XNU,INP,IPR
      DIMENSION GI(10),OME(10),GIL(10),OMEL(10),GW(2,6),R(2)
C
C  DATA FOR THE GAUSS QUADRATURE
C
      DATA GI/0.9739065285,-0.9739065285,0.8650633666,-0.8650633666
     @,0.6794095682,-0.6794095682,0.4333953941,-0.4333953941,
     @0.1488743389,-0.1488743389/
      DATA OME/0.0666713443,0.0666713443,0.1494513491,0.1494513491
     @,0.2190863625,0.2190863625,0.2692667193,0.2692667193,
     @0.2955242247,0.2955242247/
C
C  DATA FOR THE SPECIAL QUADRATURE
C
      DATA GIL/0.0090426309,0.0539712662,0.1353118246,0.2470524162
     @,0.3802125396,0.5237923179,0.6657752055,0.7941904160,
     @0.8981610912,0.9688479887/
      DATA OMEL/0.1209551319,0.1863635425,0.1956608732,0.1735771421
     @,0.1356956729,0.0936467585,0.0557877273,0.0271598109,
     @0.0095151826,0.0016381576/
C
C  SET LOCAL COORDINATES SYSTEM
C
      GOTO(1,2,3),NODO
    1 X3=XG3-XG1
      Y3=YG3-YG1
      X2=XG2-XG1
      Y2=YG2-YG1
      A1=(X3-2*X2)*0.5
      B1=X2
      A2=(Y3-2*Y2)*0.5
      B2=Y2
      GO TO 4
```

```
    2 X3=XG3-XG2
      Y3=YG3-YG2
      X1=XG1-XG2
      Y1=YG1-YG2
      A1=X1+X3
      B1=X3-X1
      A2=Y1+Y3
      B2=Y3-Y1
      GO TO 4
    3 X2=XG2-XG3
      Y2=YG2-YG3
      X1=XG1-XG3
      Y1=YG1-YG3
      A1=(X1-2*X2)*0.5
      B1=-X2
      A2=(Y1-2*Y2)*0.5
      B2=-Y2
    4 CONTINUE
C
      DO 10 I=1,2
      DO 10 J=1,6
   10 GW(I,J)=0.
      A=A1**2 + A2**2
      B=2*(A1*B1 + A2*B2)
      C=B1**2 + B2**2
      CONT1=(3-4*XNU)/(8.*3.1415926*GE*(1-XNU))
      CONT2=CONT1/(3-4*XNU)
      DO 250 I=1,10
      T1=((A1*GI(I)+B1)**2)/((A2*GI(I)+B2)**2 + (A1*GI(I)+B1)**2)
      T2=((A2*GI(I)+B2)**2)/((A2*GI(I)+B2)**2 + (A1*GI(I)+B1)**2)
      T3=(A1*GI(I)+B1)*(A2*GI(I)+B2)/((A1*GI(I)+B1)**2+(A2*GI(I)+B2)**2)
C
C   COMPUTE SHAPE FUNCTIONS FOR NUMERICAL INTEGRATION
C
      F3=0.5*GI(I)*(GI(I)+1.)
      F2=1.-GI(I)**2
      F1=0.5*GI(I)*(GI(I)-1.)
      FL3=GIL(I)*(2.*GIL(I)-1.)
      FL2=4.*GIL(I)*(1.-GIL(I))
      FL1=(GIL(I)-1.)*(2.*GIL(I)-1.)
      FLN3=0.5*GIL(I)*(GIL(I)+1.)
      FLN2=1.-GIL(I)**2
      FLN1=0.5*GIL(I)*(GIL(I)-1.)
C
C   COMPUTE GW COEFFICIENTS
C
      GO TO(50,60,70) NODO
   50 XJA1=SQRT((4*A1*GIL(I)-2*A1+0.5*X3)**2+(4*A2*GIL(I)-2*A2+0.5*Y3)**
     *2)*2
      XJA2=SQRT((A1*GI(I)*2+0.5*X3)**2+(A2*GI(I)*2+0.5*Y3)**2)
      XLO=-ALOG(2*SQRT((GI(I)*A1+B1)**2+(GI(I)*A2+B2)**2))
      S3=CONT1*(FL3*XJA1*OMEL(I)+F3*XJA2*XLO*OME(I))
      S2=CONT1*(FL2*XJA1*OMEL(I)+F2*XJA2*XLO*OME(I))
      S1=CONT1*(FL1*XJA1*OMEL(I)+F1*XJA2*XLO*OME(I))
      GO TO 200
   60 XJA1=SQRT((0.5*B1-A1*GIL(I))**2+(0.5*B2-A2*GIL(I))**2)
      XJA11=SQRT((0.5*B1+A1*GIL(I))**2+(0.5*B2+A2*GIL(I))**2)
      XJA2=SQRT((0.5*B1+A1*GI(I))**2+(0.5*B2+A2*GI(I))**2)
      XLO=-0.5*ALOG((GI(I)*A1*0.5+B1*0.5)**2+(GI(I)*A2*0.5+B2*0.5)**2)
      S3=CONT1*((FLN1*XJA1+FLN3*XJA11)*OMEL(I)+F3*XJA2*XLO*OME(I))
      S2=CONT1*(FLN2*(XJA1+XJA11)*OMEL(I)+F2*XJA2*XLO*OME(I))
      S1=CONT1*((FLN3*XJA1+FLN1*XJA11)*OMEL(I)+F1*XJA2*XLO*OME(I))
      GO TO 200
   70 XJA1=SQRT((2*A1-4*A1*GIL(I)-0.5*X1)**2+(2*A2-4*A2*GIL(I)-0.5*Y1)**
     *2)*2
      XJA2=SQRT((2*A1*GI(I)-0.5*X1)**2+(2*A2*GI(I)-0.5*Y1)**2)
      XLO=-ALOG(2*SQRT((A1*GI(I)+B1)**2+(A2*GI(I)+B2)**2))
      S3=CONT1*(FL1*XJA1*OMEL(I)+F3*XJA2*XLO*OME(I))
      S2=CONT1*(FL2*XJA1*OMEL(I)+F2*XJA2*XLO*OME(I))
      S1=CONT1*(FL3*XJA1*OMEL(I)+F1*XJA2*XLO*OME(I))
  200 GW(1,5)= GW(1,5)+S3+CONT2*F3*T1*XJA2*OME(I)
      GW(1,3)= GW(1,3)+S2+CONT2*F2*T1*XJA2*OME(I)
      GW(1,1)= GW(1,1)+S1+CONT2*F1*T1*XJA2*OME(I)
```

```
      GW(1,6)= GW(1,6)+ CONT2*T3*XJA2*F3*OME(I)
      GW(1,4)= GW(1,4)+ CONT2*T3*XJA2*F2*OME(I)
      GW(1,2)= GW(1,2)+ CONT2*T3*XJA2*F1*OME(I)
      GW(2,5)= GW(1,6)
      GW(2,3)= GW(1,4)
      GW(2,1)=GW(1,2)
      GW(2,6)= GW(2,6)+ S3+CONT2*T2*XJA2*F3*OME(I)
      GW(2,4)=GW(2,4)+S2+CONT2*T2*XJA2*F2*OME(I)
      GW(2,2)= GW(2,2)+ S1+CONT2*T2*XJA2*F1*OME(I)
C
  250 CONTINUE
C
      RETURN
      END
```

6. Routine INTEREQ

This subroutine first reorders the vectors DFI and FI in such a way that all the boundary displacements are stored in FI and all the tractions in DFI. It then computes the displacements and stresses at internal points.

The displacement at any interior point is given by

$$\mathbf{u}^i = \sum_{j=1}^{NE} \left\{ \int_{\Gamma_j} \mathbf{u}^* \boldsymbol{\phi} \, d\Gamma \right\} \mathbf{p}^j - \sum_{j=1}^{NE} \left\{ \int_{\Gamma_j} \mathbf{p}^* \boldsymbol{\phi} \, d\Gamma \right\} \mathbf{u}^j \tag{4.62}$$

where the integrals along the boundary elements are computed numerically by calling again the subroutine EXTINEQ.

Analogously, the stresses are given by

$$\boldsymbol{\sigma}^i_{kl} = \sum_{j=1}^{NE} \left\{ \int_{\Gamma_j} \mathbf{D}_{kl} \boldsymbol{\phi} \, d\Gamma \right\} \mathbf{p}^j - \sum_{j=1}^{NE} \left\{ \int \mathbf{S}_{kl} \boldsymbol{\phi} \, d\Gamma \right\} \mathbf{u}^j \tag{4.63}$$

where

$$\mathbf{D}_{kl} = [D_{1kl}, D_{2kl}]; \qquad \mathbf{S}_{kl} = [S_{1kl}, S_{2kl}]$$

The values of D_{mkl} and S_{mkl} have been given in equations (3.110) and (3.111). The integrals along the elements in equation (4.63) are computed by calling the subroutine SIGMAEQ.

The listing of INTEREQ is as follows,

```
C-----------------------------------------------------------------------
      SUBROUTINE INTEREQ(FI,DFI,KODE,CX,CY,X,Y,SSOL,DSOL)
C
C  PROGRAM 41
C
C  THIS SUBROUTINE COMPUTES THE VALUES OF STRESSES AND DISPLACEMENTS AT
C  INTERNAL POINTS.
C
      COMMON N,L,GE,XNU,INP,IPR
      DIMENSION FI(1),DFI(1),KODE(1),CX(1),CY(1)
      DIMENSION X(1),Y(1),SSOL(1),DSOL(1)
      DIMENSION HW(2,6),GW(2,6)
      DIMENSION D11(6),D12(6),D22(6),S11(6),S12(6),S22(6)
C
C  REARRANGE FI AND DFI ARRAYS TO STORE ALL THE VALUES OF THE
C  DISPLACEMENT IN FI AND ALL THE VALUES TRACTION IN DFI
C
```

```
      NE=N/2
      DO 180 I=1,NE
      DO 170 J=1,6
      IF(KODE(6*I-6+J)) 110,110,170
  110 IF((I-NE).NE.0 .OR. J.LT.5) GO TO 125
      IF(KODE(J-4)) 114,114,113
  113 CH=FI(J-4)*GE
      FI(J-4)=DFI(6*I-6+J)
      DFI(6*I-6+J)=CH
      GO TO 170
  114 DFI(6*I-6+J)=DFI(J-4)
      GO TO 170
  125 IF(I.EQ.1 .OR. J.GT.2 .OR. KODE(6*I-8+J).EQ.1) GO TO 130
      DFI(6*I-6+J)=DFI(6*I-8+J)
      GO TO 170
  130 CH=FI(4*I-4+J)*GE
      FI(4*I-4+J)=DFI(6*I-6+J)
      DFI(6*I-6+J)=CH
  170 CONTINUE
  180 CONTINUE
C
C  COMPUTE THE VALUES OF STRESSES AND DISPLACEMENTS AT INTERNAL POINTS
C
      IF(L.EQ.0) GO TO 50
      DO 240 K=1,L
      SSOL(3*K-2)=0.
      SSOL(3*K-1)=0.
      SSOL(3*K)=0.
      DSOL(2*K-1)=0.
      DSOL(2*K)=0.
      DO 230 I=1,NE
      CALL EXTINEQ(CX(K),CY(K),X(2*I-1),Y(2*I-1),X(2*I),Y(2*I),
     1X(2*I+1),Y(2*I+1),HW,GW)
      CALL SIGMAEQ(CX(K),CY(K),X(2*I-1),Y(2*I-1),X(2*I),Y(2*I),
     1X(2*I+1),Y(2*I+1),D11,D12,D22,S11,S12,S22)
      DO 220 J=1,6
      IJ4=4*I-4+J
      IF(IJ4.GT.(4*NE)) IJ4=J-4
      SSOL(3*K-2)=SSOL(3*K-2)+D11(J)*DFI(6*I-6+J)-S11(J)*FI(IJ4)
      SSOL(3*K-1)=SSOL(3*K-1)+D12(J)*DFI(6*I-6+J)-S12(J)*FI(IJ4)
      SSOL(3*K)=SSOL(3*K)+D22(J)*DFI(6*I-6+J)-S22(J)*FI(IJ4)
      DSOL(2*K-1)=DSOL(2*K-1)+GW(1,J)*DFI(6*I-6+J)-HW(1,J)*FI(IJ4)
  220 DSOL(2*K)=DSOL(2*K)+GW(2,J)*DFI(6*I-6+J)-HW(2,J)*FI(IJ4)
  230 CONTINUE
  240 CONTINUE
   50 RETURN
      END
```

7. Routine SIGMAEQ

The integrals of the S and D coefficients multiplied by the shape functions in equation (4.63) are evaluated using Gaussian quadrature. The procedure is similar to that used in routine EXTINEQ, the only difference being that now the expressions of D and S given by equation (3.110) and (3.111) are computed at the integration points instead of the values of u^* and p^* as in EXTINEQ.

The FORTRAN listing of SIGMAEQ is as follows,

```
C-----------------------------------------------------------------------
      SUBROUTINE SIGMAEQ(XP,YP,X1,Y1,X2,Y2,X3,Y3,D11,D12,D22,S11,S12
     *,S22)
C
C  PROGRAM 42
C
C  THIS SUBROUTINE COMPUTES THE VALUES OF THE S AND D MATRICES
C  USING GAUSS QUADRATURE IN ORDER TO COMPUTE THE STRESS
C  AT ANY INTERNAL POINT.
```

```
C  RA              = RADIUS
C  RD1,RD2,RDN     = RADIUS DERIVATIVES
C  ETA1,ETA2       = COMPONENTS OF THE UNIT NORMAL TO THE ELEMENT
C  XCO,YCO         = INTEGRATION POINT ALONG THE ELEMENT
C  XJA             = JACOBIAN
C
      COMMON N,L,GE,XNU,INP,IPR
      DIMENSION D11(6),D12(6),D22(6),S11(6),S12(6),S22(6)
      DIMENSION GI(10),OME(10)
      DATA GI/0.9739065285,-0.9739065285,0.8650633666,-0.8650633666
     @,0.6794095683,-0.6794095682,0.4333953941,-0.4333953941,
     @0.1488743389,-0.1488743389/
      DATA OME/0.0666713443,0.0666713443,0.1494513491,0.1494513491
     @,0.2190863625,0.2190863625,0.2692667193,0.2692667193,
     @0.2955242247,0.2955242247/
      DO 20 J=1,6
      D11(J)=0.
      D12(J)=0.
      D22(J)=0.
      S11(J)=0.
      S12(J)=0.
   20 S22(J)=0.
C
      FA=1-4*XNU
      AL=1-2*XNU
      A=X3-2*X2+X1
      B=(X3-X1)/2
      C=Y3-2*Y2+Y1
      D=(Y3-Y1)/2
      DE=4*3.141592*(1-XNU)
      DO 40 I=1,10
C
C  COMPUTE VALUES OF THE SHAPE FUNCTIONS AT THE INTEGRATION POINTS
C
      F1=GI(I)*(GI(I)-1)*0.5
      F2=1.-GI(I)**2
      F3=GI(I)*(GI(I)+1)*0.5
C
C  COMPUTE GEOMETRICAL PARAMETERS
C
      XCO=X1*F1+X2*F2+X3*F3
      YCO=Y1*F1+Y2*F2+Y3*F3
      XJA=SQRT((GI(I)*A+B)**2+(GI(I)*C+D)**2)
      ETA1=(GI(I)*C+D)/XJA
      ETA2=-(GI(I)*A+B)/XJA
      RA=SQRT((XP-XCO)**2+(YP-YCO)**2)
      RD1=(XCO-XP)/RA
      RD2=(YCO-YP)/RA
      RDN=RD1*ETA1+RD2*ETA2
C
C  COMPUTE D AND S COEFFICIENTS
C
      D11(1)=D11(1)+(AL*RD1+2*RD1**3)*OME(I)*XJA*F1/(DE*RA)
      D11(2)=D11(2)+(2*RD1**2*RD2-AL*RD2)*OME(I)*XJA*F1/(DE*RA)
      D11(3)=D11(3)+(AL*RD1+2*RD1**3)*OME(I)*XJA*F2/(DE*RA)
      D11(4)=D11(4)+(2*RD1**2*RD2-AL*RD2)*OME(I)*XJA*F2/(DE*RA)
      D11(5)=D11(5)+(AL*RD1+2*RD1**3)*OME(I)*XJA*F3/(DE*RA)
      D11(6)=D11(6)+(2*RD1**2*RD2-AL*RD2)*OME(I)*XJA*F3/(DE*RA)
      D12(1)=D12(1)+(AL*RD2+2*RD1**2*RD2)*F1/(DE*RA)*OME(I)*XJA
      D12(2)=D12(2)+(AL*RD1+2*RD1*RD2**2)*F1/(DE*RA)*OME(I)*XJA
      D12(3)=D12(3)+(AL*RD2+2*RD1**2*RD2)*F2/(DE*RA)*OME(I)*XJA
      D12(4)=D12(4)+(AL*RD1+2*RD1*RD2**2)*F2/(DE*RA)*OME(I)*XJA
      D12(5)=D12(5)+(AL*RD2+2*RD1**2*RD2)*F3/(DE*RA)*OME(I)*XJA
      D12(6)=D12(6)+(AL*RD1+2*RD1*RD2**2)*F3/(DE*RA)*OME(I)*XJA
      D22(1)=D22(1)+(2*RD1*RD2**2-AL*RD1)*F1/(DE*RA)*OME(I)*XJA
      D22(2)=D22(2)+(AL*RD2+2*RD2**3)*F1/(DE*RA)*OME(I)*XJA
      D22(3)=D22(3)+(2*RD1*RD2**2-AL*RD1)*F2/(DE*RA)*OME(I)*XJA
      D22(4)=D22(4)+(AL*RD2+2*RD2**3)*F2/(DE*RA)*OME(I)*XJA
      D22(5)=D22(5)+(2*RD1*RD2**2-AL*RD1)*F3/(DE*RA)*OME(I)*XJA
      D22(6)=D22(6)+(AL*RD2+2*RD2**3)*F3/(DE*RA)*OME(I)*XJA
      S11(1)=S11(1)+(2*RDN*(AL*RD1+XNU*2*RD1-4*RD1**3)+4*XNU*ETA1
     1*RD1**2+AL*(2*ETA1*RD1**2+2*ETA1)-FA*ETA1)*2*GE*F1/(DE*RA**2)*
     2OME(I)*XJA
```

```
      S11(2)=S11(2)+(2*RDN*(AL*RD2-4*RD1**2*RD2)+4*XNU*ETA1*RD1*RD2+
     1AL*2*ETA2*RD1**2-FA*ETA2)*2*GE*F1/(DE*RA**2)*OME(I)*XJA
      S11(3)=S11(3)+(2*RDN*(AL*RD1+XNU*2*RD1-4*RD1**3)+4*XNU*ETA1
     1*RD1**2+AL*(2*ETA1*RD1**2+2*ETA1)-FA*ETA1)*2*GE*F2/(DE*RA**2)*
     2OME(I)*XJA
      S11(4)=S11(4)+(2*RDN*(AL*RD2-4*RD1**2*RD2)+4*XNU*ETA1*RD1*RD2+
     1AL*2*ETA2*RD1**2-FA*ETA2)*2*GE*F2/(DE*RA**2)*OME(I)*XJA
      S11(5)=S11(5)+(2*RDN*(AL*RD1+XNU*2*RD1-4*RD1**3)+4*XNU*ETA1
     1*RD1**2+AL*(2*ETA1*RD1**2+2*ETA1)-FA*ETA1)*2*GE*F3/(DE*RA**2)*
     2OME(I)*XJA
      S11(6)=S11(6)+(2*RDN*(AL*RD2-4*RD1**2*RD2)+4*XNU*ETA1*RD1*RD2+
     1AL*2*ETA2*RD1**2-FA*ETA2)*2*GE*F3/(DE*RA**2)*OME(I)*XJA
      S12(1)=S12(1)+(2*RDN*(XNU*RD2-4*RD1**2*RD2)+2*XNU*(ETA1*RD2*
     1RD1+ETA2*RD1**2)+AL*(2*ETA1*RD1*RD2+ETA2))*2*GE*F1/
     2(DE*RA**2)*OME(I)*XJA
      S12(2)=S12(2)+(2*RDN*(XNU*RD1-4*RD1*RD2**2)+2*XNU*(ETA1*RD2**2
     1+ETA2*RD1*RD2)+AL*(2*ETA2*RD1*RD2+ETA1))*2*GE*F1/
     2(DE*RA**2)*OME(I)*XJA
      S12(3)=S12(3)+(2*RDN*(XNU*RD2-4*RD1**2*RD2)+2*XNU*(ETA1*RD2*
     1RD1+ETA2*RD1**2)+AL*(2*ETA1*RD1*RD2+ETA2))*2*GE*F2/
     2(DE*RA**2)*OME(I)*XJA
      S12(4)=S12(4)+(2*RDN*(XNU*RD1-4*RD1*RD2**2)+2*XNU*(ETA1*RD2**2
     1+ETA2*RD1*RD2)+AL*(2*ETA2*RD1*RD2+ETA1))*2*GE*F2/
     2(DE*RA**2)*OME(I)*XJA
      S12(5)=S12(5)+(2*RDN*(XNU*RD2-4*RD1**2*RD2)+2*XNU*(ETA1*RD2*
     1RD1+ETA2*RD1**2)+AL*(2*ETA1*RD1*RD2+ETA2))*2*GE*F3/
     2(DE*RA**2)*OME(I)*XJA
      S12(6)=S12(6)+(2*RDN*(XNU*RD1-4*RD1*RD2**2)+2*XNU*(ETA1*RD2**2
     1+ETA2*RD1*RD2)+AL*(2*ETA2*RD1*RD2+ETA1))*2*GE*F3/
     2(DE*RA**2)*OME(I)*XJA
      S22(1)=S22(1)+(2*RDN*(AL*RD1-4*RD1*RD2**2)+4*XNU*ETA2*RD1*RD2+
     1AL*2*ETA1*RD2**2-FA*ETA1)*2*GE*F1/(DE*RA**2)*OME(I)*XJA
      S22(2)=S22(2)+(2*RDN*(AL*RD2+2*XNU*RD2-4*RD2**3)+4*XNU*ETA2
     1*RD2**2+AL*(2*ETA2*RD2**2+2*ETA2)-FA*ETA2)*2*GE*F1/
     2(DE*RA**2)*OME(I)*XJA
      S22(3)=S22(3)+(2*RDN*(AL*RD1-4*RD1*RD2**2)+4*XNU*ETA2*RD1*RD2+
     1AL*2*ETA1*RD2**2-FA*ETA1)*2*GE*F2/(DE*RA**2)*OME(I)*XJA
      S22(4)=S22(4)+(2*RDN*(AL*RD2+2*XNU*RD2-4*RD2**3)+4*XNU*ETA2
     1*RD2**2+AL*(2*ETA2*RD2**2+2*ETA2)-FA*ETA2)*2*GE*F2/
     2(DE*RA**2)*OME(I)*XJA
      S22(5)=S22(5)+(2*RDN*(AL*RD1-4*RD1*RD2**2)+4*XNU*ETA2*RD1*RD2+
     1AL*2*ETA1*RD2**2-FA*ETA1)*2*GE*F3/(DE*RA**2)*OME(I)*XJA
   40 S22(6)=S22(6)+(2*RDN*(AL*RD2+2*XNU*RD2-4*RD2**3)+4*XNU*ETA2
     1*RD2**2+AL*(2*ETA2*RD2**2+2*ETA2)-FA*ETA2)*2*GE*F3/
     2(DE*RA**2)*OME(I)*XJA
      RETURN
      END
```

8. Routine OUTPTEQ

This subroutine prints the results in the following order.

(i) Displacements at boundary nodes.
(ii) Tractions at boundary nodes (tractions 'before' and 'after' each node are printed).
(iii) Internal point displacements.
(iv) Internal point stresses.

The listing is as follows:

```
C-----------------------------------------------------------------------
      SUBROUTINE OUTPTEQ(X,Y,FI,DFI,CX,CY,SSOL,DSOL)
C
C  PROGRAM 43
C
C  THIS SUBROUTINE PRINTS THE VALUES OF THE DISPLACEMENTS
```

```
C  AND TRACTIONS AT BOUNDARY NODES. IT ALSO PRINTS THE VALUES
C  OF DISPLACEMENTS AND STRESSES AT INTERNAL POINTS
C
      DIMENSION X(1),Y(1),FI(1),DFI(1)
      DIMENSION CX(1),CY(1),SSOL(1),DSOL(1)
      COMMON N,L,GE,XNU,INP,IPR
      NE=N/2
C
      WRITE(IPR,100)
  100 FORMAT(' ',79('*')//1X,'RESULTS'//2X,'BOUNDARY NODES'///10X
     1,'X',17X,'Y',11X,'DISPLACEMENT X',4X,'DISPLACEMENT Y'/)
      DO 10 I=1,N
   10 WRITE(IPR,200) X(I),Y(I),FI(2*I-1),FI(2*I)
  200 FORMAT(4(4X,E14.7))
      WRITE(IPR,150)
  150 FORMAT(///28X,'TRACTION X',3X,'TRACTION Y',3X,'TRACTION X'
     1,3X,'TRACTION Y'/7X,'X',12X,'Y',7X,'BEFORE NODE',2X,'BEFORE NODE'
     2,2X,'AFTER NODE',3X,'AFTER NODE'/)
      WRITE(IPR,250) X(1),Y(1),DFI(6*NE-1),DFI(6*NE),DFI(1),DFI(2)
      WRITE(IPR,250) X(2),Y(2),DFI(3),DFI(4),DFI(3),DFI(4)
      DO 15 I=2,NE
      WRITE(IPR,250) X(2*I-1),Y(2*I-1),DFI(6*I-7),DFI(6*I-6),
     1DFI(6*I-5),DFI(6*I-4)
   15 WRITE(IPR,250) X(2*I),Y(2*I),DFI(6*I-3),DFI(6*I-2),
     1DFI(6*I-3),DFI(6*I-2)
  250 FORMAT(6(1X,E12.5))
C
      IF(L.EQ.0) GO TO 30
      WRITE(IPR,300)
  300 FORMAT(///2X,'INTERNAL POINTS DISPLACEMENTS'//8X,'X',15X,'Y',9X
     1,'DISPLACEMENT X',2X,'DISPLACEMENT Y')
      DO 20 K=1,L
   20 WRITE(IPR,400)CX(K),CY(K),DSOL(2*K-1),DSOL(2*K)
      WRITE(IPR,350)
  350 FORMAT(//2X,'INTERNAL POINTS STRESSES'//8X,'X',15X,'Y',12X,
     1'SIGMA X',10X,'TAU XY',9X,'SIGMA Y')
      DO 25 K=1,L
   25 WRITE(IPR,450) CX(K),CY(K),SSOL(3*K-2),SSOL(3*K-1),SSOL(3*K)
  400 FORMAT(4(2X,E14.7))
  450 FORMAT(5(2X,E14.7))
   30 WRITE(IPR,500)
  500 FORMAT(' ',79('*'))
      RETURN
      END
```

Example 4.2

The circular cavity under internal pressure described in Example 4.1 is now studied using 12 quadratic elements. This discretization gives the same number of nodes as the one which used 24 constant elements and results can then be easily compared.

The input for the ELQUABE code is as follows

CIRCULAR CAVITY (DATA)

```
  CIRCULAR CAVITY UNDER INTERNAL PRESSURE  (12 QUADRATIC ELEMENTS)
12 5 94500. .1
-0.76985 -2.87295
-1.4872  -2.57575
-2.1032 -2.1032
-2.57585 -1.4872
-2.87295 -0.76985
-2.9745 0.
-2.87295 0.76985
-2.57585  1.4872
-2.1032  2.1032
-1.4872  2.57585
-0.76985  2.87295
0.       2.9743
```

```
0.76985  2.87295
1.4872   2.57575
2.1032  2.1032
2.57585  1.4872
2.87295  0.76985
2.9745 0.
2.87295 -0.76985
2.57585   -1.4872
2.1032  -2.1032
1.4872  -2.57585
0.76985   -2.87295
0.      -2.9743
1   -25.88 1 -96.59  1  -50.   1 -86.6 1   -70.71 1 -70.71
1   -70.71 1 -70.71  1  -86.6 1 -50.   1   -96.59 1 -25.88
1   -96.59 1 -25.88  1  -100. 1  0.    1   -96.59 1  25.88
1   -96.59 1  25.88  1  -86.6 1  50.   1   -70.71 1  70.71
1   -70.71 1  70.71  1  -50.   1  86.6 1   -25.88 1  96.59
1   -25.88 1  96.59  0    0.   1  100. 1 25.88 1  96.59
1 25.88 1   96.59  1 50.   1   86.6 1 70.71 1   70.71
1 70.71 1   70.71  1 86.6 1  50.   1 96.59 1  25.88
1 96.59 1  25.88  1 100. 0  0.     1 96.59 1 -25.88
1 96.59 1 -25.88  1 86.6 1 -50.   1 70.71 1 -70.71
1 70.71 1 -70.71  1 50.   1 -86.6 1 25.88 1 -96.59
1 25.88 1 -96.59  0    0.   1 -100. 1   -25.88 1 -96.59
4. 0. 2.82843 2.82843 -4. 0. 6. 0. 10. 0.
```

The output given below also presents symmetry of results, same absolute values (with different sign) for hoop and radial stresses at internal points and exponential decay of both displacements and stresses with distance from the cavity.

The comparison is as follows.

Table 4.2 12 Quadratic Elements Discretization

Distance from centre of cavity	Elasticity theory	Constant elements		Quadratic elements	
		Stress	% Error	Stress	% Error
4	−56.25	−57.234	1.75%	−55.275	1.73%
6	−25.00	−25.295	1.18%	−24.565	1.74%
10	−9.00	−9.106	1.18%	−8.844	1.73%

(Notice that only internal points with comparatively large stresses are taken into consideration as the stresses are too low for points further away from the cavity to give a meaningful error estimate.)

It is interesting to see that in this case the use of quadratic elements does not appreciably alter the error in the results on the contrary, they are overall less accurate for the same number of degrees of freedom. This is due to the problem being rather of a special character. As we will see in what follows, although constant elements behave well in problems such as this they give poor results when used to model problems with bending.

CIRCULAR CAVITY (OUTPUT)

```
******************************************************************************
 CIRCULAR CAVITY UNDER INTERNAL PRESSURE  (12 QUADRATIC ELEMENTS)

 DATA
 NUMBER OF BOUNDARY ELEMENTS= 12
 NUMBER OF INTERNAL POINTS=  5
 SHEAR MODULUS= 0.9450000E+05
 POISSON RATIO= 0.1000000E+00
```

BOUNDARY NODES COORDINATES

NODE	X	Y
1	-0.7698500E+00	-0.2872950E+01
2	-0.1487200E+01	-0.2575750E+01
3	-0.2103200E+01	-0.2103200E+01
4	-0.2575850E+01	-0.1487200E+01
5	-0.2872950E+01	-0.7698500E+00
6	-0.2974500E+01	0.0000000E+00
7	-0.2872950E+01	0.7698500E+00
8	-0.2575850E+01	0.1487200E+01
9	-0.2103200E+01	0.2103200E+01
10	-0.1487200E+01	0.2575850E+01
11	-0.7698500E+00	0.2872950E+01
12	0.0000000E+00	0.2974300E+01
13	0.7698500E+00	0.2872950E+01
14	0.1487200E+01	0.2575750E+01
15	0.2103200E+01	0.2103200E+01
16	0.2575850E+01	0.1487200E+01
17	0.2872950E+01	0.7698500E+00
18	0.2974500E+01	0.0000000E+00
19	0.2872950E+01	-0.7698500E+00
20	0.2575850E+01	-0.1487200E+01
21	0.2103200E+01	-0.2103200E+01
22	0.1487200E+01	-0.2575850E+01
23	0.7698500E+00	-0.2872950E+01
24	0.0000000E+00	-0.2974300E+01

BOUNDARY CONDITIONS

	PRESCRIBED VALUES											
	-----FIRST NODE------				-----SECOND NODE-----				-----THIRD NODE------			
ELE	X DIR.	C	Y DIR.	C	X DIR.	C	Y DIR.	C	X DIR.	C	Y DIR.	C
1	-25.880	1	-96.590	1	-50.000	1	-86.600	1	-70.710	1	-70.710	1
2	-70.710	1	-70.710	1	-86.600	1	-50.000	1	-96.590	1	-25.880	1
3	-96.590	1	-25.880	1	-100.000	1	0.000	1	-96.590	1	25.880	1
4	-96.590	1	25.880	1	-86.600	1	50.000	1	-70.710	1	70.710	1
5	-70.710	1	70.710	1	-50.000	1	86.600	1	-25.880	1	96.590	1
6	-25.880	1	96.590	1	0.000	0	100.000	1	25.880	1	96.590	1
7	25.880	1	96.590	1	50.000	1	86.600	1	70.710	1	70.710	1
8	70.710	1	70.710	1	86.600	1	50.000	1	96.590	1	25.880	1
9	96.590	1	25.880	1	100.000	1	0.000	0	96.590	1	-25.880	1
10	96.590	1	-25.880	1	86.600	1	-50.000	1	70.710	1	-70.710	1
11	70.710	1	-70.710	1	50.000	1	-86.600	1	25.880	1	-96.590	1
12	25.880	1	-96.590	1	0.000	0	-100.000	1	-25.880	1	-96.590	1

****

RESULTS

BOUNDARY NODES

X	Y	DISPLACEMENT X	DISPLACEMENT Y
-0.7698500E+00	-0.2872950E+01	-0.4072179E-03	-0.1519695E-02
-0.1487200E+01	-0.2575750E+01	-0.7866786E-03	-0.1362574E-02
-0.2103200E+01	-0.2103200E+01	-0.1112431E-02	-0.1112502E-02
-0.2575850E+01	-0.1487200E+01	-0.1362535E-02	-0.7866730E-03
-0.2872950E+01	-0.7698500E+00	-0.1519704E-02	-0.4072844E-03
-0.2974500E+01	0.0000000E+00	-0.1573262E-02	0.1928129E-08
-0.2872950E+01	0.7698500E+00	-0.1519712E-02	0.4072882E-03
-0.2575850E+01	0.1487200E+01	-0.1362557E-02	0.7866755E-03
-0.2103200E+01	0.2103200E+01	-0.1112482E-02	0.1112481E-02
-0.1487200E+01	0.2575850E+01	-0.7866530E-03	0.1362540E-02
-0.7698500E+00	0.2872950E+01	-0.4071935E-03	0.1519707E-02
0.0000000E+00	0.2974300E+01	0.0000000E+00	0.1573358E-02
0.7698500E+00	0.2872950E+01	0.4072177E-03	0.1519692E-02
0.1487200E+01	0.2575750E+01	0.7866795E-03	0.1362572E-02
0.2103200E+01	0.2103200E+01	0.1112433E-02	0.1112499E-02
0.2575850E+01	0.1487200E+01	0.1362536E-02	0.7866717E-03
0.2872950E+01	0.7698500E+00	0.1519705E-02	0.4072846E-03

```
0.2974500E+01     0.0000000E+00     0.1573262E-02     0.0000000E+00
0.2872950E+01    -0.7698500E+00     0.1519712E-02    -0.4072880E-03
0.2575850E+01    -0.1487200E+01     0.1362557E-02    -0.7866766E-03
0.2103200E+01    -0.2103200E+01     0.1112481E-02    -0.1112482E-02
0.1487200E+01    -0.2575850E+01     0.7866526E-03    -0.1362543E-02
0.7698500E+00    -0.2872950E+01     0.4071933E-03    -0.1519709E-02
0.0000000E+00    -0.2974300E+01     0.0000000E+00    -0.1573361E-02
```

```
                                TRACTION X   TRACTION Y   TRACTION X   TRACTION Y
      X            Y           BEFORE NODE  BEFORE NODE  AFTER NODE   AFTER NODE

-0.76985E+00 -0.28730E+01 -0.25880E+02 -0.96590E+02 -0.25880E+02 -0.96590E+02
-0.14872E+01 -0.25758E+01 -0.50000E+02 -0.86600E+02 -0.50000E+02 -0.86600E+02
-0.21032E+01 -0.21032E+01 -0.70710E+02 -0.70710E+02 -0.70710E+02 -0.70710E+02
-0.25759E+01 -0.14872E+01 -0.86600E+02 -0.50000E+02 -0.86600E+02 -0.50000E+02
-0.28730E+01 -0.76985E+00 -0.96590E+02 -0.25880E+02 -0.96590E+02 -0.25880E+02
-0.29745E+01  0.00000E+00 -0.10000E+03  0.00000E+00 -0.10000E+03  0.00000E+00
-0.28730E+01  0.76985E+00 -0.96590E+02  0.25880E+02 -0.96590E+02  0.25880E+02
-0.25759E+01  0.14872E+01 -0.86600E+02  0.50000E+02 -0.86600E+02  0.50000E+02
-0.21032E+01  0.21032E+01 -0.70710E+02  0.70710E+02 -0.70710E+02  0.70710E+02
-0.14872E+01  0.25759E+01 -0.50000E+02  0.86600E+02 -0.50000E+02  0.86600E+02
-0.76985E+00  0.28730E+01 -0.25880E+02  0.96590E+02 -0.25880E+02  0.96590E+02
 0.00000E+00  0.29743E+01 -0.32454E-03  0.10000E+03 -0.32454E-03  0.10000E+03
 0.76985E+00  0.28730E+01  0.25880E+02  0.96590E+02  0.25880E+02  0.96590E+02
 0.14872E+01  0.25758E+01  0.50000E+02  0.86600E+02  0.50000E+02  0.86600E+02
 0.21032E+01  0.21032E+01  0.70710E+02  0.70710E+02  0.70710E+02  0.70710E+02
 0.25759E+01  0.14872E+01  0.86600E+02  0.50000E+02  0.86600E+02  0.50000E+02
 0.28730E+01  0.76985E+00  0.96590E+02  0.25880E+02  0.96590E+02  0.25880E+02
 0.29745E+01  0.00000E+00  0.10000E+03  0.61401E-03  0.10000E+03  0.61401E-03
 0.28730E+01 -0.76985E+00  0.96590E+02 -0.25880E+02  0.96590E+02 -0.25880E+02
 0.25759E+01 -0.14872E+01  0.86600E+02 -0.50000E+02  0.86600E+02 -0.50000E+02
 0.21032E+01 -0.21032E+01  0.70710E+02 -0.70710E+02  0.70710E+02 -0.70710E+02
 0.14872E+01 -0.25759E+01  0.50000E+02 -0.86600E+02  0.50000E+02 -0.86600E+02
 0.76985E+00 -0.28730E+01  0.25880E+02 -0.96590E+02  0.25880E+02 -0.96590E+02
 0.00000E+00 -0.29743E+01  0.24495E-03 -0.10000E+03  0.24495E-03 -0.10000E+03

 INTERNAL POINTS DISPLACEMENTS

      X                 Y           DISPLACEMENT X   DISPLACEMENT Y
  0.4000000E+01     0.0000000E+00     0.1169808E-02   -0.4138201E-10
  0.2828430E+01     0.2828430E+01     0.8271520E-03    0.8271745E-03
 -0.4000000E+01     0.0000000E+00    -0.1169808E-02   -0.1176431E-08
  0.6000000E+01     0.0000000E+00     0.7798681E-03   -0.2637535E-10
  0.1000000E+02     0.0000000E+00     0.4679200E-03   -0.9115411E-09

 INTERNAL POINTS STRESSES

      X                 Y              SIGMA X             TAU XY            SIGMA Y
  0.4000000E+01     0.0000000E+00    -0.5527559E+02    -0.2949312E-03     0.5527643E+
  0.2828430E+01     0.2828430E+01     0.2465338E-02    -0.5527016E+02     0.3696308E-
 -0.4000000E+01     0.0000000E+00    -0.5527559E+02    -0.2649724E-03     0.5527644E+
  0.6000000E+01     0.0000000E+00    -0.2456595E+02    -0.1923647E-03     0.2456613E+
  0.1000000E+02     0.0000000E+00    -0.8843707E+01    -0.5847029E-04     0.8843761E+
******************************************************************************
```

Example 4.3

The following example demonstrates the use of quadratic elements for a simple problem which because of flexure, can not be solved accurately using constant elements.

Figure 4.9 shows a rectangular plate under a linear distribution of tractions in the horizontal direction, which represents two applied moments. The plate is considered to be in plane stress with shear modulus $\mu = G = 80{,}000$ MPa and Poisson's ratio $\nu = 0.25$. The boundary has been discretized into only 6 quadratic

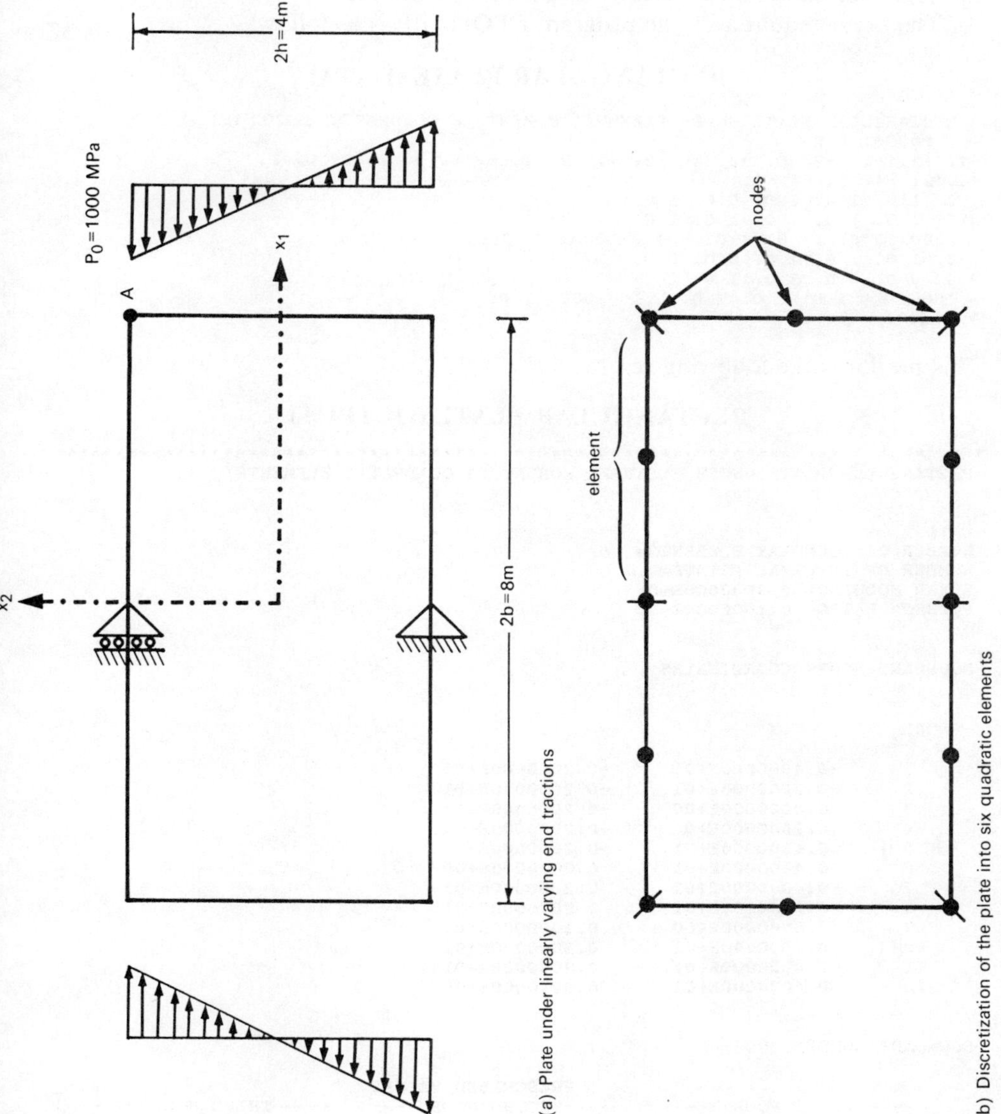

(a) Plate under linearly varying end tractions

(b) Discretization of the plate into six quadratic elements

Figure 4.9 Plate stretching elements under flexion

elements with tractions prescribed along all elements (note tractions in horizontal sides are prescribed as zero in the two directions), with the exception of the nodes located on the vertical axes where the displacements can be prescribed as indicated in the figure. These displacements are the minimum needed to avoid rigid body motions and make the problem well posed for solution.

The input required by the program ELQUABE is as follows:

RECTANGULAR PLATE (DATA)

```
  RECTANGULAR PLATE UNDER FLEXURAL MOMENT (6 QUADRATIC ELEMENTS)
6 1 80000. 0.2
-4. -2. -2. -2. 0. -2. 2. -2. 4. -2. 4. 0. 4. 2. 2. 2. 0. 2.
-2. 2. -4. 2. -4. 0.
1 0. 1 0. 1 0. 1 0. 0 0. 1 0.
0 0. 1 0. 1 0. 1 0. 1 0. 1 0.
1 1000. 1 0. 1  0. 1 0.   1  -1000.    1 0.
1 0. 1 0. 1 0. 1 0. 0 0. 0 0.
0 0. 0 0. 1 0. 1 0. 1 0. 1 0.
1 1000.     1 0. 1 0. 1 0.   1 -1000. 1 0.
0. 0.
```

This produces the following results:

RECTANGULAR PLATE (OUTPUT)

```
*****************************************************************************
 RECTANGULAR PLATE UNDER FLEXURAL MOMENT (6 QUADRATIC ELEMENTS)


 DATA
 NUMBER OF BOUNDARY ELEMENTS=  6
 NUMBER OF INTERNAL POINTS=  1
 SHEAR MODULUS= 0.8000000E+05
 POISSON RATIO= 0.2000000E+00

 BOUNDARY NODES COORDINATES


   NODE          X                  Y

     1      -0.4000000E+01     -0.2000000E+01
     2      -0.2000000E+01     -0.2000000E+01
     3       0.0000000E+00     -0.2000000E+01
     4       0.2000000E+01     -0.2000000E+01
     5       0.4000000E+01     -0.2000000E+01
     6       0.4000000E+01      0.0000000E+00
     7       0.4000000E+01      0.2000000E+01
     8       0.2000000E+01      0.2000000E+01
     9       0.0000000E+00      0.2000000E+01
    10      -0.2000000E+01      0.2000000E+01
    11      -0.4000000E+01      0.2000000E+01
    12      -0.4000000E+01      0.0000000E+00

 BOUNDARY CONDITIONS

                                  PRESCRIBED VALUES
     -----FIRST NODE------    -----SECOND NODE-----    -----THIRD NODE------
ELE   X DIR.  C   Y DIR.  C   X DIR.  C   Y DIR.  C   X DIR.  C   Y DIR.  C

  1    0.000  1    0.000  1    0.000  1    0.000  1    0.000  0    0.000  1
  2    0.000  0    0.000  1    0.000  1    0.000  1    0.000  1    0.000  1
  3 1000.000  1    0.000  1    0.000  1    0.000  1-1000.000  1    0.000  1
  4    0.000  1    0.000  1    0.000  1    0.000  1    0.000  0    0.000  0
  5    0.000  0    0.000  0    0.000  1    0.000  1    0.000  1    0.000  1
  6 1000.000  1    0.000  1    0.000  1    0.000  1-1000.000  1    0.000  1
*****************************************************************************
```

```
RESULTS
 BOUNDARY NODES
        X                  Y                DISPLACEMENT X      DISPLACEMENT Y
   -0.4000000E+01     -0.2000000E+01     -0.1999997E-01      0.1999998E-01
   -0.2000000E+01     -0.2000000E+01     -0.9999990E-02      0.5000000E-02
    0.0000000E+00     -0.2000000E+01      0.0000000E+00     -0.2648449E-08
    0.2000000E+01     -0.2000000E+01      0.1000000E-01      0.4999989E-02
    0.4000000E+01     -0.2000000E+01      0.2000001E-01      0.1999998E-01
    0.4000000E+01      0.0000000E+00      0.1396984E-07      0.1874997E-01
    0.4000000E+01      0.2000000E+01     -0.1999997E-01      0.1999997E-01
    0.2000000E+01      0.2000000E+01     -0.9999976E-02      0.4999987E-02
    0.0000000E+00      0.2000000E+01      0.0000000E+00      0.0000000E+00
   -0.2000000E+01      0.2000000E+01      0.9999998E-02      0.4999997E-02
   -0.4000000E+01      0.2000000E+01      0.2000002E-01      0.2000000E-01
   -0.4000000E+01      0.0000000E+00     -0.2517481E-08      0.1875000E-01

                                  TRACTION X   TRACTION Y   TRACTION X   TRACTION Y
      X             Y            BEFORE NODE  BEFORE NODE  AFTER NODE   AFTER NODE
-0.40000E+01 -0.20000E+01 -0.10000E+04  0.00000E+00  0.00000E+00  0.00000E+00
-0.20000E+01 -0.20000E+01  0.00000E+00  0.00000E+00  0.00000E+00  0.00000E+00
 0.00000E+00 -0.20000E+01 -0.10431E-02  0.00000E+00 -0.10431E-02  0.00000E+00
 0.20000E+01 -0.20000E+01  0.00000E+00  0.00000E+00  0.00000E+00  0.00000E+00
 0.40000E+01 -0.20000E+01  0.00000E+00  0.00000E+00  0.10000E+04  0.00000E+00
 0.40000E+01  0.00000E+00  0.00000E+00  0.00000E+00  0.00000E+00  0.00000E+00
 0.40000E+01  0.20000E+01 -0.10000E+04  0.00000E+00  0.00000E+00  0.00000E+00
 0.20000E+01  0.20000E+01  0.00000E+00  0.00000E+00  0.00000E+00  0.00000E+00
 0.00000E+00  0.20000E+01 -0.18626E-03 -0.18626E-03 -0.18626E-03 -0.18626E-03
-0.20000E+01  0.20000E+01  0.00000E+00  0.00000E+00  0.00000E+00  0.00000E+00
-0.40000E+01  0.20000E+01  0.00000E+00  0.00000E+00  0.10000E+04  0.00000E+00
-0.40000E+01  0.00000E+00  0.00000E+00  0.00000E+00  0.00000E+00  0.00000E+00

 INTERNAL POINTS DISPLACEMENTS

      X                Y             DISPLACEMENT X  DISPLACEMENT Y
  0.0000000E+00    0.0000000E+00    0.1111766E-07  -0.1250002E-02

 INTERNAL POINTS STRESSES

      X                Y                SIGMA X          TAU XY          SIGMA Y
  0.0000000E+00    0.0000000E+00    0.1640320E-03    0.6055832E-04    0.6484985E-
**************************************************************************************
```

Results for u_1 and u_2 displacements at corner A (figure 4.9(a)) are summarized in Table 4.3, where they are compared against the analytical solution and the values obtained using the program ELCONBE with 50 constant elements. It can be seen that while the accuracy of the quadratic elements is excellent, poor results are obtained using constant elements. This demonstrates that constant elements are not to be recommended for problems with flexural strains, although they give good results in other applications.

Table 4.3 Displacements at Corner A (figure 4.9)

Displacement	Analytical solution	6 Quadratic element solution		50 Constant elements	
		Values	Error (%)	Values	Error (%)
u_1	−0.02	−0.02	0%	−0.0157	21.5%
u_2	0.02	0.02	0%	+0.0165	17.5%

(All results in m.)

Example 4.4

Another flexural problem which can be solved accurately using quadratic elements is the cantilever beam shown in figure 4.10. The beam has been discretized using 12 quadratic elements and is under a transverse parabolic load distribution at its two ends. On the end on the left hand side one also considers horizontal displacement constraints as shown in the figure, with the lower corner node displacements

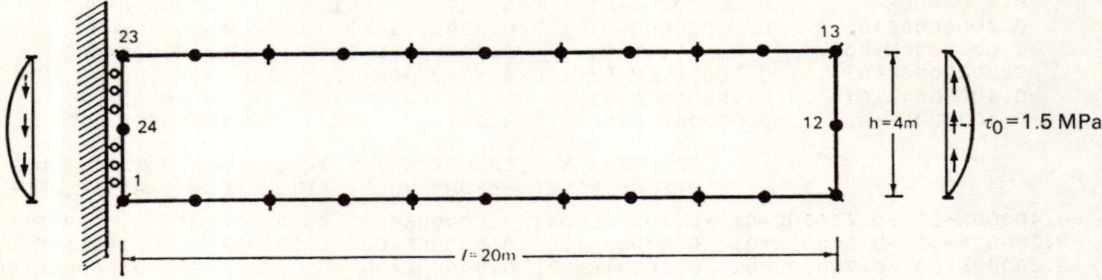

Figure 4.10 Cantilever beam under end loading

set both equal to zero. The problem is in plane stress and the material constants are assumed to be the same as those in Example 4.3.

The beam theory gives for this problem a free end deflection of,

$$u_2 = \frac{Pl^3}{3EI} = 10 \text{ mm} \tag{a}$$

where P is the total load applied at the end, l is the length of the beam, E its modulus of elasticity and I the moments of inertia of the beam.

The stress distributions along the fixed part of the boundary is linear in accordance with the beam theory and equal to

$$\sigma = \frac{Mh}{2I} = 30 \text{ MPa} \tag{b}$$

Boundary Element Values of vertical displacements at nodes 11–12–13 at one end and normal tractions at nodes 23–24–1 at the other end are given in Table 4.4. The results agree well with those of formulae (a) and (b).

Table 4.4 Vertical Displacements and Tractions for the Cantilever Beam of figure 4.10

Node	Vertical displacement u_2	Normal tractions σ_n
11	10.166	—
12	10.169	—
13	10.166	—
23	—	29.85
24	—	−0.00
1	—	−29.85

The input for this problem to run program ELQUABE is as follows:

CANTILEVER BEAM (DATA)

```
   CANTILEVER BEAM UNDER TRANSVERSAL END LOAD (12 QUADRATIC ELEMENTS)
 12 1  80000. 0.2
 0. 0. 2. 0. 4. 0. 6. 0. 8. 0. 10. 0. 12. 0. 14. 0. 16. 0. 18. 0. 20. 0.
 20. 2. 20. 4. 18. 4. 16. 4. 14. 4. 12. 4. 10. 4. 8. 4. 6. 4. 4. 4.
 2. 4. 0. 4. 0. 2.
 1 0. 0 0. 1 0. 1 0. 1 0. 1 0.
 1 0. 1 0. 1 0. 1 0. 1 0. 1 0.
 1 0. 1 0. 1 0. 1 0. 1 0. 1 0.
 1 0. 1 0. 1 0. 1 0. 1 0. 1 0.
 1 0. 1 0. 1 0. 1 0. 1 0. 1 0.
 1 0. 1 0.    1 0. 1 1500. 1 0. 1 0.
 1 0. 1 0. 1 0. 1 0. 1 0. 1 0.
 1 0. 1 0. 1 0. 1 0. 1 0. 1 0.
 1 0. 1 0. 1 0. 1 0. 1 0. 1 0.
 1 0. 1 0. 1 0. 1 0. 1 0. 1 0.
 1 0. 1 0. 1 0. 1 0. 1 0. 1 0.
 0 0. 1 0.    0 0. 1 -1500. 0 0. 1 0.
10 2.
```

The corresponding output is listed below,

CANTILEVER BEAM (OUTPUT)

```
*****************************************************************************
 CANTILEVER BEAM UNDER TRANSVERSAL END LOAD (12 QUADRATIC ELEMENTS)
DATA
NUMBER OF BOUNDARY ELEMENTS= 12
NUMBER OF INTERNAL POINTS=  1
SHEAR MODULUS= 0.8000000E+05
POISSON RATIO= 0.2000000E+00

BOUNDARY NODES COORDINATES


    NODE           X                  Y

       1      0.0000000E+00      0.0000000E+00
       2      0.2000000E+01      0.0000000E+00
       3      0.4000000E+01      0.0000000E+00
       4      0.6000000E+01      0.0000000E+00
       5      0.8000000E+01      0.0000000E+00
       6      0.1000000E+02      0.0000000E+00
       7      0.1200000E+02      0.0000000E+00
       8      0.1400000E+02      0.0000000E+00
       9      0.1600000E+02      0.0000000E+00
      10      0.1800000E+02      0.0000000E+00
      11      0.2000000E+02      0.0000000E+00
      12      0.2000000E+02      0.2000000E+01
      13      0.2000000E+02      0.4000000E+01
      14      0.1800000E+02      0.4000000E+01
      15      0.1600000E+02      0.4000000E+01
      16      0.1400000E+02      0.4000000E+01
      17      0.1200000E+02      0.4000000E+01
      18      0.1000000E+02      0.4000000E+01
      19      0.8000000E+01      0.4000000E+01
      20      0.6000000E+01      0.4000000E+01
      21      0.4000000E+01      0.4000000E+01
      22      0.2000000E+01      0.4000000E+01
      23      0.0000000E+00      0.4000000E+01
      24      0.0000000E+00      0.2000000E+01
```

BOUNDARY CONDITIONS

	PRESCRIBED VALUES											
	-----FIRST NODE------				-----SECOND NODE-----				-----THIRD NODE------			
ELE	X DIR.	C	Y DIR.	C	X DIR.	C	Y DIR.	C	X DIR.	C	Y DIR.	C
1	0.000	1	0.000	0	0.000	1	0.000	1	0.000	1	0.000	1
2	0.000	1	0.000	1	0.000	1	0.000	1	0.000	1	0.000	1
3	0.000	1	0.000	1	0.000	1	0.000	1	0.000	1	0.000	1
4	0.000	1	0.000	1	0.000	1	0.000	1	0.000	1	0.000	1
5	0.000	1	0.000	1	0.000	1	0.000	1	0.000	1	0.000	1
6	0.000	1	0.000	1	0.000	1	1500.000	1	0.000	1	0.000	1
7	0.000	1	0.000	1	0.000	1	0.000	1	0.000	1	0.000	1
8	0.000	1	0.000	1	0.000	1	0.000	1	0.000	1	0.000	1
9	0.000	1	0.000	1	0.000	1	0.000	1	0.000	1	0.000	1
10	0.000	1	0.000	1	0.000	1	0.000	1	0.000	1	0.000	1
11	0.000	1	0.000	1	0.000	1	0.000	1	0.000	1	0.000	1
12	0.000	0	0.000	1	0.000	0	-1500.000	1	0.000	0	0.000	1

****

RESULTS

BOUNDARY NODES

X	Y	DISPLACEMENT X	DISPLACEMENT Y
0.0000000E+00	0.0000000E+00	0.0000000E+00	0.0000000E+00
0.2000000E+01	0.0000000E+00	0.2831624E+00	0.1641325E+00
0.4000000E+01	0.0000000E+00	0.5382484E+00	0.5995840E+00
0.6000000E+01	0.0000000E+00	0.7601910E+00	0.1273359E+01
0.8000000E+01	0.0000000E+00	0.9555768E+00	0.2155845E+01
0.1000000E+02	0.0000000E+00	0.1118159E+01	0.3217330E+01
0.1200000E+02	0.0000000E+00	0.1254107E+01	0.4428082E+01
0.1400000E+02	0.0000000E+00	0.1357250E+01	0.5758359E+01
0.1600000E+02	0.0000000E+00	0.1433837E+01	0.7178458E+01
0.1800000E+02	0.0000000E+00	0.1477282E+01	0.8658765E+01
0.2000000E+02	0.0000000E+00	0.1492648E+01	0.1016628E+02
0.2000000E+02	0.2000000E+01	0.1201406E-05	0.1016907E+02
0.2000000E+02	0.4000000E+01	-0.1492655E+01	0.1016628E+02
0.1800000E+02	0.4000000E+01	-0.1477283E+01	0.8658760E+01
0.1600000E+02	0.4000000E+01	-0.1433836E+01	0.7178457E+01
0.1400000E+02	0.4000000E+01	-0.1357249E+01	0.5758362E+01
0.1200000E+02	0.4000000E+01	-0.1254107E+01	0.4428082E+01
0.1000000E+02	0.4000000E+01	-0.1118160E+01	0.3217330E+01
0.8000000E+01	0.4000000E+01	-0.9555784E+00	0.2155848E+01
0.6000000E+01	0.4000000E+01	-0.7601921E+00	0.1273359E+01
0.4000000E+01	0.4000000E+01	-0.5382496E+00	0.5995839E+00
0.2000000E+01	0.4000000E+01	-0.2831629E+00	0.1641333E+00
0.0000000E+00	0.4000000E+01	0.0000000E+00	-0.2929009E-06
0.0000000E+00	0.2000000E+01	0.0000000E+00	-0.4010320E-01

X	Y	TRACTION X BEFORE NODE	TRACTION Y BEFORE NODE	TRACTION X AFTER NODE	TRACTION Y AFTER NODE
0.00000E+00	0.00000E+00	-0.29853E+05	0.00000E+00	0.00000E+00	0.39985E-01
0.20000E+01	0.00000E+00	0.00000E+00	0.00000E+00	0.00000E+00	0.00000E+00
0.40000E+01	0.00000E+00	0.00000E+00	0.00000E+00	0.00000E+00	0.00000E+00
0.60000E+01	0.00000E+00	0.00000E+00	0.00000E+00	0.00000E+00	0.00000E+00
0.80000E+01	0.00000E+00	0.00000E+00	0.00000E+00	0.00000E+00	0.00000E+00
0.10000E+02	0.00000E+00	0.00000E+00	0.00000E+00	0.00000E+00	0.00000E+00
0.12000E+02	0.00000E+00	0.00000E+00	0.00000E+00	0.00000E+00	0.00000E+00
0.14000E+02	0.00000E+00	0.00000E+00	0.00000E+00	0.00000E+00	0.00000E+00
0.16000E+02	0.00000E+00	0.00000E+00	0.00000E+00	0.00000E+00	0.00000E+00
0.18000E+02	0.00000E+00	0.00000E+00	0.00000E+00	0.00000E+00	0.00000E+00
0.20000E+02	0.00000E+00	0.00000E+00	0.00000E+00	0.00000E+00	0.00000E+00
0.20000E+02	0.20000E+01	0.00000E+00	0.15000E+04	0.00000E+00	0.15000E+04
0.20000E+02	0.40000E+01	0.00000E+00	0.00000E+00	0.00000E+00	0.00000E+00
0.18000E+02	0.40000E+01	0.00000E+00	0.00000E+00	0.00000E+00	0.00000E+00
0.16000E+02	0.40000E+01	0.00000E+00	0.00000E+00	0.00000E+00	0.00000E+00
0.14000E+02	0.40000E+01	0.00000E+00	0.00000E+00	0.00000E+00	0.00000E+00
0.12000E+02	0.40000E+01	0.00000E+00	0.00000E+00	0.00000E+00	0.00000E+00
0.10000E+02	0.40000E+01	0.00000E+00	0.00000E+00	0.00000E+00	0.00000E+00

```
0.80000E+01  0.40000E+01  0.00000E+00  0.00000E+00  0.00000E+00  0.00000E+00
0.60000E+01  0.40000E+01  0.00000E+00  0.00000E+00  0.00000E+00  0.00000E+00
0.40000E+01  0.40000E+01  0.00000E+00  0.00000E+00  0.00000E+00  0.00000E+00
0.20000E+01  0.40000E+01  0.00000E+00  0.00000E+00  0.00000E+00  0.00000E+00
0.00000E+00  0.40000E+01  0.00000E+00  0.00000E+00  0.29853E+05  0.00000E+00
0.00000E+00  0.20000E+01  0.44191E-01 -0.15000E+04  0.44191E-01 -0.15000E+04


INTERNAL POINTS DISPLACEMENTS

       X                Y            DISPLACEMENT X   DISPLACEMENT Y
 0.1000000E+02   0.2000000E+01   -0.1154840E-06    0.3198673E+01

 INTERNAL POINTS STRESSES

       X                Y               SIGMA X            TAU XY           SIGMA Y
  0.1000000E+02   0.2000000E+01   0.1745605E-01   0.1438650E+04   -0.1143646E-
*******************************************************************************
```

Example 4.5

This problem represents the case of a plane-strain hollow cylinder under internal pressure as shown in figure 4.11. The pressure is assumed to be $p = 100$ N/mm^2, while the internal and external radii are $r_1 = 10$ mm and $r_2 = 25$ mm, respectively. The elastic constants of the material are; Elasticity Modulus $E = 200{,}000$ N/mm^2 and Poisson's ratio, $v = 0.25$.

Due to the symmetry of the problem only one quarter of the section needs to be discretized. The boundary conditions are shown in figure 4.10(b). Three different discretizations consisting of 4, 10 and 15 quadratic elements are used as shown in figures 4.11(c).

Table 4.5 shows the relevant displacements computed for the points A, B and C along the radius. The displacements are given in microns, i.e. 10^{-3} mm and are compared with the exact values given by the theory of elasticity. It is interesting to notice that even for the coarse discretization, results with 2% of the exact solution are obtained. When 15 elements are employed, the results are within a range of 0.1% of the exact solution.

Table 4.5 Radial Displacements for Hollow Cylinder under Internal Pressure (in 10^{-3} mm)

Node	Exact value	Discretization		
		4 elements	10 elements	15 elements
A	8.0325	7.8781	8.0246	8.0350
B	5.2912	5.1668	5.2845	5.2928
C	4.4526	4.3896	4.4520	4.4631

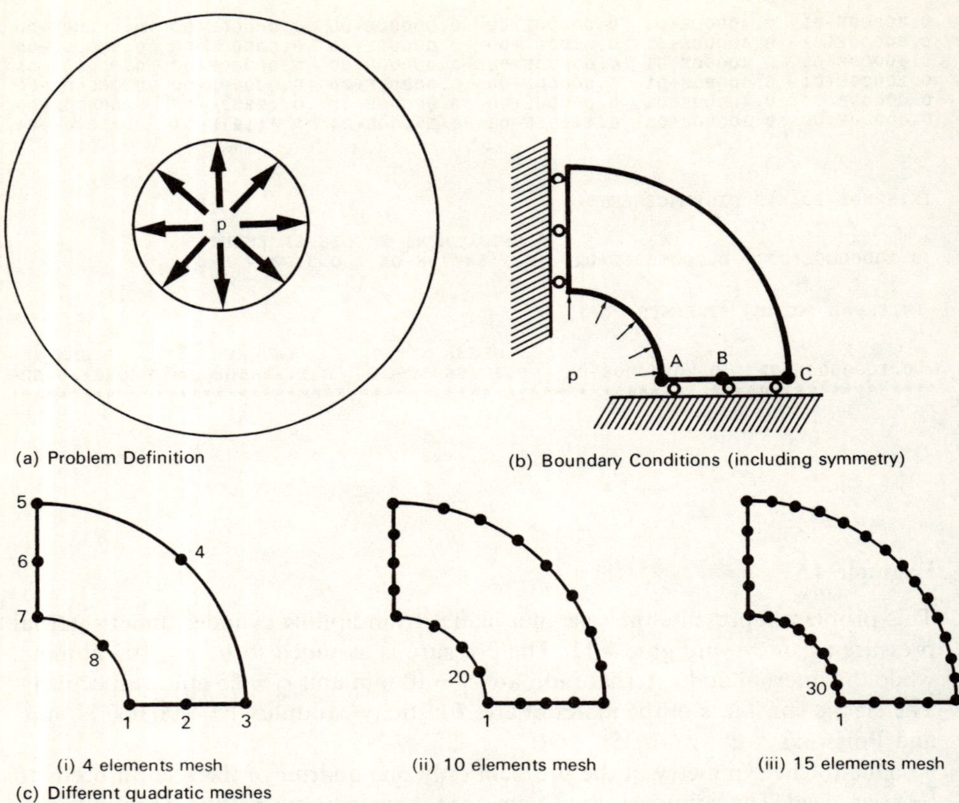

Figure 4.11 Hollow cylinder under internal pressure

Example 4.6

This example studies the stress concentration around a circular hole in a plate stretching specimen as shown in figure 4.12(a). The geometric and material constants are as follows:

$$\frac{l}{r} = 20$$

$$G = 80{,}000 \text{ MPa} \qquad \text{(a)}$$

$$\nu = 0.25$$

Figure 4.12(b) shows the boundary discretization for half the plate using only quadratic elements. Notice that only one element has been assumed over the circular section.

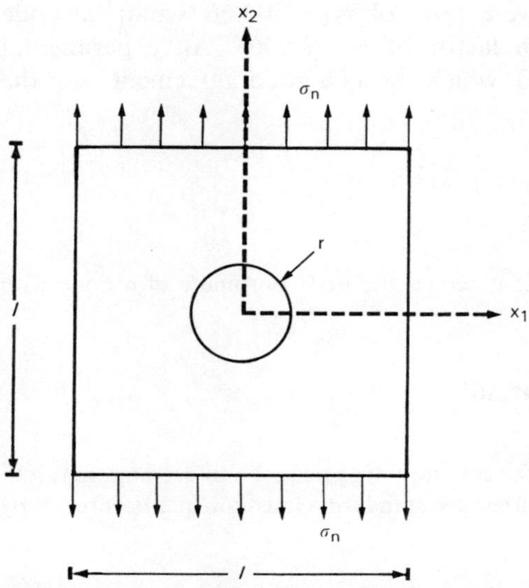

(a) Square plate with hole. Geometry and boundary conditions

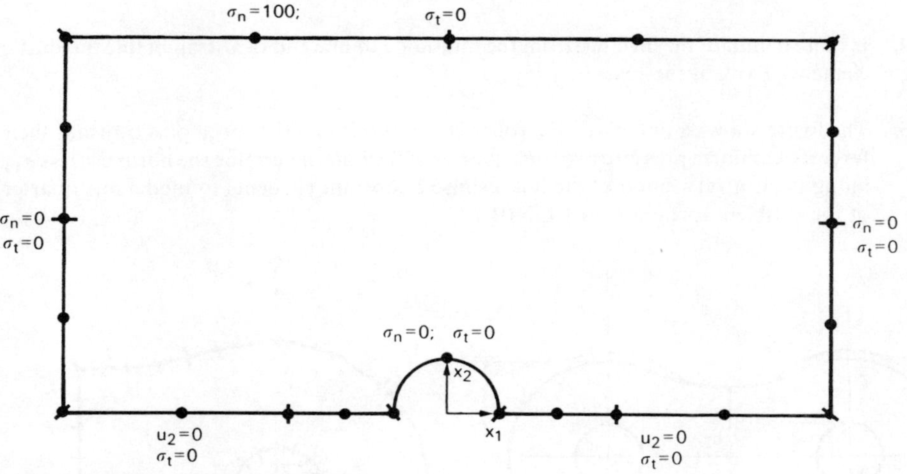

(b) Boundary element discretization

Figure 4.12 Square plate with hole. Geometrical definition and boundary discretization

In order to check the accuracy of the Boundary Element solution the stress concentration factor around the hole is computed. This factor can be defined as,

$$K = \frac{(\sigma_{22})_{\max}}{\sigma_n} \tag{b}$$

where σ_n is the applied stress at the top and bottom end of the plate.

The boundary element solution gave a value of $\sigma_{22} = 303.66$ N/mm^2 at node A, which gives a stress concentration factor of $K = 3.0366$. An experimental solution for this problem gave $K = 3.03$, which shows a good agreement with the Boundary Element solution.

Exercises

4.1. Compute analytically the term G_{11}^{ii} that relates the first coordinate of a node with itself in a linear elements program.

$$G_{11}^{ii} = \int_{\Gamma_{i-1}} \phi_2 u_{11}^* \, d\Gamma + \int_{\Gamma_i} \phi_1 u_{11}^* \, d\Gamma$$

4.2. Do the necessary transformations to the equation of exercise 4.1 to decompose it into integrals which may be computed either by standard Gaussian quadrature or by logarithmic Gaussian quadrature.

4.3. Discuss the effect of shifting the mid-node of a quadratic element from the centre on the numerical integrations (equation (4.56)).

4.4. Is there a limit of the distance from the mid-node to one end of a straight line quadratic element? Explain the reasons.

4.5. The figure shows a link plate of a roller chain. Assuming that both pins transmit their force as a uniform pressure over one quarter of the hole, determine the normal stress σ_{11} along the central section of the link using 32 constant elements to model one quarter of the problem (program ELCONBE).

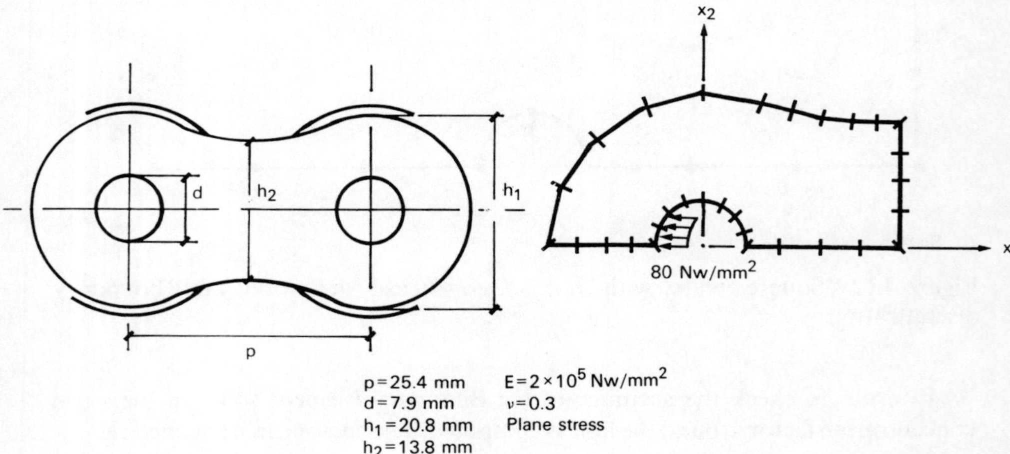

4.6. Solve problem 4.5 using program ELQUABE and the same number of nodes with 16 quadratic elements. Solve the same problem using 32 quadratic elements. Compare the three solutions obtained for the problem. (The DIMENSIONS in the main program ELQUABE must be changed to run the 32 elements example.)

4.7. The link plate of a silent chain is shown in the figure. Using constant elements and the discretization of the figure compute the normal stresses along the central section.

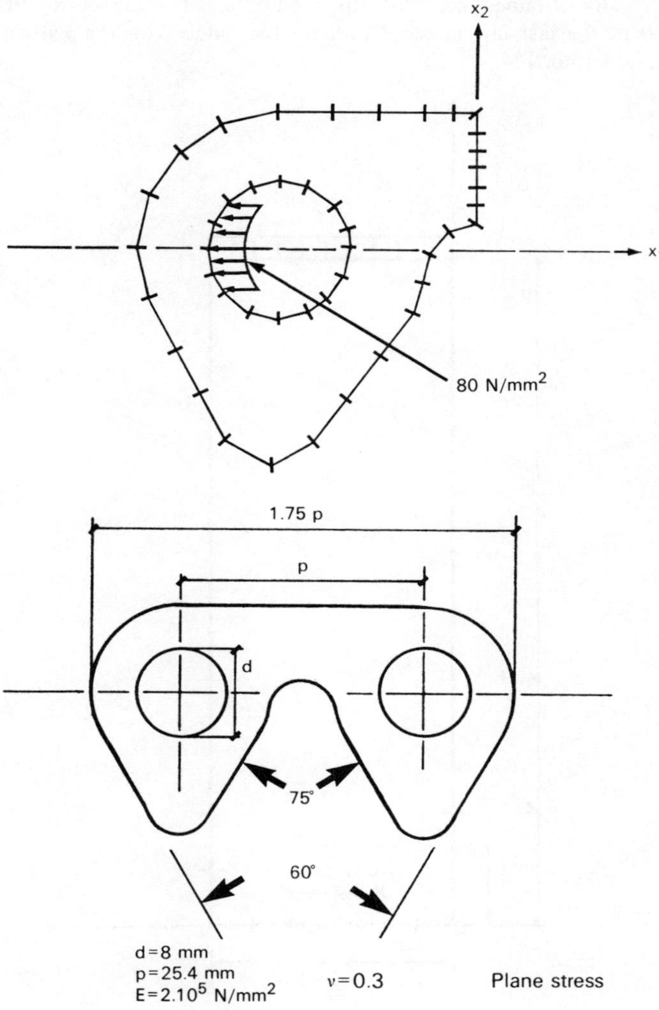

4.8. Introduce the necessary changes in program ELQUABE to solve multiboundary problems.

4.9. Solve problem 4.7 with the same number of nodes but using quadratic elements. Also solve the problem with quadratic elements and double the number of nodes. Compare the two solutions and also that of exercise 4.7.

4.10. The figure shows a centre cracked plate under traction. Because of the symmetry only one quarter of the plate is discretized. The σ_{22} stress near the crack tip for points along the line A–B is known to be of the form $\sigma_{22} = K_I/\sqrt{2\pi r}$, r being the distance to the tip and K_I the mode $-I$ stress intensity factor. Use the discretization (19 quadratic elements) shown in the figure to solve the problem and represent the value of $\sigma_{22}\sqrt{2\pi r}$ along the line A–B.

Comment on the results obtained near the tip. Compute the value of K_I by extrapolation to $r=0$ of the last few nodes. Compare the value with the known solution $K_I = 6.67$, exact within 1%.

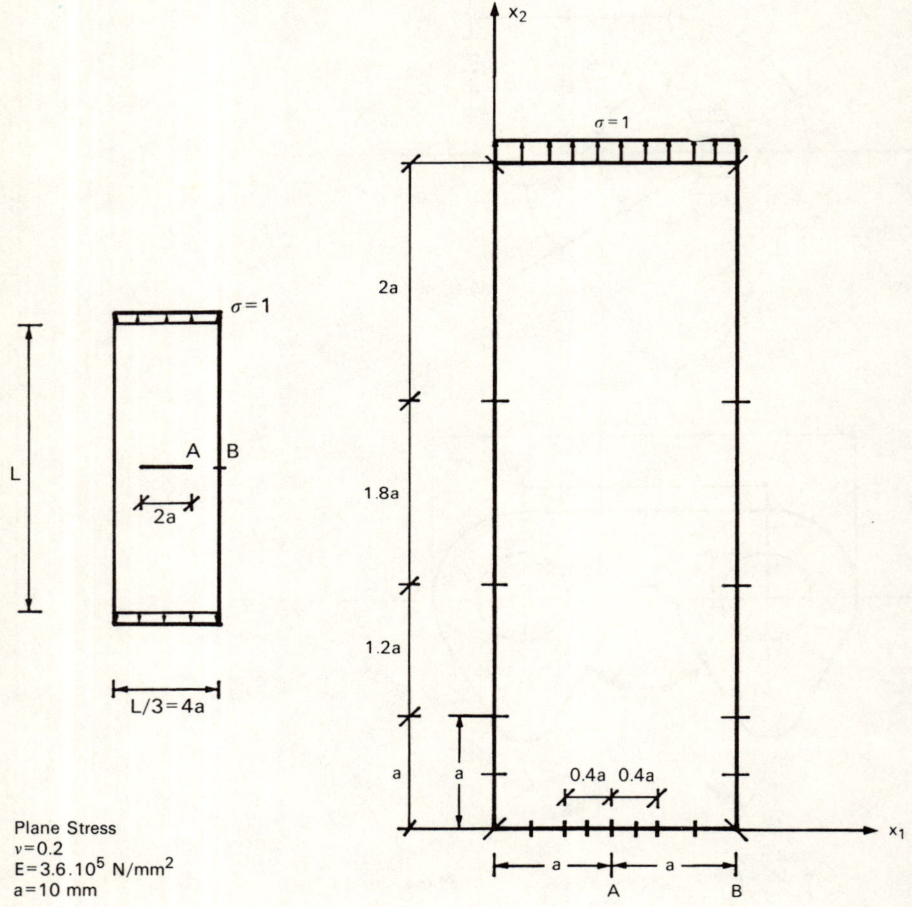

Centre cracked plate under traction and discretization of one quarter of the plate.

Chapter 5

Other Interesting Topics

5.1 Introduction

This chapter comprises a series of topics which are of general interest and extend the range of applications of boundary elements but are not as essential as those described in previous chapters to understand the method.

One of the most interesting possibilities of boundary elements is that it is easy to combine the technique with other numerical methods. The range of combinations varies from those needed in analysis where boundary layers are interfaced to boundary element regions, to simple coupling in static analysis. Sometimes for instance, matrices for potential fluids are formed using boundary elements and combined with finite element models for shells which represent a container, an aerospace structure, an offshore platform, etc. The coupling is particularly simple as boundary elements accepts discontinuity of variables and full compatibility is not required to obtain accurate answers. Ways of combining finite and boundary elements are particularly attractive in view of the widespread use of both methods. Although many papers have been published on the topic, section 5.2 only discusses comparatively simple ways of carrying out this combination. The first technique is a purely intuitive approach which is not justified mathematically. It consists of using the finite element results from a global solution as boundary conditions when focusing on a particular region of the system. The finite element boundary variables used are potentials or displacements as they are given with a higher order of accuracy than fluxes or tractions.

The first of the other two approaches consists of treating the boundary element region as a finite element and appropriately transforming the matrices. The second approach consists of treating the finite element region as a boundary element and manipulating the FE matrices in such a way that they can be implemented in the boundary element system. Both approaches give similar results and using one or the other will depend on which part of the problem (i.e. finite or boundary element parts) is predominant.

Section 5.3 discusses special types of boundary element, which are produced by asymptotic considerations when the boundary is far from the part under perturbation. Under certain hypotheses these elements are equivalent to the radiation type conditions presented by different authors. The section gives a methodology of how these conditions can be obtained from basic boundary integral considerations.

One of the first applications of boundary elements was the study of elastic fracture mechanics problems for which singularities arise at the tip of the crack.

These problems have been solved by different authors using a variety of boundary integral formulations but more recently they have been studied using a simple transformation which produces a singularity at the tip of the crack. This is achieved by using the so-called quarter point elements which were first developed for finite elements. They give a very elegant formulation in the case of boundary elements, which directly produce the stress intensity coefficients in a way that can not be done using finite elements.

The last section in this chapter explains how the technique can be expanded to study steady state elastodynamics problems. Although this is similar to what has been shown in Chapter 2 for the Helmholtz equation (section 2.14) it was decided to include this amongst the special topics as the frequency dependent formulation in elasticity is rather complex.

In spite of that, the implementation of the resulting relationships in existing elastostatics codes (including those presented in Chapter 4 of this book) can be attempted by the reader as explained in section 5.5.

5.2 Combination of Boundary and Finite Elements

There are sometimes advantages in combining finite and boundary element solutions. In many unbounded field problems for instance, boundary elements may provide the appropriate conditions to represent the infinite domain while finite elements can solve complex material properties in the near domain. Boundary elements are also of interest in regions of high stresses or potentials, but finite elements may be adequate for other parts of the boundary and may be simpler to use in cases such as layered continuum, anisotropic and non-linear materials. Hence it is important for the analyst to be able to represent a body using finite or boundary element techniques, depending on the particular geometries or boundary conditions.

There are many papers written on the combination of the two techniques, but from the viewpoint of simplicity of application as it relates to the codes already described here, we will consider only three methods, i.e.

(i) *Method* (*i*) Using the finite element solution to define the boundary conditions for a localized boundary element region.
(ii) *Method* (*ii*) Treating a boundary element region as a finite element and combining with finite elements.
(iii) *Method* (*iii*) The converse of method (ii), i.e. treating a finite element region as an equivalent boundary element and combining with the other boundary element region.

Method (*i*) This approach is a purely empirical technique and consists of having solved a problem using finite elements to 'zoom' in a particular region using as boundary conditions the finite element results for displacements or potentials.

The approach can not easily be justified from a mathematical standpoint but it is used in several codes and seems to produce reasonable results. One can apply this technique in the case of studying a region with a crack as shown in figure 5.1. First a global finite element solution is found using the mesh described in figure 5.1(a) and then the boundary element method is used to study the crack region in more detail as shown in figure 5.1(b) using as boundary conditions the displacements obtained in the finite element code. The reason why the approach works is due to the fact that the finite element results for displacements (or potentials) are usually accurate. The method would not give good results if the finite element stresses (or fluxes) were used instead.

Method (*ii*) The second approach consists in treating the boundary element region as a finite element. Consider the two regions as shown in figure 5.2 where region Ω^1 s expressed in terms of boundary solutions and Ω^2 discretized into finite elements.

The boundary element matrices for Ω^1 can be written as

$$\mathbf{HU} = \mathbf{GP} \tag{5.1}$$

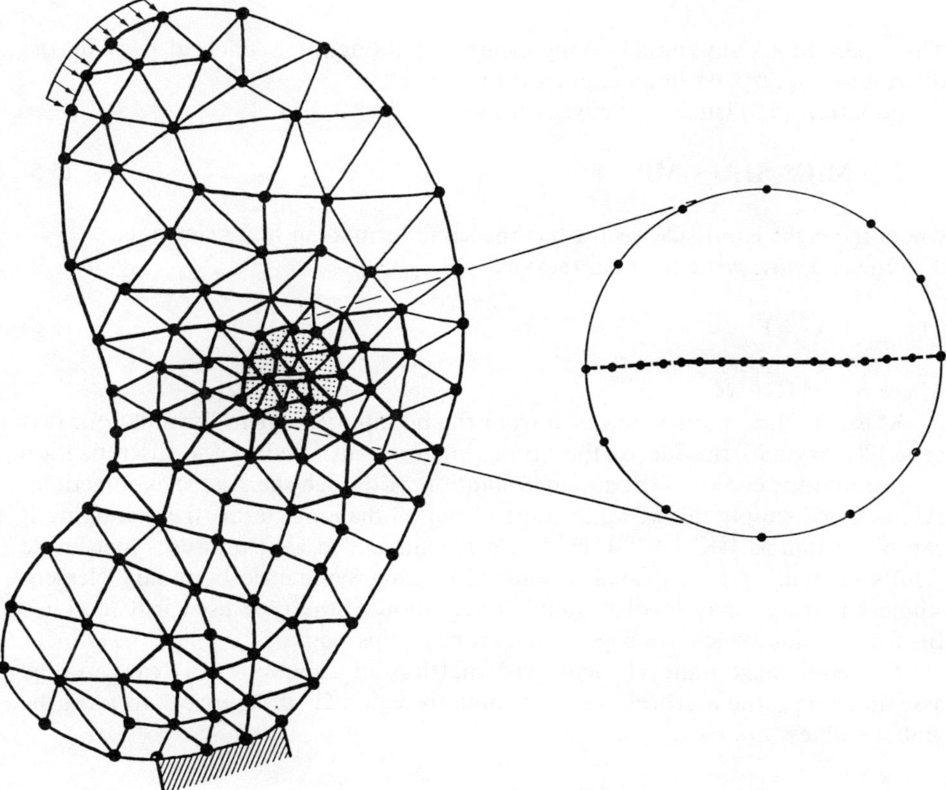

Figure 5.1. Method (i) combining global finite element solution with a localized boundary element region

while for region Ω^2 the finite element matrices are

$$\mathbf{KU} = \mathbf{F} \tag{5.2}$$

where **K** is the stiffness matrix of the problem and **F** the *equivalent* nodal forces. Note that **U** represents the displacements (or potentials) and **P** are the surface tractions (or fluxes).

In order to combine (5.1) and (5.2) one can reduce the first to a finite element form by inverting **G**, i.e.

$$\mathbf{G}^{-1}\mathbf{HU} = \mathbf{P} \tag{5.3}$$

Next one can convert the values of tractions at the nodes (as given by **P**) into an equivalent nodal force matrix of the type used in finite elements. This is done by weighting the boundary tractions by the interpolation function used for the displacements and produces a matrix M such that

$$\mathbf{F} = \mathbf{MP} \tag{5.4}$$

This operation is standard in finite elements although it is unusual to write the distribution matrix **M** in an explicit form.

Equation (5.3) can now be written as

$$\mathbf{M}(\mathbf{G}^{-1}\mathbf{H})\mathbf{U} = \mathbf{MP} = \mathbf{F}' \tag{5.5}$$

where the right hand side vector has the same form as in finite elements.

One can now write formula (5.5) as

$$\mathbf{K}'\mathbf{U} = \mathbf{F}' \tag{5.6}$$

where $\mathbf{K}' = \mathbf{MG}^{-1}\mathbf{H}$.

$\mathbf{K}'$ is a stiffness matrix obtained from the boundary element formulation. It is generally asymmetric due to the approximations involved in the discretization process and the choice of the assumed solution. Although this matrix is sometimes symmetrized simply taking an average of the off-diagonal terms (i.e. assuming it can be written as $\frac{1}{2}(\mathbf{K}' + \mathbf{K}'^{\mathrm{T}})$) this is not recommended as it produces inaccurate results in many practical applications. Obtaining symmetric boundary element stiffness matrices may involve double integration of the type used in Galerkin's BE formulation which are beyond the scope of this book.

The equivalent finite element type matrices of equation (5.6) can now be assembled with the matrices corresponding to region Ω^2 in figure 5.2 to form the global stiffness matrix.

Method (*iii*) This approach was proposed by Brebbia and Georgiou in 1979 [1] and consists in treating the finite element region as an equivalent boundary element.

Consider the two regions described in figure 5.2. For region 1 one can write the governing equations in a manner similar to that previously shown for multiregion problems, i.e.

$$[\mathbf{H}^1 \quad \mathbf{H}_I^1]\begin{Bmatrix}\mathbf{U}^1\\ \mathbf{U}_I^1\end{Bmatrix} = [\mathbf{G}^1 \quad \mathbf{G}_I^1]\begin{Bmatrix}\mathbf{P}^1\\ \mathbf{P}_I^1\end{Bmatrix} \tag{5.7}$$

where the subscript I defines the interface.

The matrices for the finite element region 2 can be written in a similar manner using the concept of distribution matrix defined in formula (5.4), i.e.

$$[\mathbf{K}^2 \quad \mathbf{K}_I^2]\begin{Bmatrix}\mathbf{U}^2\\ \mathbf{U}_I^2\end{Bmatrix} = [\mathbf{M}^2 \quad \mathbf{M}_I^2]\begin{Bmatrix}\mathbf{P}^2\\ \mathbf{P}_I^2\end{Bmatrix} \tag{5.8}$$

By writing $\mathbf{P}_I = \mathbf{P}_I^1 = -\mathbf{P}_I^2$ and $\mathbf{U}_I = \mathbf{U}_I^1 = \mathbf{U}_I^2$ one automatically satisfies the equilibrium and compatibility conditions on the interface and equation (5.7) and (5.8) can be rearranged and written together as follows.

$$\begin{bmatrix}\mathbf{H}^1 & \mathbf{H}_I^1 & -\mathbf{G}_I^1 & \mathbf{0}\\ \mathbf{0} & \mathbf{K}_I^2 & \mathbf{M}_I^2 & \mathbf{K}^2\end{bmatrix}\begin{Bmatrix}\mathbf{U}^1\\ \mathbf{U}_I\\ \mathbf{P}_I\\ \mathbf{U}^2\end{Bmatrix} = \begin{bmatrix}\mathbf{G}^1 & \mathbf{0}\\ \mathbf{0} & \mathbf{M}^2\end{bmatrix}\begin{Bmatrix}\mathbf{P}^1\\ \mathbf{P}^2\end{Bmatrix} \tag{5.9}$$

These equations will of course need to be rearranged in accordance with the boundary conditions. Notice that this approach does not require any matrix inversion.

Method (ii) – without forced symmetrization – and method (iii) are equivalent and give the same numerical results. Using one or the other depends mainly upon the problem in the sense of which part is more dominant, the finite elements or the boundary elements, in which case one can use method (ii) or (iii) respectively.

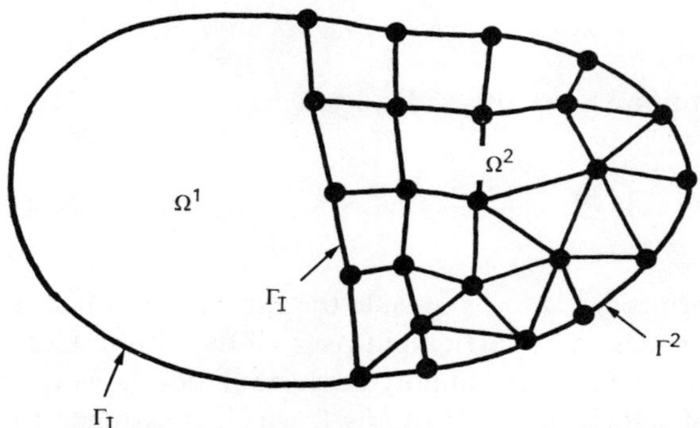

Figure 5.2 Boundary divided into a finite elements and a boundary element region

Method (ii) is essentially a stiffness method and can easily be incorporated in existing finite element packages, although it does require the inversion of the non-banded **G** matrix. In contrast, method (iii) does not require this inversion and both displacements and tractions remain unknown along the interface.

5.3 Approximate Boundary Elements

Combination of boundary elements with finite elements is particularly useful when dealing with domains tending to infinity. In these cases the near region is discretized into finite elements and the outside domain is simulated with the boundary elements on the interface of the two regions. This avoids having to discretize a larger region and at a certain distance putting some boundary conditions which try to represent the domain going to infinity. The main drawback of this approach however, is that the boundary element matrices are fully populated representing coupling of all the nodes on the boundary. In practice this coupling can be avoided by assuming that far from the region being perturbed the solution behaves in a smooth manner. This produces an approximate boundary element formulation that in many cases is equivalent to the use of radiation or similar boundary conditions. It is important to point out that these approximate boundary elements are not related to the so-called infinite elements which are based on domain rather than boundary integration.

Consider the example shown in figure 5.3 where the internal region is assumed to be subdivided into finite elements and the external region extending to infinity is modelled using boundary elements. Let us consider that the problem is governed by the Laplace's equation and hence the fundamental solution for two dimensions is

$$u^* = \frac{1}{2\pi} \ln\left(\frac{1}{r}\right) \tag{5.10}$$

For any point in the internal domain (including those near but not on the Γ_I interface) one can write,

$$\int_{\Gamma_I} u^* q \, d\Gamma = \int_{\Gamma_I} u q^* \, d\Gamma \tag{5.11}$$

Substituting the fundamental solution (5.10) into (5.11) leads to

$$\int_{\Gamma_I} \ln\left(\frac{1}{r}\right) q \, d\Gamma = \int_{\Gamma_I} u \frac{\partial}{\partial n}\left(\ln\left(\frac{1}{r}\right)\right) d\Gamma \tag{5.12}$$

Notice that the reference point is considered to be outside the external region and hence $c_i = 0$. The integration still needs to be carried out over all the Γ_I interfaces and all u and q values are interrelated. One can simplify the formulation however, if Γ_I is considered to be a circle of sufficiently large radius R which is assumed to be constant, hence $d\Gamma = R\, d\theta$, where θ is the angular coordinate. Notice also that

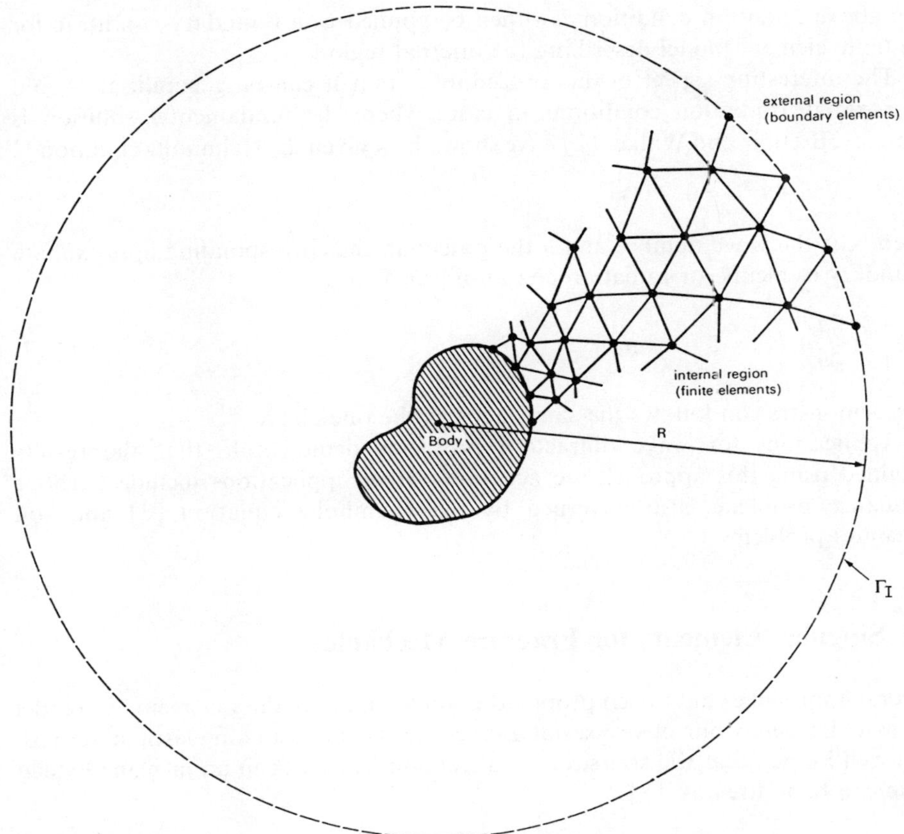

Figure 5.3 Finite and boundary element regions

$n \cong r$ if the reference point is far from the boundary Γ_I and hence equation (5.12) can be written as

$$\int_0^{2\pi} \left(\ln \frac{1}{r} \frac{\partial u}{\partial n} + u \frac{1}{r} \right) d\theta = 0 \tag{5.13}$$

$q = \partial u / \partial r$.

This is a special form of the Sommerfeld condition which can also be written as,

$$\frac{1}{r} u = (\ln r) \frac{\partial u}{\partial r} \qquad \text{on } \Gamma_I \tag{5.14}$$

Applying similar considerations for three dimensional potential problems one finds another form of this condition, i.e.

$$\frac{\partial u}{\partial r} + \frac{1}{r} u = 0 \qquad \text{on } \Gamma_I \tag{5.15}$$

The above radiation condition can then be applied as a boundary condition for the finite element model describing the internal region.

The interesting aspect of this procedure is that it can be generalized to find approximate radiation conditions in cases where the fundamental solution is complex. Brebbia and Walker [2] have shown how given the Helmholtz equation

$$\nabla^2 u + \kappa^2 u = 0 \tag{5.16}$$

where κ is the wave number and u the potential, the corresponding approximate boundary elements (or radiation condition) on Γ_I are

$$\frac{\partial u}{\partial n} + i\kappa u = 0 \qquad \text{on } \Gamma_I \tag{5.17}$$

The demonstration follows the same steps as the one above.

Applications for wave diffraction problems demonstrate that the results obtained using this approach are accurate. Other applications include harbour resonance problems also governed by the Helmholtz equation [2] and soil dynamics problems.

5.4 Singular Elements for Fracture Mechanics

Several approaches have been proposed in finite and boundary elements to model the singular behaviour of stresses at a crack tip which occurs in elastic materials. It is well known that the stresses near a traction free crack in an in-plane loaded plate can be written as, [3]

$$\begin{aligned}
\sigma_{11} &= \frac{K_I}{\sqrt{(2\pi r)}} \cos\frac{\theta}{2}\left(1 - \sin\frac{\theta}{2}\sin\frac{3}{2}\theta\right) - \frac{K_{II}}{\sqrt{(2\pi r)}} \sin\frac{\theta}{2}\left(2 + \cos\frac{\theta}{2}\cos\frac{3}{2}\theta\right) \\
\sigma_{22} &= \frac{K_I}{\sqrt{(2\pi r)}} \cos\frac{\theta}{2}\left(1 + \sin\frac{\theta}{2}\sin\frac{3}{2}\theta\right) + \frac{K_{II}}{\sqrt{(2\pi r)}} \cos\frac{\theta}{2}\sin\frac{\theta}{2}\cos\frac{3}{2}\theta \\
\sigma_{12} &= \frac{K_I}{\sqrt{(2\pi r)}} \sin\frac{\theta}{2}\cos\frac{\theta}{2}\cos\frac{3}{2}\theta + \frac{K_{II}}{\sqrt{(2\pi r)}} \cos\frac{\theta}{2}\left(1 - \sin\frac{\theta}{2}\sin\frac{3}{2}\right)
\end{aligned} \tag{5.18}$$

where r and θ are defined in figure 5.4, K_I and K_{II} are the stress intensity factors corresponding to the opening and sliding mode, respectively, and the size of r is much smaller than the crack length. The displacements near the crack tip are,

$$\begin{aligned}
u_1 &= \frac{K_I}{\mu}\sqrt{\left(\frac{r}{2\pi}\right)} \cos\frac{\theta}{2}\left(1 - 2\nu + \sin^2\frac{\theta}{2}\right) + \frac{K_{II}}{\mu}\sqrt{\left(\frac{r}{2\pi}\right)} \sin\frac{\theta}{2}\left(2 - 2\nu + \cos^2\frac{\theta}{2}\right) \\
u_2 &= \frac{K_I}{\mu}\sqrt{\left(\frac{r}{2\pi}\right)} \sin\frac{\theta}{2}\left(2 - 2\nu - \cos^2\frac{\theta}{2}\right) + \frac{K_{II}}{\mu}\sqrt{\left(\frac{r}{2\pi}\right)} \cos\frac{\theta}{2}\left(-1 + 2\nu + \sin^2\frac{\theta}{2}\right)
\end{aligned} \tag{5.19}$$

where μ is the shear modulus and v the Poisson ratio as in plane strain problems. Formulae (5.18) and (5.19) describe the stress and displacement distribution near the crack tip and have been obtained analytically. Values of K_I and K_{II} are difficult to obtain for general cases and it is then important to be able to model the behaviour of cracks in boundary element codes.

Snyder and Cruse [4], Stern *et al.* [5] and Cruse [6] presented several procedures to compute stress intensity factors using boundary elements, in particular a singular quarter-point boundary element was proposed by Blandford *et al.* [7] and by Martinez and Dominguez [8]. While some of the approaches proposed are complex to implement, the quarter point element is easy to use in boundary elements, gives accurate results and is no way sensitive to the discretization used. As with other boundary element techniques, the domain needs to be divided into subdomains by means of interfaces (figure 5.5) starting at the crack tip in order to avoid having two similar sets of equations which will produce a singular matrix. This subdivision avoids the numerical problems derived from having two displacement variables for the same geometrical point along the crack. All boundaries are discretized into elements and elements are also defined along the two faces of the crack and the interfaces between different regions. Boundary conditions are applied at external boundaries including zero traction conditions along the crack and the usual equilibrium and compatibility requirements are satisfied at the interfaces.

The quarter point element is based on the quadratic expansion. In this case any displacement, traction or coordinate such as x_1 and x_2 can be represented as

$$\mathbf{f} = \boldsymbol{\phi}\mathbf{f}^j \tag{5.20}$$

where f represents geometrical displacements or traction variables as seen in equations (4.51) to (4.54) and $\boldsymbol{\phi}$ are the quadratic shape function matrices. Any of these components can be written as,

$$f_i = \phi_1 f_i^1 + \phi_2 f_i^2 + \phi_3 f_i^3 \tag{5.21}$$

For the particular case that the quadratic element has a straight-line geometry and the mid-node is placed at a quarter of the length (figure 5.6) a simple relationship can be found between the coordinate ξ and the variable $\bar{r}$ along the element. In this case equation (5.21) gives

$$f_i = a_i^1 + a_i^2 \sqrt{\frac{\bar{r}}{l}} + a_i^3 \frac{\bar{r}}{l} \tag{5.22}$$

where

$$\begin{aligned} a_i^1 &= f_i^1 \\ a_i^2 &= -f_i^3 + 4f_i^2 - 3f_i^1 \\ a_i^3 &= 2f_i^3 - 4f_i^2 + 2f_i^1 \end{aligned} \tag{5.23}$$

Equation (5.22) ensures that for this position of the mid-point, the $\sqrt{\bar{r}}$ behaviour

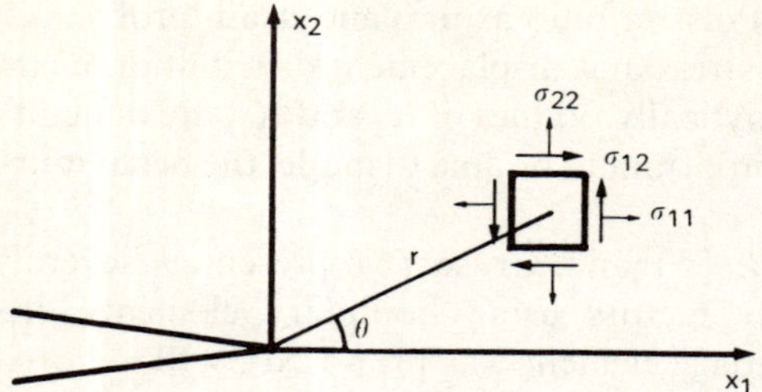

Figure 5.4 Coordinates near the tip

of the displacement near the crack tip as given by equation (5.19) is reproduced by the boundary element. This type of element is usually known as the 'quarter-point element'.

Since in the BEM displacements and tractions are represented independently, a correct representation of the displacements is compatible with an incorrect representation of the tractions. However, the singularity may be included in the representation of the tractions by using modified shape functions. Assume, for instance, that the crack tip is at node 1 (figure 5.6). One may write,

$$p_i = \phi_1 \bar{p}_i^1 \sqrt{\frac{l}{\bar{r}}} + \phi_2 \bar{p}_i^2 \sqrt{\frac{l}{\bar{r}}} + \phi_3 \bar{p}_i^3 \sqrt{\frac{l}{\bar{r}}} \tag{5.24}$$

or

$$p_i = \bar{\phi}_1 \bar{p}_i^1 + \bar{\phi}_2 \bar{p}_i^2 + \bar{\phi}_3 \bar{p}_i^3 \tag{5.25}$$

where $\bar{\phi}_1$, $\bar{\phi}_2$ and $\bar{\phi}_3$ are the modified shape functions which include the $r^{-1/2}$ singularity. Now $\bar{p}_i^j$ stands for the value of p_i at node j divided by the value of $\bar{\phi}_i$ at that node; i.e.,

$$\bar{p}_i^3 = p_i^3$$

$$\bar{p}_i^2 = p_i^2/2$$

$$\bar{p}_i^1 = \lim_{\bar{r} \to 0} p_i^1 \sqrt{\frac{\bar{r}}{l}}$$

Equation (5.24) for p_i can now be written as

$$p_i = \bar{a}_i^1 \sqrt{\frac{l}{\bar{r}}} + \bar{a}_i^2 + \bar{a}_i^3 \sqrt{\frac{\bar{r}}{l}} \tag{5.26}$$

where $\bar{a}_i^1 = \bar{p}_i^1$; $\bar{a}_i^2 = -\bar{p}_i^3 + 4\bar{p}_i^2 - 3\bar{p}_i^1$ and $\bar{a}_i^3 = 2\bar{p}_i^3 - 4\bar{p}_i^2 + 2\bar{p}_i^1$.

Using the quarter-point element with the shape functions of equation (5.25) for the tractions, both displacements and tractions will be correctly represented. The element including this kind of representation is known as the traction singular quarter-point element.

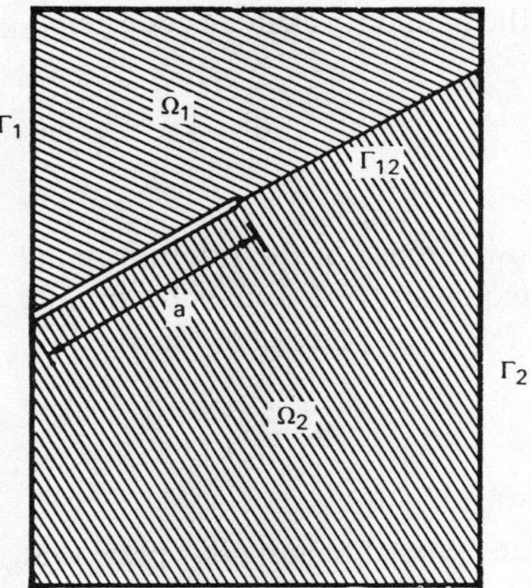

Figure 5.5 Edge cracked plate

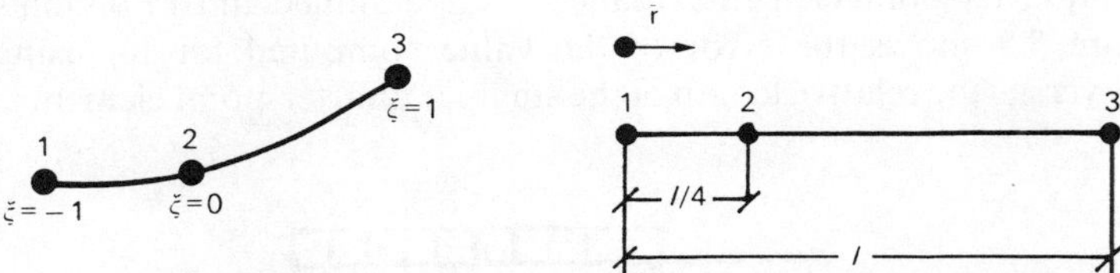

Figure 5.6 Quadratic and quadratic quarter point elements

The first and second mode stress intensity factors can be defined by the following limits (figure 5.4).

$$K_I = \lim_{x_1 \to 0} \{\sqrt{2\pi x_1}\, \sigma_{22}\}$$

$$K_{II} = \lim_{x_1 \to 0} \{\sqrt{2\pi x_1}\, \sigma_{12}\} \tag{5.27}$$

If the boundary discretization is done in such a way that the first interface element from the crack tip has $\theta = 0°$ and this element is a singular quarter-point boundary element, then for this element, $\bar{r} \equiv x_1$, $p_1 \equiv \sigma_{12}$, $p_2 \equiv \sigma_{22}$ and the nodal values for the tractions at the tip node K are:

$$\bar{p}_1^k = \lim_{\bar{r} \to 0} \{ p_1^k \sqrt{\bar{r}/l}\} = \lim_{x_1 \to 0} \{\sigma_{12} \sqrt{x_1/l}\}$$

$$\bar{p}_2^k = \lim_{\bar{r} \to 0} \{ p_1^k \sqrt{\bar{r}/l}\} = \lim_{x_1 \to 0} \{\sigma_{22} \sqrt{x_1/l}\} \tag{5.28}$$

Thus, the stress intensity factors coincide with the tractions nodal values except for a constant and may be computed directly with the boundary element code, i.e.

$$K_I = \bar{p}_2^k (2\pi l)^{1/2}$$
$$K_{II} = \bar{p}_1^k (2\pi l)^{1/2} \qquad (5.29)$$

Martinez and Dominguez [8] have shown how the use of the traction nodal values of the singular element at the crack tip (equation (5.29)) is substantially less sensitive to he discretization than any of the displacement correlation procedures.

Example 5.1

As an example, figure 5.7 shows the case of a centre cracked rectangular plate that has been studied by several authors [7], [8]. Because of the symmetry only one quarter of the plate is discretized. The total number of elements is nine, two of them being singular quarter-point elements. Plane stress is assumed and a Poisson's ratio $\nu = 0.2$. Figure 5.8 shows the error of the value computed for K_I using boundary elements versus the relative length of the singular quarter-point elements.

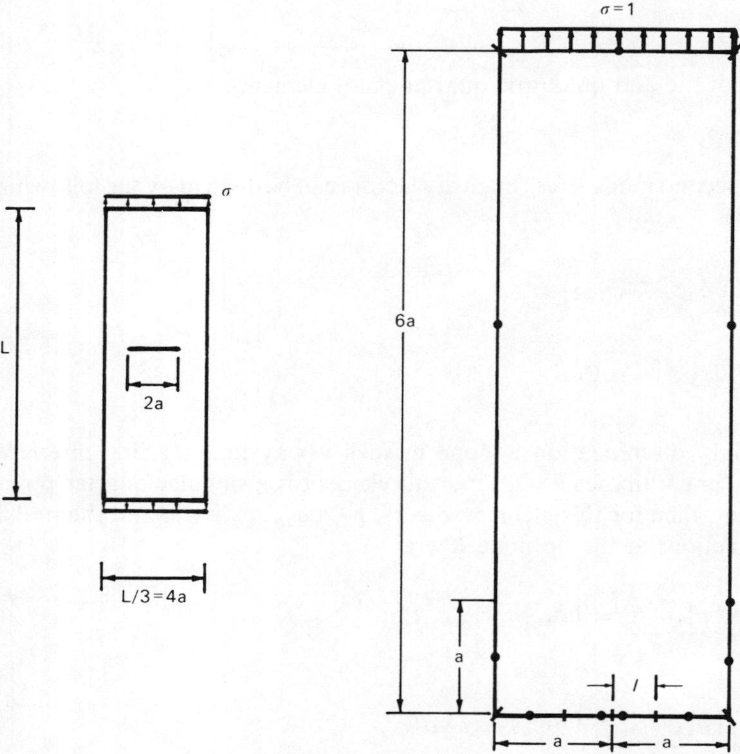

Figure 5.7 Centre cracked plate under traction. Discretization of one quarter of the plate.

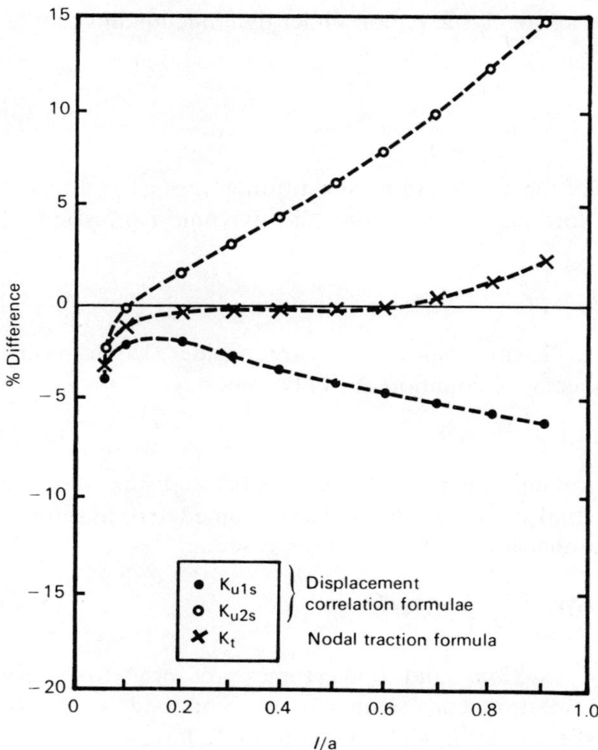

Figure 5.8 K_I stress intensity factor computed using quarter-point singular boundary elements

The value given by Bowie and Neal [8] is taken as reference as this value is accurate within one percent. The results obtained by the quarter-point boundary element procedure are given by K_t. Results computed using singular quarter-point elements and two different displacement correlation formulae (K_{u1s} and K_{u2s}) are given also for comparison. As can be seen in the figure, the nodal traction procedure with singular quarter-point elements (K_t) gives accurate results for a wide range of element sizes.

The implementation of a singular quarter-point element in the ELQUABE code of Chapter 4, requires only some minor changes in the integration routine.

5.5 Steady State Elastodynamics

Another interesting application of Boundary Elements which can be solved by modifying the codes presented in the previous chapters is the case of steady state elastodynamics. This is in a certain way similar to the harmonic wave propagation equation – or Helmholtz equation – discussed in section 2.14.

The equilibrium equations for an elastic region under dynamic loading can be written as,

$$\frac{\partial \sigma_{ij}}{\partial x_j} + b_i = \rho \ddot{u}_i \tag{5.30}$$

where $\ddot{u}_i$ are the components of the acceleration. Substituting stresses in terms of displacements in the above formulae one obtains the dynamic expression of Navier's equation, i.e.

$$(\lambda + \mu)u_{j,ji} + \mu u_{i,jj} + b_i = \rho \ddot{u}_i \tag{5.31}$$

where λ and μ are the Lamé's coefficients. If all variables are considered as harmonic functions in time, with a frequency ω, equation (5.31) becomes

$$(\lambda + \mu)u_{j,ji} + \mu u_{i,jj} + b_i + \rho\omega^2 u_i = 0 \tag{5.32}$$

The boundary element formulation can be obtained as before using weighted residuals and when both the actual and weighting field are assumed to be harmonic, the integral equations become the same as for the static case, i.e.

$$c^i_{kl} u^i_k + \int_\Gamma p^*_{kl} u_k \, d\Gamma = \int_\Gamma u^*_{kl} p_k \, d\Gamma + \int_\Omega u^*_{kl} b_k \, d\Omega \tag{5.33}$$

where all the displacements, tractions and body forces are now frequency dependent. The fundamental solution corresponds to a point load with time variation $\exp(-i\omega t)$ and satisfies equation (5.32) without body forces, i.e.

$$(\lambda + \mu)u^*_{k,kl} + \mu u^*_{l,kk} + \rho\omega^2 u^*_l + \Delta_i(\omega) = 0 \tag{5.34}$$

For three dimensional problems this solution is given by

$$u^*_{kl} = \frac{1}{\alpha\pi\rho C_s^2} [\psi\delta_{kl} - \chi r_{,k} r_{,l}] \tag{5.35}$$

where $\alpha = 4$, C_s is the shear wave velocity, C_p the P-wave velocity, and the functions ψ and χ are

$$\begin{aligned}\psi &= \left(1 - \frac{C_s^2}{\omega^2 r^2} + \frac{C_s}{i\omega r}\right)\frac{\exp(-i\omega r/C_s)}{r} \\ &\quad - \left(\frac{C_s^2}{C_p^2}\right)\left(-\frac{C_p^2}{\omega^2 r^2} + \frac{C_p}{i\omega r}\right)\frac{\exp(-i\omega r/C_p)}{r}\end{aligned} \tag{5.36}$$

$$\begin{aligned}\chi &= -\left(\frac{3C_s^2}{\omega^2 r^2} + \frac{3C_s}{i\omega r} + 1\right)\frac{\exp(-i\omega r/C_s)}{r} \\ &\quad - \left(\frac{C_s^2}{C_p^2}\right)\left(-\frac{3C_p^2}{\omega^2 r^2} + \frac{3C_p}{i\omega r} + 1\right)\frac{\exp(-i\omega r/C_p)}{r}\end{aligned} \tag{5.37}$$

where

$$C_s = \sqrt{\frac{\mu}{\rho}} \quad \text{and} \quad C_p = \sqrt{\frac{(\lambda + 2\mu)}{\rho}}, \qquad i = \sqrt{-1}$$

The tractions are given by the following relationships

$$p^*_{kl} = \frac{1}{\alpha\pi}\left\{\left(\frac{d\psi}{dr} - \frac{1}{r}\chi\right)\left(\delta_{kl}\frac{\partial r}{\partial n} + r_{,k}n_l\right) - \frac{2}{r}\chi\left(n_k r_{,l} - 2r_{,k}r_{,l}\frac{\partial r}{\partial n}\right)\right.$$
$$\left. - 2\frac{d\chi}{dr}r_{,k}r_{,l}\frac{\partial r}{\partial n} + \left(\frac{C_p^2}{C_s^2} - 2\right)\left(\frac{d\psi}{dr} - \frac{d\chi}{dr} - \frac{\alpha}{2r}\chi\right)r_{,l}n_k\right\} \tag{5.38}$$

The steady-state fundamental solution for two-dimensions is also given by equations (5.35) and (5.38) for the case $\alpha = 2$, and the following ψ and χ functions

$$\psi = K_0\left(\frac{i\omega r}{C_s}\right) + \frac{C_s}{i\omega r}\left[K_1\left(\frac{i\omega r}{C_s}\right) - \frac{C_s}{C_p}K_1\left(\frac{i\omega r}{C_p}\right)\right] \tag{5.39}$$

and

$$\chi = K_2\left(\frac{i\omega r}{C_s}\right) - \frac{C_s^2}{C_p^2}K_2\left(\frac{i\omega r}{C_p}\right) \tag{5.40}$$

Functions K_0, K_1 and K_2 are the modified Bessel functions of the second kind and order 0, 1 and 2 respectively.

Integral equation (5.33) can then be discretized into boundary elements in the same form as for elastostatics and the codes previously studied can be extended to elastodynamics simply by changing the fundamental solutions and the solutions to compute internal stresses and set all variables as complex. For instance codes ELCONBE and ELQUABE of Chapter 4 would only require changes in subroutines EXTINEC and LOCINEC or EXTINEQ and LOCINEQ in addition to the general changes in the definition of variables as complex.

It is worth pointing out that as the fundamental solution is frequency dependent the system $\mathbf{AX} = \mathbf{F}$ has to be formed and solved for each frequency. The numerical treatment of Bessel function in the two dimensional formulation also requires more care than the logarithm of the static problem, in particular for high and very low frequency values.

Some applications of the use of boundary element methods in steady state elastodynamics can be seen in the work of Domínguez and Alarcón [10] and Domínguez [11].

References

[1] Brebbia, C. A. and Georgiou, P. Combination of Boundary and Finite Elements in Elastostatics, *Appl. Math. Modell.*, **3**(2), June 1979.

[2] Brebbia, C. A. and Walker, S. Simplified Boundary Elements for Radiation Problems, *Res. Note Appl. Math. Modell.*, **2**(2), June 1978.

[3] Irwin, G. R. Fracture, in *Encyclopaedia of Physics* (Ed. S. Flugge), Vol. VI, Springer-Verlag, 1958.
[4] Snyder, M. D. and Cruse, T. A. Boundary Integral Equation Analysis of Anisotropic Cracked Plates, *Int. J. Fracture*, **11**, 315–328, 1975.
[5] Stern, M., Becker, E. B. and Dunham, R. S. A Contour Integral Computation of Mixed-mode Stress Intensity Factors, *Int. J. Fracture*, **12**, 359–368, 1976.
[6] Cruse, T. A. Two-dimensional BIE Fracture Mechanics Analysis, *Appl. Math. Modell.*, **2**, 287–293, 1978.
[7] Blandford, G. E., Ingraffea, A. R. and Liggett, J. A. Two-dimensional Stress Intensity Factor Computations using the Boundary Element Method, *Int. J. Num. Meth. Eng.*, **17**, 387–404, 1981.
[8] Martinez, J. and Dominguez, J. On the Use of Quarter-point Boundary Elements for Stress Intensity Factor Computations, *Int. J. Num. Meth. Eng.*, **20**, 1941–1950, 1985.
[9] Bowie, O. L. and Neal, D. M. A Note on the Central Crack in a Uniformly Stressed Strip, *Eng. Fracture Mech.*, **2**, 181, 1970.
[10] Domínguez, J. and Alarcón, E. Elastodynamics, in *Progress in Boundary Element Methods, Vol. 1* (C. A. Brebbia, Ed.), Pentech Press, London, 1981.
[11] Domínguez, J. Dynamic Stiffness of Rectangular Foundations, *M.I.T. Research Report No. 1278-20,* Civil Eng. Dept., 1978.

Appendix A

Numerical Integration

A.1 Introduction

There is a large number of numerical integration schemes in the literature, many of them designed for special problems. Gaussian integration formulae are general, simple and very accurate and these formulae have been used for potential and elasticity programs in Chapters 2 and 4.

In the following, Gaussian integral formulae for non-singular functions are presented first. They can be used for integration over elements and internal cells. Formulae for one, two and three-dimensional domains are given. A second group refers to integrals in which a logarithmic singularity is located at one end of the integration domain. This formula has been used for elements that include the source point in the two-dimensional potential and elasticity programs of Chapters 2 and 4. In these cases the fundamental solutions present a logarithmic singularity and the special integration formula is used for elements or cells in which the source is located.

Numerical integration formulae for functions with $1/r$ singularity and finite part integrals for singular functions may be seen in references [1] and [2].

A.2 One-dimensional Gaussian Quadrature [3]

The integrals in this case can be written as,

$$I = \int_{-1}^{+1} f(\xi)\, d\xi = \sum_{i=1}^{n} w_i f(\xi_i) + E_n \tag{A1.1}$$

where n is the number of integration points, ξ_i is the coordinate of the ith integration point, w_i is the associated weighting factor and E_n is the error or residual, i.e.

$$E_n = \frac{2^{2N+1}(n!)^4}{(2n+1)[(2n!)]^3} \frac{d^{2n}f(\xi)}{d\xi^{2n}}; \qquad (-1 < \xi < 1) \tag{A1.2}$$

Formulae (A1.1) is based on the representation of $f(\xi)$ by means of Legendre polynomials $P_n(\xi)$. The ξ_i value is the coordinate at a point i where P_n is zero and for which the weights are given by

$$w_i = 2/(1 - \xi_i^2)\left[\frac{dP_n(\xi)}{d\xi}\right]^2_{\xi = \xi_i}$$

Values of ξ_i and w_i are listed in Table A1.1. Notice that ξ_i values are symmetric with respect to $\xi = 0$, $w_i$being the same for two symmetric values.

Table A1.1

$\pm\xi_i$	w_i
$n=2$	
0.57735 02691 89626	1.00000 00000 00000
$n=3$	
0.00000 00000 00000	0.88888 88888 88888
0.77459 66692 41483	0.55555 55555 55555
$n=4$	
0.33998 10435 84856	0.65214 51548 62546
0.86113 63115 94053	0.34785 48451 37454
$n=5$	
0.00000 00000 00000	0.56888 88888 88889
0.53846 93101 05683	0.47862 86704 99366
0.90617 98459 38664	0.23692 68850 56189
$n=6$	
0.23861 91860 83197	0.46791 39345 72691
0.66120 93864 66265	0.36076 15730 48139
0.93246 95142 03152	0.17132 44923 79170
$n=7$	
0.00000 00000 00000	0.41795 91836 73469
0.40584 51513 77397	0.38183 00505 05119
0.74153 11855 99394	0.27970 53914 89277
0.94910 79123 42759	0.12948 49661 68870
$n=8$	
0.18343 46424 95650	0.36268 37833 78362
0.52553 24099 16329	0.31370 66458 77887
0.79666 64774 13627	0.22238 10344 53374
0.96028 98564 97536	0.10122 85362 90376
$n=9$	
0.00000 00000 00000	0.33023 93550 01260
0.32425 34234 03809	0.31234 70770 40003
0.61337 14327 00590	0.26061 06964 02935
0.83603 11073 26636	0.18064 81606 94857
0.96816 02395 07626	0.08127 43883 61574
$n=10$	
0.14887 43389 81631	0.29552 42247 14753
0.43339 53941 29247	0.26926 67193 09996
0.67940 95682 99024	0.21908 63625 15982
0.86506 33666 88985	0.14945 13491 50581
0.97390 65285 17172	0.06667 13443 08688
$n=12$	
0.12523 34085 11469	0.24914 70458 13403
0.36783 14989 98180	0.23349 25365 38355
0.58731 79542 86617	0.20316 74267 23066
0.76990 26741 94305	0.16007 83285 43346
0.90411 72563 70475	0.10693 93259 95318
0.98156 06342 46719	0.04717 53363 86512

A.3 Two and Three-dimensional Quadrature for Rectangles and Rectangular Hexahedra

Two and three-dimensional formulae are obtained simply by combination of (A1.1), i.e.

$$I = \int_{-1}^{1} \int_{-1}^{1} f(\xi, \eta)\, d\xi\, d\eta \cong \sum_{j=1}^{n} \sum_{i=1}^{n} f(\xi_i, \eta_j) w_i w_j \qquad \text{(A1.3)}$$

and

$$I = \int_{-1}^{1} \int_{-1}^{1} \int_{-1}^{1} f(\xi, \eta, \zeta)\, d\xi\, d\eta\, d\zeta \cong \sum_{j=1}^{n} \sum_{i=1}^{n} \sum_{k=1}^{n} f(\xi_i, \eta_j, \zeta_k) w_i w_j w_k \qquad \text{(A1.4)}$$

where integration point coordinates and weighting factors are given in Table A1.1.

A.4 Two and Three-dimensional Quadrature for Triangular and Pentahedral Domains [4]

Numerical integration over a triangle can be carried out using triangular coordinates as shown in figure A1.1. This gives

$$I = \int_0^1 \left(\int_0^{1-2} f(\xi_1, \xi_2, \xi_3)\, d\xi_1 \right) d\xi_2 = \sum_{i=1}^{n} w_i f(\xi_1^i, \xi_2^i, \xi_3^i) \tag{A1.5}$$

where n is the number of integration points; ξ_1^i, ξ_2^i and ξ_3^i ar the coordinates of the i integration point and w_i the associate weighting factor. Values of ξ_1^i, ξ_2^i, ξ_3^i and w_i compiled from Hammer *et al.* [4] are given in Table A1.2.

Table A1.2

n	i	ξ_1^i	ξ_2^i	ξ_3^i	w_i
1 (linear)	1	1/3	1/3	1/3	1
2 (quadratic)	1	1/2	1/2	0	1/3
	2	0	1/2	1/2	1/3
	3	1/2	0	1/2	1/3
4 (cubic)	1	1/3	1/3	1/3	−9/16
	2	3/5	1/5	1/5	25/48
	3	1/5	3/5	1/5	25/48
	4	1/5	1/5	3/5	25/48
7 (quintic)	1	0.333 333 33	0.333 333 33	0.333 333 33	0.225 000 00
	2	0.797 426 99	0.101 286 51	0.101 286 51	0.125 939 18
	3	0.101 286 51	0.797 426 99	0.101 286 51	0.125 939 18
	4	0.101 286 51	0.101 286 51	0.797 426 99	0.125 939 18
	5	0.059 715 87	0.470 142 06	0.470 142 06	0.132 394 15
	6	0.470 142 06	0.059 715 87	0.470 142 06	0.132 394 15
	7	0.470 142 06	0.470 142 06	0.059 715 87	0.132 394 15

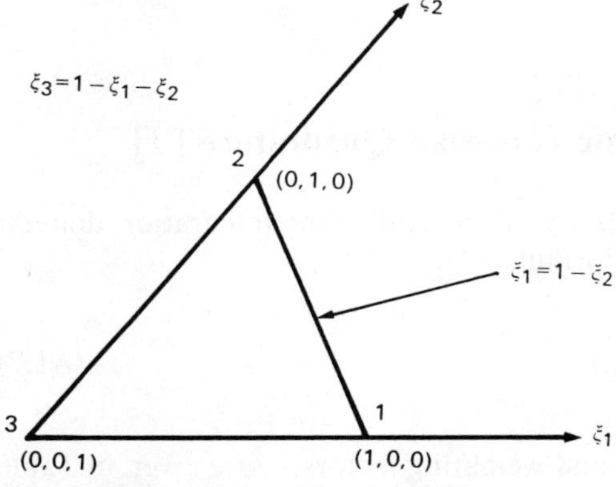

Figure A1.1 Triangular coordinates

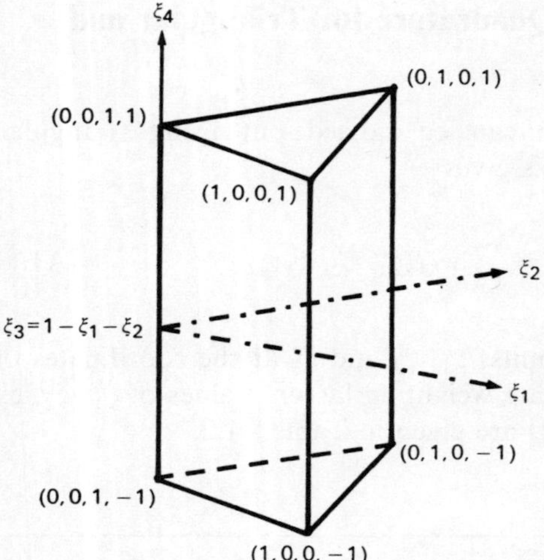

Figure A1.2 Pentahedral cell

Integration over a pentahedral cell (figure A1.2) can be done by combination of (A1.5) and the one-dimensional Gaussian quadrature (A1.1), i.e.

$$I = \int_{-1}^{1} \left[\int_{0}^{1} \left(\int_{0}^{1-\xi} f(\xi_1, \xi_2, \xi_3, \xi_4)\, d\xi_1 \right) d\xi_2 \right] d\xi_4$$

$$\cong \sum_{i=1}^{n} \sum_{j=1}^{m} w_i^{(1)} w_j^{(2)} f(\xi_1^i, \xi_2^i, \xi_3^i, \xi_4^j) \tag{A1.6}$$

where n is the number of integration points over the base and m the number of integration points along the height; ξ_1^i, ξ_2^i, ξ_3^i and $w_i^{(1)}$ are given in Table A1.2 and ξ_4^j and $w_j^{(2)}$ are shown in Table A1.1.

A.5 One-dimensional Logarithmic Gaussian Quadrature [3]

Kernels including a logarithmic singularity at one end of the integration domain can be integrated using the following formula [3].

$$I = \int_{0}^{1} \ln\left(\frac{1}{\xi}\right) f(\xi)\, d\xi \cong \sum_{i=1}^{n} w_i f(\xi_i) \tag{A1.7}$$

where integration point coordinates ξ_i and weighting factors w_i are given in Table A1.3.

Table A1.3

n	ξ_i	w_i	n	ξ_i	w_i
2	0.11200880	0.71853931	8	0.13320243 (−1)	0.16441660
	0.60227691	0.28146068		0.79750427 (−1)	0.23752560
3	0.63890792 (−1)	0.51340455		0.19787102	0.22684198
	0.36899706	0.39198004		0.35415398	0.17575408
	0.76688030	0.94615406 (−1)		0.52945857	0.11292402
4	0.41448480 (−1)	0.38346406		0.70181452	0.57872212 (−1)
	0.24527491	0.38687532		0.84937932	0.20979074 (−1)
	0.55616545	0.19043513		0.95332645	0.36864071 (−2)
	0.84898239	0.39225487 (−1)	9	0.10869338 (−1)	0.14006846
5	0.29134472 (−1)	0.29789346		0.64983682 (−1)	0.20977224
	0.17397721	0.34977622		0.16222943	0.21142716
	0.41170251	0.23448829		0.29374996	0.17715622
	0.67731417	0.98930460 (−1)		0.44663195	0.12779920
	0.89477136	0.18911552 (−1)		0.60548172	0.78478879 (−1)
6	0.216344005 (−1)	0.23876366		0.75411017	0.39022490 (−1)
	0.12958339	0.30828657		0.87726585	0.13867290 (−1)
	0.31402045	0.24531742		0.96225056	0.24080402 (−2)
	0.53865721	0.14200875	10	0.90425944 (−2)	0.12095474
	0.75691533	0.55454622 (−1)		0.53971054 (−1)	0.18636310
	0.92266884	0.10168958 (−1)		0.13531134	0.19566066
7	0.16719355 (−1)	0.19616938		0.24705169	0.17357723
	0.10018568	0.27030264		0.38021171	0.13569597
	0.24629424	0.23968187		0.52379159	0.93647084 (−1)
	0.43346349	0.16577577		0.66577472	0.55787938 (−1)
	0.63235098	0.88943226 (−1)		0.79419019	0.27159893 (−1)
	0.81111862	0.33194304 (−1)		0.89816102	0.95151992 (−2)
	0.94084816	0.59327869 (−2)		0.96884798	0.16381586 (−2)

Note: Numbers are to be multiplied by the power of 10 in parentheses.

A.6 Variable Transformation Methods

To deal with singular and nearly singular integrals, there are methods using transformation of variables in order to weaken or cancel out the singularity by the Jacobian of the transformation before applying the ordinary Gauss rule. This approach has been proposed by Telles [5] using quadratic and cubic transformation and by Hayami and Brebbia [6] who developed the idea of transforming in the radial direction working in polar coordinates.

References

[1] Brebbia, C. A., Telles, J. C. F. and Wrobel, L. C. *Boundary Element Techniques*, Springer-Verlag, Berlin, N.Y., 1984.

[2] Kutt, H. R. Quadrature Formulae for Finite Part Integrals, *Report WISK 178*, The National Research Institute for Mathematical Sciences, Pretoria, 1975.

[3] Stroud, A. H. and Secrest, D. *Gaussian Quadrature Formulas*, Prentice-Hall, N.Y., 1966.

[4] Hammer, P. C., Marlowe, O. J. and Stroud, A. H. Numerical Integration over Simplexes and Cones, *Math. Tables and Other Aids to Computation*, *Vol. 10,* 1956.
[5] Telles, J. C. F. A Self-Adaptive Coordinate Transformation for Efficient Numerical Evaluation of General Boundary Element Integrals, *International Journal for Numerical Methods in Engineering*, 1989, **24**, 959–973.
[6] Hayami, K. and Brebbia, C. A. A New Coordinate Transformation Method for Singular and Nearly Singular Integrals over General Curved Boundary Elements, *Boundary Elements IX*, *Vol. 1*, 375–399, 1987.

Appendix B

Diskette Contents

A diskette is available from Computational Mechanics Institute. It contains six FORTRAN codes and nine data files. Each FORTRAN file contains one main program and all the associate subroutines in such a way that each file only needs to be compiled and linked with the FORTRAN library to be ready to run.

The FORTRAN files are:

POCONBE.FOR:	Potential Constant Boundary Elements Program
POLINBE:FOR:	Potential Linear Boundary Elements Program
POQUABE.FOR:	Potential Quadratic Boundary Elements Program
POMCOBE.FOR:	Potential Multiboundary Constant Boundary Elements Program
ELCONBE.FOR:	Elastic Constant Boundary Elements Program
ELQUABE.FOR:	Elastic Quadratic Boundary Elements Program

The data files are the following:

PC12HF.DAT:	Heat Flow Problem using 12 Constant Elements (Example 2.1)
PL12HF.DAT:	Heat Flow Problem using 12 Linear Elements (Example 2.2)
PL04HF.DAT:	Heat Flow Problem using 4 Linear Elements (Example 2.3)
PQ10ES.DAT:	Elliptical Section under Torsion using 10 Quadratic Elements (Example 2.4)
PQ05ES.DAT:	Elliptical Section under Torsion using 5 Quadratic Elements (Example 2.5)
EC24CC.DAT:	Circular Cavity under Internal Pressure using 24 Constant Elements (Example 4.1)
EQ12CC.DAT:	Circular Cavity under Internal Pressure using 12 Quadratic Elements (Example 4.2)
EQ06RP.DAT:	Rectangular Plate under Flexural Moment using 6 Quadratic Elements (Example 4.3)
EQ12CB.DAT:	Cantilever Beam under Flexure using 12 Quadratic Elements (Example 4.4)

For further information contact Computational Mechanics Institute, Ashurst Lodge, Ashurst, Southampton, SO4 2AA, U.K. Telephone 0 42129 3223, FAX 0 42129 2853.

Appendix C

References for Further Reading

[1] Pian, T. H. H. and Tong, P. Basis of Finite Element Method for Solid Continua, *Int. Jnl. Numerical Methods Engng.*, 1969, **1**, 3–28.

[2] Washizu, K. *Variational Methods in Elasticity and Plasticity, 2nd edn.* Pergamon Press, New York, 1975.

[3] Muskhelishvili, N. I. *Some Basic Problems of the Mathematical Theory of Elasticity*, P. Noordhoff Ltd., Groningen, 1953.

[4] Mikhlin, S. G. *Integral Equations*, Pergamon, New York, 1957.

[5] Kupradze, O. D. *Potential Methods in the Theory of Elasticity*, Daniel Davey and Co., New York, 1965.

[6] Smirnov, V. J. Integral Equations and Partial Differential Equations, in *A Course in Higher Mathematics, Vol. IV*, Addison-Wesley, 1964.

[7] Kellogg, O. D. *Foundations of Potential Theory*, Dover, New York, 1953.

[8] Fredholm, I. Sur Une Classe D'Equations Fonctionelles, *Acta Math.* 1963, **27**, 365–390.

[9] Jaswon, M. A. Integral Equation Methods in Potential Theory, I, *Proc. R. Soc. Ser. A*, 1963, **275**, 23–32.

[10 Symm, G. T. Integral Equation Methods in Potential Theory, II, *Proc. R. Soc. Ser. A*, 1963, **275**, 33–46.

[11] Massonnet, C. E. Numerical Use of Integral Procedures in Stress Analysis, *Stress Analysis* (O. C. Zienkiewicz and G. S. Holister, Eds.), Wiley, 1966.

[12] Shaw, R. Diffraction of Plane Shock Waves by Obstacles of Arbitrary Shape, *Ph.D. Thesis*, Columbia University, 1960.

[13] Friedman, M. B. and Shaw, R. Diffraction of a Plane Shock Wave by an Arbitrary Rigid Cylindrical Obstacle, *Journ. Appl. Mech.*, 1962, **29**, 40–46.

[14] Shaw, R. Retarded Potential Approach to the Scattering of Elastic Waves by Rigid Obstacles of Arbitrary Shape, *J.A.S.A.*, 1968, **44**, 745–748.

[15] Shaw, R. Integral Equation Formulation of Dynamic Acoustic Fluid-Elastic Solid Interaction Problems, *J.A.S.A.*, 1973, **53**, 514–520.

[16] Tai, R. C. and Shaw, R. Helmholtz Equation Eigenvalues and Eigennodes for Arbitrary Domains, *J.A.S.A.*, 1974, **56**, 796–804.

[17] Shaw, R. P. An Integral Equation Approach to Diffusion, *Int. Jnl. Heat Transfer*, 1974, **17**, 693–699.

[18] Shaw, R. and Tai, R. C. Time Harmonic Acoustic Radiation from a Non-Concentric Circular Cylinder, *J.A.S.A.*, 1974, **56**, 1354–1360.

[19] Jaswon, M. A. and Ponter, A. R. S. An Integral Equation Solution of the Torsion Problem, *Proc. Roy. Soc.*, 1963, **A273**, 237–246.

[20] Symm, G. T. External Thermal Resistance of Buried Cables and Troughs, *Proc. I.E.E.*, 1969, **116**(10), 1695–1698.

[21] Symm, G. T. Capacitance of Coaxial Lines with Steps and Tapes, in *Recent Advances in BEM*, Proc. 1st Int. Conf. on BEM, Southampton University (C. A. Brebbia, Ed.), Pentech Press London, 1978.

[22] Symm, G. T. An Integral Equation Method in Conformal Mapping, *Num. Math.*, 1966, **9**, 250–258.

[23] Symm, G. T. Conformal Mapping of Doubly Connected Domains, *Num. Math.*, 1969, **13**, 448–457.

[24] Jaswon, M. A., Maiti, M. and Symm, G. T. Numerical Biharmonic Analysis and Some Applications, *Int. J. Solids Structures*, 1967, **3**, 309–332.

[25] Jaswon, M. A. and Maiti, M. 'An Integral Formulation of Plate Bending Problems, *Jnl. Eng. Math.*, 1968, **2**(1), 83–93.

[26] Cruse, T. A. Numerical Solutions in Three Dimensional Elastostatics, *Int. Jnl. of Solids Structures*, 1969, **5**, 1259–1274.

[27] Brebbia, C. A. (Ed.) *Variational Methods in Engineering*, Proc. of the 1st Int. Conf on Variatonal Methods in Engng., Southampton University, 1972, 2 volumes, Southampton University Press, 1973.

[28] Cruse, T. A. An Improved Boundary-Integral Equation Method for Three Dimensional Elastic Stress Analysis, *Computers and Structures*, 1974, **4**, 741–754.

[29] Cruse, T. A. Boundary-Integral Equation Method for Three Dimensional Elastic Fracture Mechanics Analysis, AFOSR-TR-75-0813, 1975.

[30] Cruse, T. A., Snow, D. W. and Wilson, R. B. Numerical Solutions in Axisymmetric Elasticity, *Computers and Structures*, 1977, **7**, 445–451.

[31] Snyder, M. D. and Cruse, T. A. Boundary-Integral Analysis of Anisotropic Cracked Plates, *Int. Jnl. of Fracture Mechanics*, 1975, **11**, 315–328.

[32] Symm, G. Practical Applications of an Integral Equation Method for the Solution of Laplace's Equation, *Comput. Electr. Eng.*, 1977, **4**, 167–170.

[33] Liu, P. L-F. and Liggett, J. A. Boundary Integral Solutions to Groundwater Problems, *Proc. Int. Conf. on Appl. Num. Mod.*, Southampton, 1977, 559–569.

[34] Liggett, J. A. and Liu, P. L-F. An Efficient Numerical Method of Two Dimensional Steady Groundwater Problems, *Water Resources Research*, 1978, **14**(3), 385–390.

[35] Blandford, G. E., Ingraffea, A. R. and Liggett, J. A. Mixed-Mode Stress Intensity Factor Calculation using the Boundary Element Method' *Proc. ASCE Eng. Mech. Div. Specialty Conf. Austin, Texas*, 1979, 797–800.

[36] Liggett, J. A. and Liu, P. L-F. Unsteady Interzonal Flow in Porous Media, *Water Resources Research*, 1979, **15**(2), 240–246.

[37] Jaswon, M. A. and Symm, G. T. *Integral Equation Methods in Potential Theory and Elastostatics*, Academic Press, London, 1977.

[38] Brebbia, C. A. *The Boundary Element Method for Engineers*, Pentech Press, London, 1978.

[39] Brebbia, C. A. and Dominguez, J. The Boundary Element Method for Potential Problems, *Applied Mathematical Modelling Jnl.*, 1977, **2**, 7.

[40] Brebbia, C. A. and Walker, S. *The Boundary Element Technique in Engineering*, Newnes-Butterworths, London, 1979.

[41] Brebbia, C. A. Weighted Residual Classification of Approximate Methods, *Applied Mathematical Modelling Journal*, Sept. 1978, **2** 3.

[42] Brebbia, C. A. and Connor, J. J. Boundary Integral Formulations, Chapter in *Topics in BE Research, Vol. 1*, Basic Principles and Applications, Springer-Verlag, Berlin and New York, 1984.

[43] Brebbia, C. A. (Ed.) *Recent Advances in Boundary Element Methods*, Pentech Press, London, 1978.

[44] Brebbia, C. A. (Ed.) *New Developments in Boundary Element Methods*, CML Publications, Southampton, 1980.

[45] Brebbia, C. A. (Ed.) *Boundary Element Methods*, Springer-Verlag, Berlin and New York and CML Publications, Southampton, 1981.

[46] Brebbia, C. A. (Ed.) *Boundary Element Methods in Engineering*, Springer-Verlag, Berlin and New York and CML Publications, Southampton, 1982.

[47] Brebbia, C. A., Futagami, T. and Tanaka, M. (Eds.) *Boundary Elements*, Springer-Verlag, Berlin and New York and CML Publications, Southampton, 1983.

[48] Brebbia, C. A. (Ed.) *Boundary Elements VI*, Springer-Verlag, Berlin and New York and CML Publications, Southampton, 1984.

[49] Brebbia, C. A. and Maier, G. (Eds.) *Boundary Elements VII*, Springer-Verlag, Berlin and New York and CML Publications, Southampton, 1985.

[50] Tanaka, M. and Brebbia, C. A. *Boundary Elements VIII*, Springer-Verlag, Berlin and New York and CML Publications, Southampton, 1986.
[51] Brebbia, C. A. and Wendland, W. *Boundary Elements IX*, Springer-Verlag, Berlin and New York and CML Publications, Southampton, 1987.
[52] Brebbia, C. A., Telles J. C. and Wrobel, L. C. *Boundary Element Techniques. Theory and Applications in Engineering*, Springer-Verlag, Berlin and New York, 1984.
[53] Brebbia, C. A. and Noye, B. J. *BETECH/85*, Springer-Verlag, Berlin and New York and CML Publications, Southampton, 1985.
[54] Brebbia, C. A. and Connor, J. J. *BETECH/86*, Computational Mechanics Publications, Southampton and Boston, 1986.
[55] Brebbia, C. A. and Venturini, W. *BETECH/87*, CM Publications, Southampton and Boston, 1987.
[56] Kermanidis, A Numerical Solution for Axially Symmetrical Elasticity Problems, *Int. Jnl. of Solids and Structures*, 1975, **11**, 493–500.
[57] Mayr, M. The Numerical Solution of Axisymmetric Elasticity Problems using an Integral Equation Approach, *Mechanics Research Communications*, 1976, **3**, 393–398.
[58] Kuhn, G. and Mohrmann, W. Boundary Element Method in Elastostatics: Theory and Applications, *Appl. Math. Mod.*, 1983, **7**, 565–572.
[59] Nageswaran, S. and Brebbia, C. A. BEASY – Boundary Element Analysis System as a CIM Tool, *Computer Aided Engineering Systems Handbook*, *Vol. 1* (J. Puig-Pey and C. A. Brebbia, Eds), Computational Mechanics Publications, Southampton, 1987.
[60] Danson, D. J. Linear Isotropic Elasticity with Body Forces, Chapter 4 in *Progress in Boundary Element Methods*, *Vol. 2*, Pentech Press, London, Springer-Verlag, NY, 1983.
[61] Bui, H. D. An Integral Equation Method for Solving the Problem of a Plane Crack of Arbitrary Shape, *Jnl. of the Mechanics and Physics of Solids*, 1977, **25**, 29–39.
[62] Cruse, T. A. Two Dimensional BIE Fracture Mechanics Analysis, *Appl. Math. Modeling*, 1978, **2**, 287–293.
[63] Tan, C. L. and Fenner, R. T. Elastic Fracture Mechanics Analysis of the Boundary Integral Equation Method, *Proc. Royal Society of London*, 1979, **A369**, 243–260.
[64] Rudolphi, T. A Boundary Element Solution of the Edge Crack Problem, *Int. Jnl. of Fracture*, 1982, **18**(3), 179–190.
[65] Atkinson, C. Fracture Mechanics Stress Analysis, Chapter 3 in *Progress of Boundary Element Methods*, *Vol. 2*, Pentech Press, London and Springer-Verlag, NY, 1983.
[66] Sato, Y., Tanaka, M. and Nakamura M. Stress Intensity Factor Computation in 3-D Elastostatics by Boundary Element Method, *Engineering Analysis*, 1984, **1**(4), 200–205.
[67] Martinez, J. and Dominguez, J. On the Use of Quarter Point Boundary Elements for Stress Intensity Factor Computations, *Int. Jnl. for Numerical Methods in Engineering*, 1984, **20**, 1941–1950.
[68] Dominguez, J. and Chirino, F. BEM for Dynamic Crack Problems in Two Dimensions, *BETECH/86* (C. A. Connor and C. A. Brebbia, Eds.), Computational Mechanics Publications, 1986, 619–631.
[69] Fan, T. Y. and Hahn, H. G. An Application of the Boundary Integral Equation Method to Dynamic Fracture Mechanics, *Engineering Fracture Mechanics*, 1985, **21**(2), 307–313.
[70] Riccardella, P. C. An Implementation of the Boundary Integral Technique for Planar Problems in Elasticity and Elastoplasticity, *Report N. SM-73-10*, Dept. Mech. Engng., Carnegie Mellon University, Pittsburg, 1973.
[71] Mendelson, A. Boundary Integral Methods in Elasticity and Plasticity, *Report No. NASA TN D-7418*, NASA, 1973.
[72] Mendelson, A. and Albers, L. U. Application of Boundary Integral Equations to Elastoplastic Problems, in *Boundary Integral Equation Method. Computational Applications in Applied Mechanics* (Cruse and Risso, Eds.), ASME, New York, 1975, 47–84.

[73] Mukherjee, S. Corrected Boundary Integral Equation in Planar Thermoelasto-plasticity, *Int. Jnl. Solids and Structures*, 1977, **13**, 331–335.

[74] Kumar, V. and Mukherjee, S. A Boundary Integral Equation Formulation for Time Dependent Inelastic Deformation in Metal, *Int. Jnl. Mech. Sci.*, 1977, **19**, 713–724.

[75] Chaudonneret, M. Methode des Equations Integrales Appliquees a la Resolution de Problemes de Viscoplasticite, *Jnl. Mecanique Appliquee*, 1988, **1**, 113–132.

[76] Bui, H. A. Some Remarks about the Formulation of Three Dimensional Thermo-elastoplastic Problems by Integral Equations, *Int. Jnl. Solids and Structures*, 1978, **14**, 935–939.

[77] Mukherjee, S. and Kumar, V. Numerical Analysis of Time Dependent Inelastic Deformation in Metallic Media using the Boundary Integral Equation Method, *Trans. ASME, Jnl. Appl. Mech.*, 1978, **45**, 785–790.

[78] Telles, J. C. F. and Brebbia, C. A. On the Applicaton of the Boundary Element Method to Plasticity, *Appl. Math. Modelling*, 1979, **3**, 466–470.

[79] Telles, J. C. F. and Brebbia, C. A. The Boundary Element Method in Plasticity, in *New Developments in Boundary Element Methods* (Bebbia, C. A. Ed.), CML Publications, Southampton, 1980, 295–317.

[80] Kobayashi, S. and Nishimura, N. Elastoplastic Analysis by the Integral Equation Method, Memo Faculty of Eng., Kyoto University, 1980, 42, Py. 3, 324–334.

[81] Telles J. C. F. and Brebbia, C. A. Elastoplastic Boundary Element Analysis, *Proc. Europe – U.S. Workshop on Nonlinear Finite Element Analysis in Structural Mechanics* (Wunderlich *et al.*, Eds.), Ruhr University Bochum, Germany, Springer-Verlag, Berlin, 1980, 403–434.

[82] Telles, J. C. F. and Brebbia, C. A. New Developments in Elastoplastic Analysis, *Appl. Math. Modelling*, 1981, **5**, 376–382.

[83] Mukherjee, S. and Morjaria, M. A Boundary Element Formulation for Planar, Time Dependent Inelastic Deformation of Plates with Cutouts, *Int. Jnl. Solids and Structures*, 1981, **17**, 115–126.

[84] Morjaria, M. and Mukherjee, S. Numerical Analysis of Planar Time Dependent Inelastic Deformation of Plates with Cracks by the Boundary Element Method, *Int. Jnl. Solids and Structures*, 1981, **17**, 127–143.

[85] Telles, J. C. F. and Brebbia, C. A. Elastic-Viscoplastic Problems using Boundary Elements, *Int. Jnl. Mech. Sci.*, 1982, **24**(1), 605–618.

[86] Brunet, M. Numerical Analysis of Cyclic Plasticity using the Boundary Integral Equation Method, in *Boundary Element Methods*, Springer-Verlag, Berlin and CML Publications, Southampton, 1981.

[87] Tanaka, M. New Boundary Element Methods for Viscoelastic Problems, in *BETECH/85* (C. A. Brebbia and B. J. Noye, Eds.), Springer-Verlag, Berlin and CML Publications, Southampton, 1985.

[88] Chandra, A. and Mukherjee, S. Applications of the Boundary Element Method to Large Strain Large Deformation Problems of Viscoplasticity, *Jnl. Strain Analysis*, October 1983, **18**, 261–270.

[89] Chandra, A. and Mukherjee, S. A Boundary Element Formulation for Sheet Metal Forming, *Applied Math. Modelling*, 1983, **9**, 175–182.

[90] Chandra, A. and Mukherjee, S. A Boundary Element Formulation for Large Strain Problems of Compressible Plasticity, *Engineering Analysis*, 1986, **3**, 71–78.

[91] Mukherjee, S. and Poddar, B. An Integral Equation for Elastic and Inelastic Shell Analysis, *Proc. of the Int. Conf. on BEM in Engineering, Beijing, China, October*, Pergamon Press, Oxford, UK, 1986.

[92] Chang, Y. P., Kang, C. S. and Chen, D. J. The Use of Fundamental Green's Functions for the Solution of Problems of Heat Conduction in Anisotropic Media, *Int. Jnl. Heat Mass Transfer*, 1973, **16**, 1905–1918.

[93] Wrobel, L. C. and Brebbia, C. A.The Boundary Element Method for Steady-State and Transient Heat Conduction, *Proc. First Conf. on Numerical Methods in Thermal Problems* (R. W. Lewis and K. Morgan, Eds.), Pineridge Press, Swansea, 1979.

[94] Brebbia, C. A. and Wrobel, L. C. Steady and Unsteady Potential Problems using the Boundary Element Method, in *Recent Advances in Numerical Methods in Fluids*, Pineridge Press, Swansea, Wales 1979.

[95] Wrobel, L. C. and Brebbia, C. A. A Formulation of the Boundary Element Method for Axisymmetric Transient Heat Conduction, *Int. Jnl. Heat Mass Transfer*, 1981, **24**, 843–850.

[96] Onishi, K. Convergence of the Boundary Element Method for Heat Equation, *T.R.U. Mathematics*, 1981, **17**(2), 213–225.

[97] Tanaka, M. and Tanaka, K. Transient Heat Conduction Problems in Inhomogeneous Media Discretized by Means of Boundary-Volume Elements, *Nuclear Eng. and Design*, 1980, **60**, 381–387.

[98] Yoshikawa, F. and Tanaka, M. A Boundary Element Analysis of Steady-State Heat Conduction Problems in Axisymmetric Body Heat Transfer, *Japan Research*, 184, **13**, 51–75.

[99] Kikuta, M., Togoh, H. and Tanaka, M. A Boundary Element Method for Non-linear Transient Heat Conduction Problems, *Boundary Element VIII* (M. Tanaka and C. A. Brebbia, Eds.), CM Publications, Southampton and Springer-Verlag, Berin, 1986.

[100] Bialeci, R. and Nowa, A. J. Boundary Value Problems in Heat Conduction with Non-linear Material and Non-linear Boundary Condition, *Appl. Math. Modelling*, 1981, **5**, 417–421.

[101] Skerget, P. and Brebbia, C. A. Non-linear Potential Problems, Chapter 1 in *Progress in Boundary Element Methods, Vol. 2*, Pentech Press, London and Springer-Verlag, NY, 1982.

[102] Skerget, P. and Brebbia, C. A. Time Dependent Non-linear Potential Problems, Chapter 3 in *Topics in Boundary Element Research, Vol. 2*, Springer-Verlag, Berlin and NY, 1985.

[103] Nardini, D. and Brebbia, C. A. The Solution of Parabolic and Hyperbolic Problems using an Alternative Boundary Element Formulation, *Boundary Elements VII* (C. A. Brebbia and G. Maier, Eds.), Springer-Verlag, Berlin and NY, 1985.

[104] Wrobel, L. C., Brebbia, C. A. and Nardini, D. Analysis of Transient Thermal Problems in the BEASY System, *BETECH/86* (J. J. Connor and C. A. Brebbia, Eds.), Computational Mechanics Publications, Southampton, 1986.

[105] Liggett, J. A. Location of Free Surface in Porous Media, *Jnl. Hydraulics Div., ASCE*, 1977, **103**, 353–365.

[106] Liu, P. L-F. and Liggett, J. A. Boundary Solutions to Two Problems in Porous Media, *Jnl. Hydralics Div., ASCE*, 1979, **105**, 171–183.

[107] Lennon, G. P., Liu, P. L-F. and Liggett, J. A. Boundary Integral Equation Solution to Axisymmetric Potential Flows: Part I Basic Formulation, – Part II, Recharge and Well Problems in Porous Media, *Water Resources Res.*, 1979, **15**, 1102–1115.

[108] Lennon, G. P., Liu, P. L-F. and Liggett, J. A. Boundary Integral Solutions to Three Dimensional Unconfined Darcy's Flow, *Water Resources Research*, 1980, **6**, 651–658.

[109] Liu, P. L-F., Cheng, A. H. D., Liggett, J. A. and Lee, J. H. Boundary Integral Equation Solutions to Moving Interface between Two Fluids in Porous Media, *Water Resources Research*, 1981, **17**, 1445–1452.

[110] Shaw, R. P. Diffracton of Acoustic Pulses by Obstacles of Arbitrary Shape with a Robin Boundary Condtion, Part A, *J.A.S.A.*, 1967, **41**, 855–859.

[111] Shaw, R. P. Singularities in Acoustic Pulse Scattering by Free Surface Obstacles with Sharp Corners, *Jnl. Appl. Mech.* 1971, **38**, 526–528.

[112] Shaw, R. P. Transient Scattering by a Circular Cylinder, *Jnl. Sound Vibration*, 1975, **42**, 295–304.

[113] Mitzner, R. M. Numerical Solution for Transient Scattering from a Hard Surface of Arbitrary Shape-Retarded Potential Technique, *Jnl. Acoust. Soc. Amer.*, 1967, **42**, 391–397.

[114] Groenenboom, P. H. L. Wave Propagation Phenomena, Chapter 2 in *Progress in Boundary Element Methods* (C. A. Bebbia, Ed.), Pentech Press London, Springer-Verlag, NY, 1983.

[115] Groenenboom, P. H. L., de Jong, J. J. and Brebbia, C. A. BEWAVE – Pressure Wave Propagation by the Boundary Element Program, *Structural Analysis Systems, Vol. 3* (A. Niku-Lari, Ed.), Pergamon Press, NY, 1986.

[116] Cole, D. M., Kosloff, D. D. and Minster, J. B. A Numerical Boundary Integral Equation Method for Elastodynamics I, *Bull. Seis. Soc. America*, 1978, **68**(5), 1331–1357.

[117] De Mey, G. M. Calculation of Eigenvalues of Helmholtz Equation by an Integral Equation, *Int. Jnl. Num. Meth. Engng.*, 1976, **10**, 59–66.

[118] Hutchinson, J. R. Determination of Membrane Vibrational Characteristics by the Boundary Integral Equation Methods, *Recent Advances in Boundary Element Methods* (C. A. Brebbia, Ed.), Pentech Press, London, 1978, 301–316.

[119] Mansur, W. J. and Brebbia, C. A. Formulation of the Boundary Element Method for Transient Problems Governed by the Scalar Wave Equation, *Appl. Math. Modelling*, 1982, **6**, 307–311.

[120] Mansur, W. J. and Brebbia, C. A. Numerical Implementation of the Boundary Element Method for Two Dimensional Transient Scalar Wave Propagation Problems, *Appl. Math. Modelling*, 1982, **6**m, 299–306.

[121] Mansur, W. J. and Brebbia, C. A. Transient Elastodynamics using a Time-Stepping Technique, *Boundary Elements* (C. A. Brebbia, *et al.* Eds.), Springer-Verlag, Berlin and CML Publications, Southampton, 1983.

[122] Mansur, W. J. and Brebbia, C. A. Transient Elastodynamics, Chapter 5 in *Topics in Boundary Element Research, Vol. 2, Time Dependent and Vibration Problems* (C. A. Brebbia, Ed.), Springer-Verlag, Berlin and NY, 1985.

[123] Manolis, G. D. and Beskos, D. E. Dynamics Stress Concentration Studies by Boundary Integrals and Laplace Transform, *Int. Jnl. Num. Meth. Engng.*, 1981, **17**, 573–599.

[124] Niwa, Y., Fukui, T., Kato, S. and Fujiki, K. An Application of the Integral Equation Method to Two Dimensional Elastodynamics, *Theoret. and Appl. Mech.*, University of Tokyo Press, 1980, **28**, 281–290.

[125] Niwa, Y., Kobayashi, S. and Azuma, N. An Analysis of Transient Stresses Produced around Cavities of an Arbitrary Shape during the Passage of Travelling Wave, *Memo. Fac. Eng.*, Kyoto University, 1975, **37**, 28–46.

[126] Niwa, Y., Kobayashi, S. and Fukui, T. Application of Integral Equation Method to Solve Some Geomechanical Problems, *Proc. 2nd Int. Conf. Numerical Meth. Geomechanics, ASCE,* 1976, 120–131.

[127] Kobayash, S. and Nishimura, N. Dynamics Analysis of Underground Structures by the Integral Equation Method, *Proc. 4th Int. Conf. Num. Method. Geomechanics, Balkema, 1982,* **1**, 401–409.

[128] Kobayashi, S. and Nishimura N. Green's Tensor for Elastic Half-Space – An Application of Boundary Integral Equation Method, *Memo. Faculty of Eng.*, Kyoto University, 1980, **42** Pt. 2, 228–241.

[129] Kobayashi, S. and Nishimura, N. *Analysis of Dynamic Soil-Structure Interaction by Boundary Integral Equation Method, Vol.* (P. Lascaux, Ed.), Pluralis, Paris, 1983, 353–362.

[130] Kitahara, M. Boundary Integral Equation Methods in Eigenvalue Problems of Elastodynamics and Thin Plates, *Studies in Appl. Maths.*, Elsevier, Amsterdam, 1985, 10.

[131] Niwa, Y., Kobayashi, S. and Kitahara M. Eigenfrequency Analysis of Plates by Integral Equation Method, in *Innovative Numerical Analysis for the Applied Engineering Sciences, Supplement* (R. P. Shaw *et al.* Eds.), University of Virginia Press, 1980.

[132] Vivoli, J. and Filippi, P. Eigenfrequencies of Thin Plates and Layer Potentials, *Jnl. Acoust. Soc. Am.*, 1974, **55**, 562–567.

[133] Niwa, Y., Kobayashi, S. and Kitahara, M. Applications of Integral Equation Method to Eigenvalue Problems of Elasticity, *Proc. Japan Soc. Civil Eng.* (in Japanese), 1979, **285**, 17–28.

[134] Niwa, Y., Kobayashi, S. and Kitahara, M. Eigenfrequency Analysis of a Plate by the Integral Equation Method, *Theoret. Appl. Mech.*, University of Tokyo Press, 1981, **29**, 287–307.

[135] Niwa, Y., Kobayashi, S. and Kitahara, M. Determination of Eigenvalues by Boundary Element Methods, Chapter 6 of *Developments in Boundary Element Methods – Vol. 2* (P. K. Banerjee and R. P. Shaw, Eds.), Applied Science, 1982.

[136] Dominguez, J. and Alarcon, E. Elastodynamics, Chapter 7 of *Progress in Boundary Element Methods – Vol. 1* (C. A. Brebbia, Ed.), Pentech Press, 1981.

[137] Brebbia, C. A. and Nardini, D. Dynamic Analysis in Solid Mechanics by an Alternative Boundary Element Procedure, *Soil Dynamics and Earthquake Engineering Journal*, 1983, **2**(4).

[138] Nardini, D. and Brebbia, C. A. Boundary Integral Formulation of Mass Matrices for Dynamic Analysis, Chapter 7 in *Topics in Boundary Element Research, Vol. 2, Time Dependent and Vibration Problems*, Springer-Verlag, Berlin and NY, 1985.

[139] Wu. J. C. and Thompson, J. F. Numerical Solutions of Time Dependent Incompressible Navier-Stokes Equation using an Integro-Differential Formulation, *Comput. Fluids*, 1973, **1**, 197–215.

[140] Bratanow, T. and Sphehert, T. Computational Flow Development for Unsteady Viscous Flow, *NASA CR-2995*, 1978.

[141] Wu, J. C. Numerical Boundary Conditions for Viscous Flow Problems, *AIAA Jnl.*, 1976, **14**, 1042–1049.

[142] Wu, J. C. and Wahbah, M. M. Numerical Solution of Viscous Flow Equations using Integral Representations, *Lecture Notes in Physics*, Springer-Verlag, Berlin, 1976, 59.

[143] Wu, J. C. and Rizk, Y. M. Integral-Representation Approach for Time Dependent Viscous Flow, *Lecture Notes in Physics*, Springer-Verlag, NY, 1978, **90**, 558–564.

[144] Wu, J. C. and Wahbah, M. M. and Sugavanam, A. Some Numerical Solutions of Turbulent Flow Problems by the Use of Integral Reprsentations, *Proc. Symposium on Applications of Computer Methods in Engineering*, University of Southern California, Los Angeles, 1977.

[145] Wu, J. C. and Sugavanam, A. Method for the Numerical Solutions of Turbulent Flow Problems, *AIAA Jnl.*, 1978, **16**, 948–955.

[146] Brebbia, C. A. and Wrobel, L. C. Application of Boundary Elements in Fluid Flow, *Proc. of the 2nd Int. Conf. on Finite Elements in Water Resources, Imperial College, London, July 1978*, Pentech Press, London, 1978.

[147] Skerget, P., Alujevic, A. and Brebbia, C. A. The Solution of Navier-Stokes Equations in Vorticity-Velocity Variables by Boundary Elements, in *Boundary Elements VI* (C. A. Brebbia, Ed.), Springer-Verlag, Berlin and CML Publications, Southampton, 1984, 4/41–4/56.

[148] Skerget, P., Alujevic, A. and Brebbia, C. A. Analysis of Laminar Flows with Separation using BEM, in *Boundary Elements VII* (C. A. Brebbia and G. Maier, Eds.), Springer-Verlag, Berlin and CML Publications, Southampton, 1985, 9/23–9/36.

[149] Onishi, K., Kuroki, T. and Tanaka, M. Boundary Element Method for Laminar Viscous Flow and Convective Diffusion Problems, *Topics in Boundary Element Research, Vol. 2*, Springer-Verlag, Berlin, 1985, 209–229.

[150] Bush, M. B. and Tanner, R. I. Numerical Solution of Viscous Flows using Integral Equation Methods, *Int. Jnl. of Viscous Flows using Integral Equation Methods, Int. Jnl. of Num. Meth. in Fluids*, 1983, **3**, 71–92.

[151] Bush, M. B. Modelling Two Dimensional Flow Past Arbitrary Cylindrical Bodies using Boundary Element Formulations, *Appl. Math. Modelling*, 1983, **7**, 386–394.

[152] Tosaka, N. and Kakuda K. Numerical Solutions of Steady Incompressible Viscous Flow Problems by the Integral Equation Method, *Proc. of 4th Int. Conf. on Numerical Methods for Engineers,* 1986, 211–222.

[153] Tosaka, N. and Fukushima, N. Integral Equation Analysis of Laminar Natural Convection Problem, in *Boundary Elements VIII* (M. Tanaka and C. A. Brebbia, Eds.), CM Publications, Southampton, Springer-Verlag, Berlin, 1986, 803–812.

[154] Kitagawa, K., Brebbia, C. A., Wrobel, L. C. and Tanaka, M. Boundary Element Analysis of Viscous Flow by Penalty Function Formulation, *Eng. Analysis*, 1986, **3**(4), 194–200.

[155] Kitagawa, K., Wrobel, L. C., Brebbia, C. A. and Tanaka, M. Modelling Thermal Transport Problems using the Boundary Element Method, *Proc. of the Int. Conf. on Development and Application of Computer Techniques to Environmental Studies*, CM Publications, Southampton, 1986, 715–731.

[156] Onishi, K., Kuroki, T. and Tanaka, M. An Application of Boundary Element Method to Incompressible Laminar Viscous Flows, *Eng. Analysis*, 1984, **1**(3), 122–127.

[157] Onishi, K., Kuroki, T. and Tanaka, M. An Application of Boundary Element Method to Natural Convection, *Appl. Math. Modelling*, 1984, **8**, 383–390.

[158] Bush, M. D. and Tanner, R. I. The Boundary Element Method Applied to the Warping Motion of a Sphere, *Proc. 2nd Int. Conf. on Numerical Methods in Laminar and Turbulent Flow*, Pineridge, Press, Swansea, 1981.

[159] Kelmanson, M. A. An Integral Equation Method for the Solution of Singular Slow Flow Problems, *Jnl. Compt. Phys.*, 1983, **51**, 139–324.

[160] Kelmanson, M. A. Modified Integral Equation Solution of Viscous Flow near Sharp Corners, *Computers and Fluids*, 1983, **11**, 307–324.

[161] Kelmanson, M. A. Boundary Integral Equation Solution of Viscous Flows with Free Surfaces, *Jnl. Engg. Math.*, 1983, **17**, 329–343.

[162] Ingham, D. B. and Kelmanson, M. A. A Boundary Integral Equation Method for the Study of Slow Flow within Bearing Geometries, *Boundary Elements* (C. A. Brebbia *et al.*,Ed.), Springer-Verlag, Berlin and CML Publications, Southampton, 1983.

[163] Brebbia, C. A. and Wrobel, L. C. Viscous Flow Problems by the Boundary Element Method, Chapter 1 in *Computational Techniques for Fluid Flow* (C. Taylor *et al.*, Ed.), Pineridge Press, Swansea, UK, 1986.

[164] Niwa, Y., Kobayashi, S. and Fukui, T. An Application of the Integral Equation Method to Plate-Bending Problems, *Memo. Faculty of Eng.*, Kyoto University, 1974, **36**, Pt. 2, 140–158.

[165] Bezine, G. P. and Gamby, D. A. A New Integral Equation Formulation for Plate Bending Problems, in *Recent Advances in Boundary Element Methods* (C. A. Brebbia, Ed.), Pentech Press, London, 1978.

[166] Bezine, G. P. Application of Similarity of Results of New Boundary Integral Equations for Plate Failure Problems, *Appl. Math. Model*, 1985, **5**.

[167] Bezine, G. P. Boundary Integral Formulation for Plate Flexure with Arbitrary Boundary Conditions, *Mech. Res. Comm.*, 1978, **5**.

[168] Stern, M. Boundary Integral Equations for Bending of Thin Plates, Chapter 6 in *Vol. 2, Progress in Boundary Elements* (C. A. Brebbia, Ed.), Pentech Press, London, Springer-Verlag, NY, 1983.

[169] Stern, M. A Choice of Fundamental Solutions, Chapter 18 in *Boundary Element Techniques in Computer-Aided Engineering* (C. A. Brebbia, Ed.), NATO ASI Series, Series E: Applied Sciences – No. 84, Martinus Nijhoff Publishers, Dordrecht, 1984.

[170] Stern, M. Formulating Nonsingular Boundary Integral Equations in Linear Elasticity, in *Advanced Topics in Boundary Element Analysis* (T. A. Cruse, A. B. Pifko and H. Armen, Eds.), ASME, New York, 1985.

[171] Weeen, Van Der F. Application of the Direct Boundary Element Method to Reissner's Plate Model, in *Boundary Elements in Engineering* (C. A. Brebbia, Ed.), Springer-Verlag, Berlin, 1982.

[172] Weeen, Van Der F. Application of the Boundary Integral Equation Method to Reissner's Plate Model, *Int. Jnl. Numer. Meth. Engng.*, 1982, **18**, 1–10.

[173] Long, S. Y., Brebbia, C. A. and Telles, J. C. F. Boundary Element Bending Analysis of Moderately Thick Plates, *Engng. Analysis*, 1987, **4**(3).

[174] Kamiya, N., Sawaki, Y. and Nakamura, J. Boundary Element Nonlinear Bending Analysis of Clamped Sandwich Plates and Shells, in *Boundary Elements in Engineering* (C. A. Brebbia, Ed.), Springer-Verlag, Berlin and CML Publications, Southampton, 1982.

[175] Tanaka, M. Integral Equation Approach to Small and Large Displacements of Thin Elastic Plates, in *Boundary Elements in Engineering* (C. A. Brebbia, Ed.), Springer-Verlag, Berlin and CML Publications, Southampton, 1982.

[176] Kim, J. W. On the Computation of the Stress Intensity Factors in Elastic Plate Flexure via Boundary Integral Equations, in *Boundary Elements in Engineering* (C. A. Brebbia, Ed.), Springer-Verlag, Berlin and CML Publications, Southampton, 1982.

[177] Tanaka, M. and Miyazaki, K. A Direct BEM for Elastic-Plate Structures subjected to Arbitrary Loadings, in *Boundary Elements VII* (C. A. Brebbia and G. Maier, Eds.), Springer-Verlag, Berlin and CML Publications, Southampton, 1985.

[178] Costa, J. A. and Brebbia, C. A. Elastic Buckling of Plates using the Boundary Element Method, in *Boundary Elements VII* (C. A. Brebbia and G. Maier, Eds.), Springer-Verlag, Berlin and CML Publications, Southampton, 1985.

[179] Antes, H. On Boundary Integral Equations for Circular Cylindrical Shells, in *Boundary Element Methods* (C. A. Brebbia, Ed.), Springer-Verlag, Berlin and NY, 1981.

[180] Kamita, N., Sawaki, Y. and Nakamura J. Finite and Postbuckling Deformation of Heated Plates and Shallow Shells, in *Boundary Elements* (C. A. Brebbia *et al.*, Eds.), Springer-Verlag, Berlin and CML Publications, Southampton, 1982.

[181] Tosaka, N. and Miyake, S. Non-linear Analysis of Elastic Shallow Shells by Boundary Element Methods, in *Boundary Element Methods VII* (C. A. Brebbia and G. Maier, Eds.), Springer-Verlag, Berlin and CML Publications, Southampton, 1985.

[182] McDonald, B. H. and Wexler, A. Finite Element Solutions of Unbounded Field Problems, *IEEE Trans. Microwave Theory MTT-20*, 1972, 841–847.

[183] Shaw, R. P. and Falby, W. FIBIE: A Combined Finite Element – Boundary Integral Equation Method, *Proc. 1st Symposium on Innovative Numerical Analysis in Applied Engineering Science, Versailles*, CETIM, 1977.

[184] Osias, J. W., Wilson, R. B. and Seitelman, L. A. Combined Boundary Integral Equation Finite Element Analysis of Solids, *Proc. 1st Symposium on Innovative Numerical Analysis in Applied Engineering Science, Versailles*, CETIM, 1979.

[185] Brebbia, C. A. and Georgiou, P. Combination of Boundary and Finite Elements for Elastostatics, *Appl. Math. Modelling*, 1979, **3**, 212–220.

[186] Margulies, M. Combination of the Boundary Element and Finite Element Methods, in *Progress in Boundary Element Methods, Vol. 1* (C. A. Brebbia, Ed.), Pentech Press, London; Halstead Press, NY, 1981.

[187] Beer, G. and Meek, J. L. Coupling of Boundary and Finite Element Methods for Infinite Domain Problems in Elasto-Plasticity, in *Boundary Element Methods* (C. A. Brebbia, Ed.), Springer-Verlag, Berlin and NY, 1981.

[188] Garrison, C. J. and Chow, P. Y. Wave Forces on Submerged Bodies, *J. Waterways, Harbours and Coastal Eng. Div., ASCE*, 1972, **98**, 375–392.

[189] Eatock-Taylor, R. Generalized Hydrodynamic Forces on Vibrating Offshore Structures by Wave Diffraction Technique, *Offshore Structures Engg.* (E. L. L. B. Cerneiro *et al.*, Eds.), Pentech Press, London, 1977, 249–274.

[190] Au, M. C. and Brebbia, C. A. Numerical Prediction of Wave Forces using the Boundary Element Method, *Applied Math. Modelling*, 1982, **6**, 218–228.

[191] Au, M. C. and Brebbia, C. A. Computation of Wave Forces on Three Dimensional Offshore Structures, *Boundary Element Methods in Engineering* (C. A. Brebbia, Ed.), Springer-Verlag, Berlin and CML Publications, Southampton, 1982.

[192] Au, M. C. and Brebbia, C. A. Water Wave Analysis, Chapter 5 in *Topics in Boundary Element Research, Vol. 1*, Springer-Verlag, Berlin and NY, 1984.

[193] Wrobel, L. C., Spahier, S. H. and Esperanca, P. T. T. Propagation of Surface Waves, Chapter 6 in *Topics in Boundary Element Research, Vol. 2*, Springer-Berlin, and NY, 1985.

[194] Masuda, K. and Kato, W. Hybrid B.E.M. for Calculating Nonlinear Wave Forces on Three Dimensional Bodies, *Boundary Elements* (C. A. Brebbia *et al.*, Eds.), Springer-Verlag, Berlin and CML Publications, Southampton, 1983.

[195] Liu, P. L-F. Integral Equation Solutions to Nonlinear Free-Surface Flows, *2nd Int. Conf. on Finite Elements in Water Resources*, Pentech Press, London, 1978.

[196] Nakayama, T. Boundary Element Analysis of Nonlinear Water Wave Problems, *Int. Jnl. Num. Meth. Engng.*, 1983, **19**, 953–970.

[197] Kim, S. K., Liu, P. L-F. and Liggett, J. A. Boundary Integral Equation Solutions for Solitary Wave Generation, Propagation and Run-up, *Coastal Engineering*, 1987, **7**, 299–317.

[198] Starfield, A. M. and Crouch, S. L. Elastic Analysis of Single Seam Extraction, in *New Horizons in Rock Mechanics* (H. R. Hardy Jr. and R. Stefhanko, Eds.), ASCE, New York, 1973, 421–439.

[199] Brady, B. H. G. and Bray, J. W. The Boundary Element Method for Elastic Analysis of Tabular Orebody Extraction, assuming Complete Plane Strain, *Int. Jnl. Rock Mech. Min. Sci. and Geomech. Abstr.*, 1978, **15**, 29–37.

[200] Hocking, G. Stress Analysis of Underground Excavations incorporating Slip and Separation along Discontinuities, in *Recent Advances in Boundary Element Methods* (C. A. Brebbia, Ed.), Pentech Press, London, 1978, 195–214.

[201] Wardle, L. J. and Crotty, J. M. Two Dimensional Boundary Integral Equation Analysis for Non-homogeneous Mining Applications, in *Recent Advances in Boundary Element Methods* (C. A. Brebbia, Ed.), Pentech Press, London, 1978, 233–251.

[202] Beer, G. *et al.* Efficient Analysis in Geomechanics, *4th Int. Conf. on Numerical Methods in Geomechanics* (A. A. Eisenstein, Ed.), Balkema, Rotterdam, 1982, 1, 5–13.

[203] Beer, G. and Meek, J. L. Coupled Finite Element – Boundary Element Analysis of Infinite Domain Problems, *Int. Conf. on Num. Meth. for Coupled Problems* (Bettess *et al.*, Eds.), Pineridge Press, Swansea, Wales, 1981.

[204] Beer, G. and Meek, J. L. Applications in Mining, Chapter 8 in *Topics in Boundary Element Research, Vol. 1*, Springer-Verlag, Berlin and NY, 1984.

[205] Venturini, W. S. and Brebbia, C. A. Application in Geomechanics, Chapter 7 in *Topics in Boundary Element Research, Vol. 1*, Springer-Verlag, Berlin and NY, 1984.

[206] Rudolphi, T. J. An Implementation of the Boundary Element Method for Zoned Media with Stress Discontinuities, *Int. Jnl. for Numerical Meth. in Engng.*, 1983, **19**, 1–15.

[207] Andersson, T. and Allan-Persson, B. G. The Boundary Element Method Applied to Two Dimensional Contact Problems, Chapter 5 in *Progress in Boundary Element Methods, Vol. 1* (C. A. Brebbia, Ed.), Pentech Press, London, 1981.

[208] Bezine, G. and Fortune, D. Contact between Plates by a New Direct Boundary Integral Equation Formulation, *Int. Jnl. Solids and Structures* 1984, **20**(8), 739–746.

[209] Abdul-Mihsein, M. J., Bakr, A. A. and Parker, A. P. A Boundary Integral Equation Method for Axisymmetric Elastic Contact Problems, *Computer and Structures*, 1986, **23**(6), 787–793.

[210] Paris, F. and Garrido, J. A. On the Use of Discontinuous Elements in Two Dimensional Contact Problems, in *Boundary Elements VII* (C. A. Brebbia and G. Maier, Eds.), Springer-Verlag, Berlin and CML Publications, Southampton, 1985.

[211] Danson, D. J. and Warne, M. A. Current Density/Voltage Calculation using Boundary Element Techniques, *NACE 1983 Conf. Proc.*, Los Angeles, USA, 1983.

[212] Adey, R. A., Brebbia, C. A. and Niku, S. M. BEASY-CP. A System for Analysis of Galvanic Corrosion and Cathodic Protection using Boundary Elements, *BETECH/86*, Computational Mechanics Publications, Southampton, 1986.

[213] Ancelle, R. and Sabonnadiere, J. C. A Numerical Solution of 3-D Magnetic Field Problems using Boundary Integral Equations, *IEEE Transactions on Magnetics*, September 1980, MAG-16, No. 5, 1089–1091.

[214] Salon, S. J., Schneider G. M. and UDA, S. Boundary Element Solution to the Eddy Current Problem, *Proc. of the 3rd Int. Conf. on BEM, Irvine, CA, 1981*, CML Publications, Southampton and Springer-Verlag, Berlin, 1981.

[215] Magureanu, R., Vasle, N. and Tiba, M. Calculation of the Magnetic Field and Parameters of the Synchronous Motors with Ceramic-Permanent Magnets using the Boundary Element Mehod, *Engineering Analysis, Jnl.*, 1987, **4**(2).

[216] Tsuboi, H. and Misaki, T. An Analysis of Three Dimensional Electromagnetic Field by using Boundary Element Method, *BEM VIII* (C. A. Brebbia *et al.*, Eds.), CM Publications, Southampton and Springer-Verlag, Berlin and NY, 1986.

[217] Brebbia, C. A. and Magureanu, R. The Boundary Element Method for Electromagnetic Problems, *Engineering Analysis, Jnl.*, 1987, **4**(4).

[218] Enokizono, M., Kagawa, R. and Nakamura T. Non Linear Analysis of Magnetic Field by Boundary Element Method taking into Account External Power Sources, *BEM VII* (C. A. Brebbia *et al.*,Eds.), CM Publications, Southampton and Springer-Verlag, Berlin, 1986.

[219] Wendland, W. Asymptotic Accuracy and Convergence, in *Progress in Boundary Element Methods, Vol. 1* (C. A. Brebbia, Ed.), Pentech Press, London, 1981, 289–312.

[220] Wendland, W. On Asymptotic Error Analysis and Underlying Mathematical Principles for Boundary Element Methods, in *Boundary Element Techniques in Computer Aided Engineering* (C. A. Brebbia, Ed.), NATO ASI Series E-84 Martinus Nijhoff Publ., Dordrecht/Boston/Lancaster, 1984, 417–436.

[221] Wendland, W. Asymptotic Accuracy and Convergence for Point Collocation Methods, *Topics in Boundary Element Research 2* (C. A. Brebbia, Ed.), Springer-Verlag, Berlin, 1985, 230–257.

[222] Wendland, W. On Some Mathematical Aspects of Boundary Element Methods for Elliptic Problems, in *The Mathematics of Finite Elements and Applications V* (J. Whiteman, Ed.), Academic Press, London, 1985, 193–227.

[223] Wendland, W., Hsiao, G. C. and Kopp, P. The Synthesis of the Collocation and the Galerin Method Applied to Some Integral Equations of the First Kind, in *New Developments in Boundary Element Methods* (C. A. Brebbia, Ed.), CML Publications, Southampton, 1980, 122–136.

[224] Wendland, W. and Schwab, Ch. 3-D BEM and Numerical Integration, in *Boundary Elements in Engineering VII* (C. A. Brebbia and G. Maier, Eds.), Springer-Verlag, Berlin and CML Publications, Southampton, 1985, Vol. II, 13.85–13.102.

[225] Wendland, W. On Asymptotic Error Estimates for the Combined Boundary and Finite Element Method, in *Innovative Numerical Methods in Engineering* (R. P. Shaw *et al.*, Eds.), Springer-Verlag, Berlin and CML Publications, Southampton, 1986, 55–70.

[226] Wendland, W. On the Asymptotic Convergence of Boundary Integral Methods, in *Boundary Element Methods* (C. A. Brebbia, Ed.), Springer-Verlag, Berlin and CML Publications, Southampton, 1981.

[227] Wendland, W. and Saranan, J. On the Asymptotic Convergence of Collocation Methods with Spline Functions of Even Degree', *Math. Comp.* 1985, **45**, 91–108.

[228] Wendland, W. and Arnold, D. N. Collocation versus Galerkin Procedures for Boundary Integral Methods, in *Boundary Element Methods in Engineering* (C. A. Brebbia, Ed.), Springer-Verlag, Berlin and CML Publications, Southampton, 1982, 18–33.

[229] Wendland, W., Hsiao, G. C. and Stephan , H. On the Integral Equation Method for the Plane Mixed Boundary Value Problem with the Laplacian, *Mathematical Methods in the Applied Sciences*, 1979, **1**, 265–321.

[230]Wendland, W. and Stephan, E. Boundary Element Method for Membrane and Torsion Crack Problems, *Computer Meth. Appl. Mech. Eng.*, 1983, **36**, 331–358.

[231] Mota Soares, C. A., Rodriguez, H. C., Oliveira Farias, L. M. and Hang, E. J. Optimization of the Shape of Solids and Hollow Shafts using Boundary Elements, in *Boundary Elements* (C. A. Brebbia *et al.*, Eds.), Springer-Verlag, Berlin and CML Publications, Southampton, 1983.

[232] Miyamoto, Y., Iwasaki, S. and Sugimoto, H. On Study of Shape Optimization of 2-Dimensional Elastic Bodies by BEM, in *Boundary Elements VIII* (M. Tanaka and C. A. Brebbia, Eds.), Springer-Verlag, Berlin and CML Publications, Southampton, 1986.

[233] Kane, J. H. Shape Optimization Utilizing a Boundary Element Formulation, in

BETECH/86 (C. A. Brebbia and J. J. Connor, Eds.), Computational Mechanics Publications, Southampton, 1986.

[234] Novati, G. and Brebbia, C. A. Boundary Element Formulation for Geometrically Nonlinear Elastostatics, *Appl. Math. Mod.*, 1982, **6**.

[235] Brebbia, C. A. and Niku, S. M. State of the Art: Computational Applications of Boundary Element Methods for Cathodic Protection of Offshore Structures. *OMAE' 88* Conference Proceedings, ASME, 1988.

[236] Brebbia, C. A. and Trevelyan, J. On the Accuracy and Convergence of Boundary Element Results for the Floyd Pressure Vessel Problem, *Technical Note in Computers and Structures*, *Vol. 24, No. 3,* 513–516, Pergamon Press, 1986.

[237] Brebbia, C. A. Coupled Systems, Chapter 5 in *Finite Element Handbook* (H. Kardestuncer *et al.*, Eds.), McGraw-Hill Book Co., NY, 1987.

[238] Kitagawa, K., Wrobel, L. C., Brebbia, C. A. and Tanaka, M. A Boundary Element Formulation for Natural Convection Problems, *Int. Jnl. for Numerical Methods in Fluids*, **3**, 1988.

[239] Brebbia, C. A. and Tang, W. On the Treatment of Boundary Force Integrals in Boundary Elements, *CMWR Conference, Morocco*, Computational Mechanics Publications, Southampton and Springer-Verlag, Berlin and NY, 1988.

[240] Brebbia, C. A. and Umetani, M. Convergence and Accuracy Studies of Adaptive BEM Solutions, in *Proc. Reliability and Robustness Conference, Como*, Computational Mechanics Publications, Southampton, 1987.

[241] Brebbia, C. A. BEASY: A Boundary Element System of Structural Analysis, *Structural Analysis Systems, Vol. 1* (A. Niku-Lari, Ed.), Pergamon Press, Oxford, 1986.

[242] Brebbia, C. A. and DeFigueriredo, T. Critical Comparison of Finite and Boundary Element Methods, *NAFEMS, Brighton International Conf. on Quality Assurance and Standards in Finite Element Analysis*, 1987.

[243] Brebbia, C. A. and Wrobel, L. C. The Solution of Parabolic Problems using the Dual Reciprocity Boundary Element Method, *Proc. of IUTAM Symposium, San Antio, Texas, 1987*, Springer-Verlag, Berlin and NY, 1988.

[244] Wrobel, L. C. and Brebbia, C. A. The Dual Reciprocity Boundary Element Formulation for Nonlinear Diffusion Problems, *Computer Methods in Applied Mechanics and Engg.*, 1987, **65**, 147–164.

[245] Alessandri, C. and Brebbia, C. A. Strength of Masonry Walls under Static Horizontal Loads: Boundary Element Analysis and Experimental Tests, *Int. Jnl. Eng. Analysis*, **4**(3), 1987.

Appendix D

Answers to Selected Exercises

Chapter 1

1.1. $u = \frac{2}{7}x(1-x)$

1.2. $u = 0.4471x(x-1)$

1.3. $u = x(1-x)\exp(-\sqrt{8}\,y)$

1.4. $u = x(1-x)\exp(-\sqrt{10}\,y)$

1.5. $u = 1.333x - 0.333x^2$

1.6. $q_0 = \frac{1}{4};\ q = -\frac{1}{4}$

1.7. $u(\frac{1}{2}) = 7/32$

1.8. $u(\frac{1}{2}) = 7/32$

1.9. $a_1 = \dfrac{1}{\pi^2};\ a_2 = \dfrac{1}{3\pi^2}$

Chapter 2

2.2. No, because the only difference between the equations for two different scales would be the term

$$\int_\Gamma (u^*_{\text{scale 1}} - u^*_{\text{scale 2}})q\,d\Gamma = \text{constant} \int_\Gamma q\,d = 0$$

2.6. The solution is obtained locating two point sources in the complete space; one at the collocation point and the other at the mirror image of the first with respect to the free surface

$$u^* = \frac{1}{2\pi}\ln\frac{r'}{r}$$

where r' is the distance from the point image of the collocation point to the point of interest.

2.7. $u = \dfrac{1}{2\pi}\ln rr'$. The image source is now negative.

2.8. $$\left(\frac{\partial u^*}{\partial x_l}\right)^i = -\frac{\partial u^*}{\partial x_l} = \frac{1}{2\pi r} r_{,l}$$

$$\left(\frac{\partial q^*}{\partial x_l}\right)^i = -\frac{\partial q^*}{\partial x} = \frac{1}{2\pi r^2}\left(n_l - 2r_{,l}\frac{\partial r}{\partial n}\right)$$

2.11. $$q = \frac{\partial p}{\partial n} = -\frac{\partial p}{\partial x}$$

$$\begin{Bmatrix} q_1 \\ q_2 \\ q_3 \\ q_4 \end{Bmatrix} = \mathbf{A}\cdot\mathbf{B}^{-1}\begin{Bmatrix} p_1 \\ p_2 \\ p_3 \\ p_4 \end{Bmatrix}$$

where

$$\mathbf{A} = -\frac{\pi}{2H}\begin{bmatrix} \cos\frac{7\pi}{16} & 3\cos\frac{21\pi}{16} & 5\cos\frac{35\pi}{16} & 7\cos\frac{49\pi}{16} \\ \cos\frac{5\pi}{16} & 3\cos\frac{15\pi}{16} & 5\cos\frac{25\pi}{16} & 7\cos\frac{35\pi}{16} \\ \cos\frac{3\pi}{16} & 3\cos\frac{9\pi}{16} & 5\cos\frac{15\pi}{16} & 7\cos\frac{21\pi}{16} \\ \cos\frac{\pi}{16} & 3\cos\frac{3\pi}{16} & 5\cos\frac{5\pi}{16} & 7\cos\frac{7\pi}{16} \end{bmatrix}$$

$$\mathbf{B} = \begin{bmatrix} \cos\frac{7\pi}{16} & \cos\frac{21\pi}{16} & \cos\frac{35\pi}{16} & \cos\frac{49\pi}{16} \\ \cos\frac{5\pi}{16} & \cos\frac{15\pi}{16} & \cos\frac{25\pi}{16} & \cos\frac{35\pi}{16} \\ \cos\frac{3\pi}{16} & \cos\frac{9\pi}{16} & \cos\frac{15\pi}{16} & \cos\frac{21\pi}{16} \\ \cos\frac{\pi}{16} & \cos\frac{3\pi}{16} & \cos\frac{5\pi}{16} & \cos\frac{7\pi}{16} \end{bmatrix}$$

Chapter 3

3.2. u^*_{lk} is of the order of $\frac{1}{\varepsilon}$ when $\varepsilon \to 0$ and p_k is finite. Since the area of the surface of integration is of the order of ε^2, the integral will always have a zero limit when $\varepsilon \to 0$.

3.3. For an edge along the x_1 axis defined by the semi-planes $x_1, x_2 \geqslant 0$ and $x_1, x_3 \geqslant 0$

$$\mathbf{C}^i = \begin{bmatrix} \frac{1}{4} & 0 & 0 \\ 0 & \frac{1}{4} & \dfrac{1}{4\pi(1-\nu)} \\ 0 & \dfrac{1}{4\pi(1-\nu)} & \frac{1}{4} \end{bmatrix}$$

3.5. $\mathbf{u}^*$ is symmetric. This fact can be interpreted as a consequence of the reciprocity relation. It can also be verified by simple inspection of equations (3.51) and (3.54). $\mathbf{p}^*$ is not symmetric.

3.6. When a system of Cartesian coordinates with two axes parallel to the sides of the element is adopted all the off diagonal terms of $\mathbf{H}^{ii}$ and $\mathbf{G}^{ii}$ are zero.

3.7.

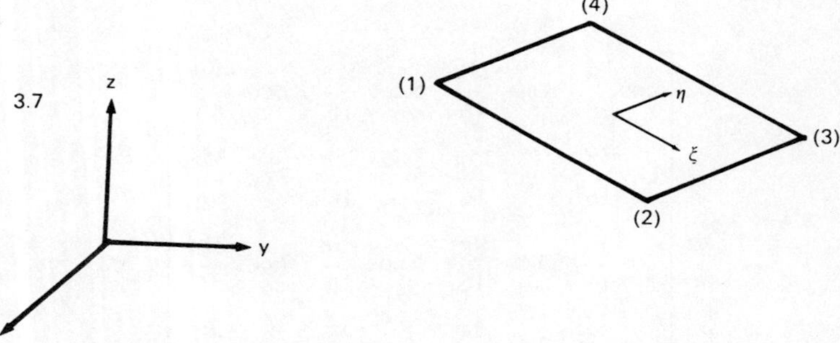

$$x = \frac{x_2 - x_1}{2}\xi + \frac{x_3 - x_2}{2}\eta + \frac{x_1 + x_2 + x_3 + x_4}{4}$$

$$y = \frac{y_2 - y_1}{2}\xi + \frac{y_3 - y_2}{2}\eta + \frac{y_1 + y_2 + y_3 + y_4}{4}$$

$$z = \frac{z_2 - z_1}{2}\xi + \frac{z_3 - z_2}{2}\eta + \frac{z_1 + z_2 + z_3 + z_4}{4}$$

$$d\Gamma = \frac{\sqrt{(x_2 - x_1)^2 + (y_2 - y_1)^2 + (z_2 - z_1)^2}}{2} \cdot \frac{\sqrt{(x_3 - x_2)^2 + (y_3 - y_2)^2 + (z_3 - z_2)^2}}{2}\, d\xi\, d\eta$$

$$d\Gamma = |G|\, d\xi\, d\eta$$

$$\int_{\Gamma_j} \mathbf{u}^*(x, y, z)\, d\Gamma = \int_{-1}^{1}\int_{-1}^{1} \mathbf{u}^*(\xi, \eta)|G|\, d\xi\, d\eta$$

$$= \sum_{l=1}^{3}\sum_{m=1}^{3} u^*(\xi_l, \eta_m) w_l w_m |G|$$

Similarly,

$$\int_{\Gamma_j} \mathbf{p}^*(x, y, z)\, d\Gamma = \int_{-1}^{1}\int_{-1}^{1} \mathbf{p}^*(\xi, \eta)|G|\, d\xi\, d\eta$$

$$= \sum_{l=1}^{3}\sum_{m=1}^{3} \mathbf{p}^*(\xi_l, \eta_m)|G| w_l w_m$$

where ξ_l, η_m are the coordinates given in table A1.1 of Appendix A and $w_l w_m$ the corresponding weighting factors.

3.8. A rigid body motion would change the term of the basic equation

$$\int_{\Gamma} u^*_{lk} p_k\, d\Gamma$$

by adding a constant to u^*_{lk}. Thus an additional term of the form

$$\int_{\Gamma} \text{constant } p_k\, d\Gamma = \text{constant} \int_{\Gamma} p_k\, d\Gamma$$

would be introduced.

For any bounded region in equilibrium,

$$\int_{\Gamma} p_k\, d\Gamma = 0$$

The above answer corresponds to problems with zero body forces, but the same can easily be shown to be true for nonzero body forces.

3.9. When the tractions over the internal boundaries have a zero resultant. Thus,

$$\int_{\Gamma_{\text{int}}} p_k\, d\Gamma = 0$$

where Γ_{int} are the internal boundaries.

3.10.

$$\mathbf{u}^*_c = \begin{bmatrix} -u^*_{21}\cos\theta - u^*_{22}\sin\theta & u^*_{21}\sin\theta - u^*_{22}\cos\theta & -u^*_{23} \\ u^*_{11}\cos\theta + u^*_{12}\sin\theta & -u^*_{11}\sin\theta + u^*_{12}\cos\theta & u^*_{13} \\ u^*_{31}\cos\theta + u^*_{32}\sin\theta & -u^*_{31}\sin\theta + u^*_{32}\cos\theta & u^*_{32} \end{bmatrix}$$

all the r are measured from points with $\theta' = -\dfrac{\pi}{2}$.

3.11.

$$\frac{1}{4}\begin{Bmatrix} u_l' \\ u_\theta' \\ u_z' \end{Bmatrix} + \begin{Bmatrix} \int_0^\pi (T_{11}^* c^2 + T_{12}^* sc) & \int_0^\pi (T_{11}^* s^2 - T_{12}^* sc) & \int_0^\pi (T_{13}^* c) \\ \int_{-\pi/2}^{\pi/2} (\hat{T}_{11}^* c^2 + \hat{T}_{12}^* sc) & \int_{-\pi/2}^{\pi/2} (\hat{T}_{11}^* s^2 - \hat{T}_{12}^* sc) & \int_{-\pi/2}^{\pi/2} (\hat{T}_{13}^* c) \\ \int_0^\pi (T_{31}^* c^2 + T_{32}^* sc) & \int_0^\pi (T_{31}^* s^2 - T_{32}^* sc) & \int_0^\pi (T_{33}^* c) \end{Bmatrix} d\theta \begin{Bmatrix} u_l' \\ u_\theta' \\ u_z' \end{Bmatrix}$$

$$= \begin{Bmatrix} \int_0^\pi (u_{11}^* c^2 + u_{12}^* sc) & \int_0^\pi (u_{11}^* s^2 - u_{12}^* sc) & \int_0^\pi (u_{13}^* c) \\ \int_{-\pi/2}^{\pi/2} (\hat{u}_{11}^* c^2 + \hat{u}_{12}^* sc) & \int_{-\pi/2}^{\pi/2} (\hat{u}_{11}^* s^2 - \hat{u}\,_{12}^* sc) & \int_{-\pi/2}^{\pi/2} (\hat{u}_{13}^* c) \\ \int_0^{2n} (u_{31}^* c^2 + u_{32}^* sc) & \int_0^\pi (u_{31}^* s^2 - u_{32}^* sc) & \int_0^\pi (u_{33}^* c) \end{Bmatrix} d\theta \begin{Bmatrix} t_\rho' \\ t_\theta' \\ t_z' \end{Bmatrix}$$

where

$$c = \cos\theta$$

$$s = \sin\theta$$

$$c^2 = \cos^2\theta$$

$$s^2 = \sin^2\theta$$

$\hat{u}_{ij}^*$, $\hat{t}_{ij}^*$ fundamental solution with the collocation point located at $\theta' = -\pi/2$.

Chapter 4

4.1.

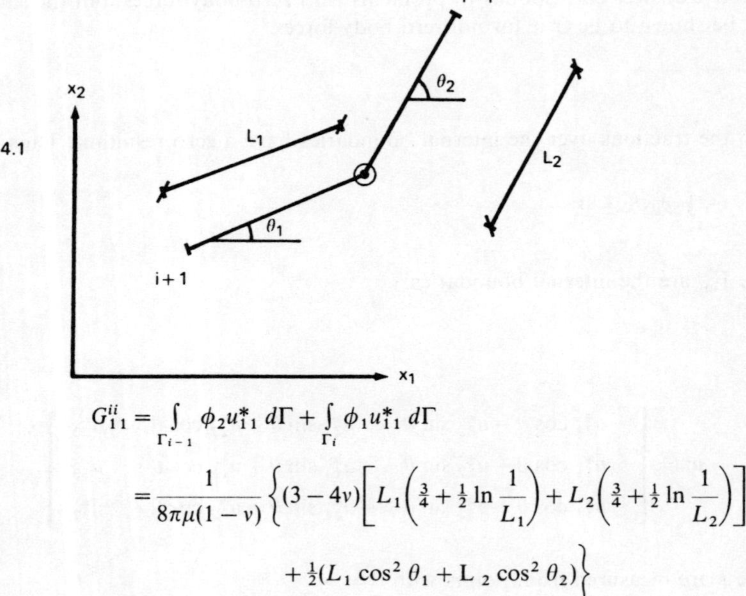

$$G_{11}^{ii} = \int_{\Gamma_{i-1}} \phi_2 u_{11}^* \, d\Gamma + \int_{\Gamma_i} \phi_1 u_{11}^* \, d\Gamma$$

$$= \frac{1}{8\pi\mu(1-\nu)} \left\{ (3-4\nu)\left[L_1\left(\tfrac{3}{4} + \tfrac{1}{2}\ln\frac{1}{L_1}\right) + L_2\left(\tfrac{3}{4} + \tfrac{1}{2}\ln\frac{1}{L_2}\right)\right] + \tfrac{1}{2}(L_1\cos^2\theta_1 + \mathrm{L}_2\cos^2\theta_2) \right\}$$

4.2.

$$G^{ii}_{11} = \frac{1}{8\pi\mu(1-\nu)}\left\{(3-4\nu)(L_1+L_2)\int_0^1 (1-\eta)\ln\frac{1}{r}\,d\eta + \tfrac{1}{4}\left[(3-4\nu)\left(L_1 \ln\frac{1}{L_1} + L_2 \ln\frac{1}{L_2}\right) + (L_1\cos^2\theta_1 + L_2\cos^2\theta_2)\right]\int_{-1}^{1}(1-\xi)\,d\xi\right\}$$

4.4. The distance must be $\geqslant L/4$; where L is the length of the element. Otherwise the geometry of the element would not be a segment going from node 1 to node 3 but a segment starting from node 1 would go in the opposite direction to node 3 and after a short distance would turn back towards node 3.

4.5 and **4.6**

Nodal Coordinates		Normal Traction		
X	Y	32 Quadr.	16 Quadr.	32 Const.
0.12700E + 02	0.00000E + 00	0.19068E + 02	0.19577E + 02	—
0.12700E + 02	0.57500E + 00	0.19537E + 02	—	0.20389E + 02
0.12700E + 02	0.11500E + 01	0.20711E + 02	0.21116EE + 02	—
0.12700E + 02	0.17250E + 01	0.22778E + 02	—	0.22627E + 02
0.12700E + 02	0.23000E + 01	0.25493E + 02	0.25605E + 02	—
0.12700E + 02	0.28750E + 01	0.28975E + 02	—	0.28570E + 02
0.12700E + 02	0.34500E + 01	0.32979E + 02	0.33133E + 02	—
0.12700E + 02	0.40250E + 01	0.37576E + 02	—	0.36346E + 02
0.12700E + 02	0.46000E + 01	0.42561E + 02	0.42425E + 02	—
0.12700E + 02	0.51750E + 01	0.47899E + 02	—	0.44667E + 02
0.12700E + 02	0.57500E + 01	0.53376E + 02	0.53303E + 02	—
0.12700E + 02	0.63250E + 01	0.58769E + 02	—	0.59331E + 02
0.12700E + 02	0.69000E + 01	0.63560E + 02	0.63044E + 02	—

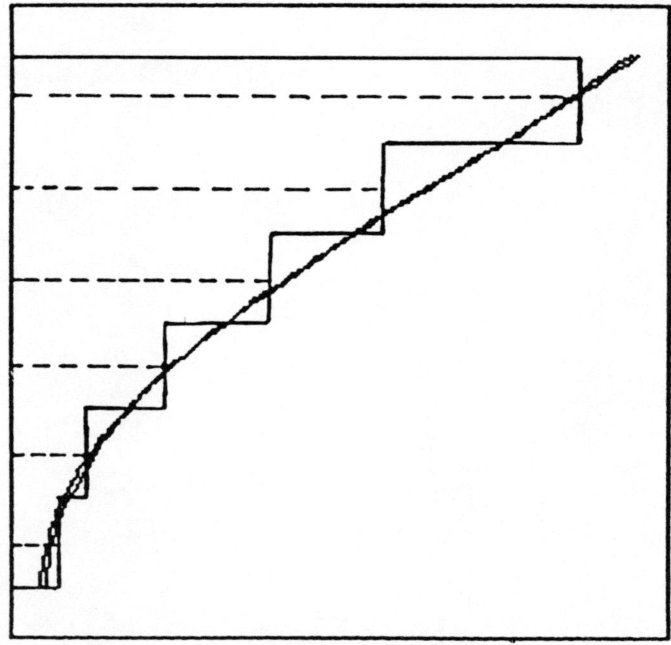

Normal Traction

4.10. The values computed for the six nodes closest to the crack tip are:

r	$t_2\sqrt{2\pi r}$
1	4.46
2	10.
3	7.61
4	8.5
5.5	8.77
7	9.18

The values at the nodes that belong to the first element are distorted because of the singularity at the tip. Using the other four values one may obtain $K_I \simeq 7$.

The stress intensity factor can also be computed from the displacement of the nodes along the crack, i.e.

r	$u_2\sqrt{\frac{2\pi}{r}}\mu(2-2\nu)$
1	6.89
2	7.12
3	6.97
4	6.81
5.5	6.51
7	6.18

$$K_I = \lim_{r\to 0} u_2\sqrt{\frac{2\pi}{r}}\mu(2-2\nu)$$